# MASS TRANSFER

## Principles and Applications

# MASS TRANSFER

## Principles and Applications

## DIRAN BASMADJIAN

CRC

## CRC PRESS

Boca Raton   London   New York   Washington, D.C.

## Library of Congress Cataloging-in-Publication Data

Basmadjian, Diran.
    Mass transfer : principles and applications / Diran Basmadjian.
        p.    cm.
    Includes bibliographical references and index.
    ISBN 0-8493-2239-1
    1. Mass transfer. I. Title

QC318.M3B37 2003
660'.28423--dc22                                                2003060755

### Visit the CRC Press Web site at www.crcpress.com

# *Preface*

The topic of mass transfer has a long and distinguished history dating to the 19th century, which saw the development and early applications of the theory of diffusion. Mass transfer operations such as distillation, drying, and leaching have an even earlier origin, although their practice was at that time an art rather than a science, and remained so well into the 20th century. Early textbook publications of that era dealt mainly with the topic of diffusion and the mathematics of diffusion.

The development of mass transfer theory based on the film concept, which began in the 1920s and continued during two decades of intense activity, brought about a shift in emphasis. The first tentative treatments of mass transfer processes dealing primarily with distillation and gas absorption began to appear, culminating with the publication, in 1952, of Robert Treybal's *Mass Transfer Operations*. It was to serve generations of students as the definitive text on the subject.

The 1950s and the decades that followed saw a second shift in emphasis, signaling a return to a more fundamental approach to the topic. Mass transfer was now seen as part of the wider basin of transport phenomena, which became the preferred topic of serious authors. The occasional text on mass transfer during this period viewed the topic on a high plane and mainly within the context of diffusion. For the most part, mass transport was seen as one of three players on the field of transport phenomena, and often a minor player at that. In the 1980s and 1990s, it became fashionable to treat mass transfer as part of the dual theme of heat and mass transfer. In these treatments, heat transfer, as the more mature discipline, predominated and mass transfer was usually given short shrift, or relegated to a secondary role. This need not be and ought not to be.

The author has felt for some time that mass transfer is a sufficiently mature discipline, and sufficiently distinct from other transport processes, to merit a separate treatment. The time is also ripe for a less stringent treatment of the topic so that readers will approach it without a sense of awe.

In other words, we do not intend to include, except in a peripheral sense, the more profound aspects of transport theory. The mainstays here are Fick's law of diffusion, film theory, and the concept of the equilibrium stage. These have been, and continue to be, the preferred tools in everyday practice. What we bring to these topics compared to past treatments is a much wider, modern set of applications and a keener sense that students need to learn how to simplify complex problems (often an art), to make engineering estimates (an art as well as a science), and to avoid common pitfalls. Such exercises, often dismissed for lacking academic rigor, are in fact a constant necessity in the engineering world.

Another departure from the norm is the organization of the material according to *mode of operation* (staged or continuous contact), rather than the type of separation process (e.g., distillation or extraction). Phase equilibria, instead of being dispersed among different operations, are likewise brought together in a single chapter. The reader will find that this approach unifies and strengthens the treatment of these topics and enables us to accommodate, under the same umbrella, processes that share the same features but are of a different origin (environmental, biological, etc.).

The readership at this level is broad. The topic of separation processes taught at all engineering schools is inextricably linked to mass transport, and students will benefit from an early introductory treatment of mass transfer combined with the basic concepts of separation theory. There is, in fact, an accelerating trend in this direction, which aims for students to address later the more complex operations, such as multicomponent and azeotropic distillation, chromatography, and the numerical procedures to simulate these and other processes.

Mass transport also plays a major role in several other important disciplines. Environmental processes are dominated by the twin topics of mass transfer and phase equilibria, and here again an early and separate introduction to these subject areas can be immensely beneficial. This text provides detailed treatments of both phase equilibria and compartmental models, which are all-pervasive in the environmental sciences. Transport, where it occurs, is almost always based on Fickian diffusion and film theory. The same topics are also dominant in the biological sciences and in biomedical engineering, and the text makes a conscious effort to draw on examples from these disciplines and to highlight the idiosyncrasies of biological processes.

Further important applications of mass transport theory are seen in the areas of materials science and materials processing. Here the dominant transport mode is one of diffusion, which in contrast to other disciplines often occurs in the solid phase. The reader will find numerous examples from these fascinating fields as well as a considerable amount of preparatory material of benefit to materials science students.

The text starts in an unconventional way by introducing the reader at an early stage to diffusion rates and Fick's law and to the related concepts of film theory and mass transfer coefficients. This is done in Chapter 1, but the topics are deemed of such importance that we return to them repeatedly in Chapters 3 and 4, and again in Chapter 5. In this manner, we develop the subject matter and our grasp of it in successive and complementary stages. The intervening Chapter 2 is entirely devoted to the art of setting up mass balances, a topic that is all too often given little attention. Without a good grasp of this subject we cannot set about the task of modeling mass transfer, and the many pitfalls we encounter here are alone sufficient reason for a separate treatment. The balances include algebraic and ordinary differential equations (ODEs). The setting up of partial differential equations (PDEs) is also discussed, and some time is spent in examining the general conservation equations in vector form. We do not attempt solutions of PDEs but instead provide the reader with known solutions and solution charts, which we use in Chapter 3 to solve a range of important problems. That chapter also considers the simultaneous occurrence of mass transfer and chemical reaction.

Chapter 6 deals with phase equilibria, which are mainly composed of topics not generally covered in conventional thermodynamics courses. These equilibria are used in Chapter 7 to analyze compartmental models and staged processes. Included in this chapter is a unique treatment of percolation processes, which should appeal to environmental and chemical engineers. Chapter 8 takes up the topic of modeling continuous-contact operations, among which the application to membrane processes is given particular prominence. Finally, in Chapter 9 we conclude the text with a brief survey of simultaneous mass and heat transfer.

The text is suitable for a third-year course addressed to engineering students, particularly those in the chemical, civil, mechanical, environmental, biomedical, and materials disciplines. Biomedical and environmental engineers will find topics of interest in almost all chapters, while materials science students may wish to concentrate on the earlier portions of the text (Chapters 1 to 5). The entire text can, with some modest omissions, be covered in a single term. The professional with a first-time interest in the topic or a need for a refresher will find this a useful and up-to-date text.

# *Acknowledgments*

The author is much obliged to his colleague, Professor Olev Trass, who was kind enough to make his course notes and problems available. Illustration 1.6, which deals with the analysis of hypothetical concentration profiles, was drawn from this source.

We were, as usual, immensely aided by the devoted efforts of Arlene Fillatre, who typed the manuscript, and Linda Staats, who produced impeccable drawings from rough sketches, which defy description. My wife, Janet, and granddaughter, Sierra, provided an oasis away from work.

# *Author*

**Diran Basmadjian** is a graduate of the Swiss Federal Institute of Technology, Zurich, and received his M.A.Sc. and Ph.D. degrees in chemical engineering from the University of Toronto. He was appointed assistant professor of chemical engineering at the University of Ottawa in 1960, moving to the University of Toronto in 1965, where he subsequently became professor of chemical engineering.

He has combined his research interests in the separation sciences, biomedical engineering, and applied mathematics with a keen interest in the craft of teaching. His current activities include writing, consulting, and performing science experiments for children at a local elementary school. Professor Basmadjian has authored four books and some fifty scientific publications.

# *Notation*

| | |
|---|---|
| $a$ | specific surface area, $m^2/m^3$ |
| $A$ | area, $m^2$ |
| $A$ | raffinate solvent, kg or kg/s |
| $B$ | extract solvent, kg or kg/s |
| $C$ | concentration, $mol/m^3$ |
| $C$ | number of components |
| $C_p$ | heat capacity at constant pressure, J/kg K or J/mol K |
| $d$ | diameter, m |
| $D$ | diffusivity, $m^2/s$ |
| $D$ | distillate, mol/s |
| $D'$ | cumulative distillate, mol |
| $D_e$ | effective diffusivity, $m^2/s$ |
| **erf** | error function |
| **erfc** | complementary error function |
| $E$ | effectiveness factor, dimensionless |
| $E$ | extract, kg or kg/s |
| $E$ | extraction ratio, dimensionless |
| $E$ | stage efficiency, dimensionless |
| $E_a$ | activation energy, J/mol |
| $E_h$ | enhancement or enrichment factor, dimensionless |
| $f$ | fraction distilled |
| $F$ | degrees of freedom |
| $F$ | feed, kg or mol, kg/s or mol/s |
| $F$ | force, N |
| $\mathfrak{F}$ | Faraday number, C/mol |
| $G$ | gas or vapor flow rate, kg/s or mol/s |
| $G_s$ | superficial carrier flow rate, $kg/m^2\,s$ |
| $h$ | heat transfer coefficient, $J/m^2\,s\,K$ |
| $h$ | height, m |
| $H$ | Henry's constant, Pa $m^3$ $mol^{-1}$ or kg solvent/kg adsorbent |
| $H$ | enthalpy, J/kg or J/mol |

| Ha | Hatta number, dimensionless |
| HETP(S) | height equivalent to a theoretical plate or stage, m |
| HTU | height of a transfer unit, m |
| $i$ | electrical current, A |
| $J_w$ | water flux, $m^3/m^2\,s$ |
| $k$ | thermal conductivity, $J/m\,s\,K$ |
| $k_C,\ k_G,\ k_L,\ k_x,\ k_y,\ k_Y$ | mass transfer coefficient, various units |
| $k_e$ | elimination rate constant, $s^{-1}$ |
| $k_r$ | reaction rate constant, $s^{-1}$ |
| $K$ | partition coefficient, various units |
| $K$ | permeability, m/s or $m^2$ |
| $K_o$ | overall mass transfer coefficient, various units |
| $\ell$ | length, m |
| $L$ | length, m |
| $L$ | liquid flow rate, kg/s or mol/s |
| $L$ | liquid mass, kg |
| $L_s$ | superficial solvent flow rate, $kg/m^2\,s$ |
| $m$ | distribution coefficient, various units |
| $m$ | mass, kg |
| $M$ | mass of emissions, kg, kg/s, or $kg/m^2\,s$ |
| $M$ | molar mass, dimensionless |
| $N$ | mass fraction (leaching), dimensionless |
| $N$ | molar flow rate, mol/s |
| $N$ | number of stages or plates |
| $N_T$ | number of mass transfer units |
| NTU | number of transfer units |
| $p$ | pressure, Pa |
| $P$ | number of phases |
| $P^o$ | vapor pressure, Pa |
| $P_T$ | total pressure, Pa |
| $P_w$ | water permeability, $mol/m^2\,s\,Pa$ |
| $p_{BM}$ | log-mean pressure difference, Pa |
| Pe | Peclet number, dimensionless |
| $q$ | heat flow, J/s |
| $q$ | thermal quality of feed, dimensionless |
| $Q$ | volumetric flow rate, $m^3/s$ |
| $r$ | radial variable, m |

| | |
|---|---|
| $r$ | recovery, dimensionless |
| $R$ | gas constant, J/mol K |
| $R$ | radius, m |
| $R$ | raffinate, kg or kg/s |
| $R$ | reflux ratio, dimensionless |
| $R$ | residue factor, dimensionless |
| $R$ | resistance, $\Omega$ |
| RO | reverse osmosis |
| $S$ | amount of solid, kg or kg/s |
| $S$ | shape factor, m |
| $S$ | solubility, cm$^3$ STP/cm$^3$ Pa |
| Sc | Schmidt number, dimensionless |
| Sh | Sherwood number, dimensionless |
| St | Stanton number, dimensionless |
| $t$ | time, s |
| $T$ | dimensionless time (adsorption) |
| $T$ | temperature, K or °C |
| $u$ | dependent variable |
| $u$ | velocity, m/s |
| $U$ | overall heat transfer coefficient, J/m$^2$ s K |
| $v$ | velocity, m/s |
| $v_H$ | specific volume, m$^3$/kg dry air |
| $V$ | voltage, V |
| $V$ | volume, m$^3$ or m$^3$/mol |
| $W$ | bottoms, mol or mol/s |
| $W$ | weight, kg |
| $x$ | liquid weight or mole fraction, dimensionless |
| $x$ | raffinate weight fraction, dimensionless |
| $x$ | solid-phase weight fraction (leaching), dimensionless |
| $X$ | adsorptive capacity, kg solute/kg solid |
| $X$ | liquid-phase mass ratio, dimensionless |
| $y$ | extract weight fraction, dimensionless |
| $y$ | vapor mole fraction, dimensionless |
| $Y$ | humidity, kg water/kg dry air |
| $Y$ | gas-phase mass ratio, dimensionless |
| $z$ | distance, m |
| $z_{FH}$ | heat transfer film thickness, m |

| $z_{FM}$ | mass transfer film thickness, m |
| $Z$ | dimensionless distance (adsorption) |
| $Z$ | flow rate ratio (dialysis) |

---

## Greek Symbols

| $\alpha$ | relative volatility, dimensionless |
| $\alpha$ | selectivity, dimensionless |
| $\alpha$ | separation factor, dimensionless |
| $\alpha$ | thermal diffusivity, m²/s |
| $\gamma$ | activity coefficient, dimensionless |
| $\dot{\gamma}$ | shear rate, s⁻¹ |
| $\delta$ | film or boundary layer thickness, m |
| $\varepsilon$ | porosity, dimensionless |
| $\lambda$ | mean free path, m |
| $\mu$ | viscosity, Pa s |
| $\nu$ | kinematic viscosity, m²/s |
| $\pi$ | osmotic pressure, Pa |
| $\rho$ | density, kg/m³ |
| $\sigma$ | liquid film thickness, m |
| $\sigma^{st}$ | length of stomatal pore, m |
| $\tau$ | shear stress, Pa |
| $\tau$ | tortuosity, dimensionless |
| $\phi$ | pressure ratio, dimensionless |

---

## Subscripts

| as | adiabatic saturation |
| $b$ | bed, bulk |
| $c$ | cold, molar concentration units ($k_c$) |
| $C$ | cross section, condenser |
| db | dry bulb |
| $D$ | distillate, dialysate |

| | |
|---|---|
| *e* | effective |
| *f, F* | feed |
| *g, G* | gas |
| *h* | hot |
| *i* | initial |
| *i* | inside |
| *i* | impeller |
| *L* | liquid |
| *m* | mean |
| *o* | outside |
| OW | octanol–water |
| *p* | particle, pellet |
| *p* | permeate |
| *p* | pore |
| *v* | vessel |
| *w* | bottoms |
| *w* | water |

---

## Superscripts

| | |
|---|---|
| * | equilibrium |
| *o* | initial |
| *o* | pure component |
| ' | cumulative |

# Table of Contents

**Chapter 1**  Some Basic Notions: Rates of Mass Transfer ............. 1
1.1   Gradient-Driven Transport ................................................................. 2
        Illustration 1.1: Transport in Systems with Vanishing
                Gradients ................................................................. 6
        Illustration 1.2: Diffusion through a Hollow Cylinder ............. 8
        Illustration 1.3: Underground Storage of Helium: Diffusion
                through a Spherical Surface ......................................... 10
1.2   Transport Driven by a Potential Difference: The Film Concept
        and the Mass Transfer Coefficient ............................................. 12
1.3   Units of the Potential and of the Mass Transfer Coefficient ............. 16
        Illustration 1.4: Conversion of Mass Transfer Coefficients ...... 17
1.4   Equimolar Counterdiffusion and Diffusion through a Stagnant Film:
        The Log-Mean Concentration Difference ................................. 18
        1.4.1   Equimolar Counterdiffusion ..................................... 19
        1.4.2   Diffusion through a Stagnant Film ........................... 20
        Illustration 1.5: Estimation of Mass Transfer Coefficients
                and Film Thickness. Transport in Blood Vessels ........ 22
1.5   The Two-Film Theory ................................................................. 24
1.6   Overall Driving Forces and Mass Transfer Coefficients ..................... 27
        Illustration 1.6: Qualitative Analysis of Concentration
                Profiles and Mass Transfer ........................................ 29
        Illustration 1.7: Drying with an Air Blower:
                A Fermi Problem ............................................................ 31
1.7   Conclusion .............................................................................. 33
Practice Problems ............................................................................. 33

**Chapter 2**  Modeling Mass Transport: The Mass Balances ........ 39
2.1   The Compartment and the One-Dimensional Pipe ..................... 40
        Illustration 2.1: Evaporation of a Solute
                to the Atmosphere ....................................................... 42
        Illustration 2.2: Reaeration of a River .................................. 47
2.2   The Classification of Mass Balances ........................................ 49
        2.2.1   The Role of Balance Space ...................................... 50
        2.2.2   The Role of Time ................................................... 50
                2.2.2.1   Unsteady Integral Balance ........................ 50
                2.2.2.2   Cumulative (Integral) Balance ................... 50
                2.2.2.3   Unsteady Differential Balances ................. 51

       2.2.2.4   Dependent and Independent Variables ......................51
       Illustration 2.3: The Countercurrent Gas Scrubber:
           Genesis of Steady Integral and Differential Mass
           Balances ........................................................................53
       Illustration 2.4: Two Examples from Biology:
           The Quasi-Steady-State Assumption ............................57
       Illustration 2.5: Batch Distillation: An Example
           of a Cumulative Balance.............................................. 62
2.3   The Information Obtained from Model Solutions ................64
2.4   Setting Up Partial Differential Equations ............................66
       Illustration 2.6: Unsteady Diffusion in One Direction:
           Fick's Equation.............................................................. 67
       Illustration 2.7: Laminar Flow and Diffusion in a Pipe:
           The Graetz Problem for Mass Transfer .......................70
       Illustration 2.8: A Metallurgical Problem: Microsegregation
           in the Casting of Alloys and How to Avoid PDEs ....73
2.5   The General Conservation Equations ....................................79
       Illustration 2.9: Laplace's Equation, Steady-State
           Diffusion in Three-Dimensional Space:
           Emissions from Embedded Sources ............................81
       Illustration 2.10: Lifetime of Volatile Underground
           Deposits ........................................................................84
Practice Problems...........................................................................85

**Chapter 3   Diffusion through Gases, Liquids, and Solids ..........91**
3.1   Diffusion Coefficients ............................................................91
     3.1.1   Diffusion in Gases ........................................................91
          Illustration 3.1: Diffusivity of Cadmium Vapor in Air ...........93
     3.1.2   Diffusion in Liquids ....................................................95
          Illustration 3.2: Electrorefining of Metals. Concentration
           Polarization and the Limiting Current Density .........98
     3.1.3   Diffusion in Solids ....................................................101
          3.1.3.1   Diffusion of Gases through Polymers
              and Metals ....................................................102
          Illustration 3.3: Uptake and Permeation of Atmospheric
           Oxygen in PVC ..........................................................104
          Illustration 3.4: Sievert's Law: Hydrogen Leakage
           through a Reactor Wall ..............................................106
          Illustration 3.5: The Design of Packaging Materials ............108
          3.1.3.2   Diffusion of Gases through Porous Solids ...............110
          Illustration 3.6: Transpiration of Water from Leaves.
           Photosynthesis and Its Implications for Global
           Warming ..................................................................... 112
          Illustration 3.7: Diffusivity in a Catalyst Pellet .....................115
          3.1.3.3   Diffusion of Solids in Solids ....................................116

Illustration 3.8: Diffusivity of a Dopant in a Silicon
              Chip ........................................................................ 117
Practice Problems ........................................................................ 117

**Chapter 4** **More about Diffusion: Transient Diffusion and**
**Diffusion with Reaction** ........................................................ **121**
4.1   Transient Diffusion ........................................................... 122
      4.1.1   Source Problems ..................................................... 123
              4.1.1.1   Instantaneous Point Source Emitting into Infinite
                        Space ................................................................ 123
              4.1.1.2   Instantaneous Point Source on an Infinite Plane
                        Emitting into Half Space ................................. 125
              4.1.1.3   Continuous Point Source Emitting into Infinite
                        Space ................................................................ 127
              Illustration 4.1: Concentration Response to an Instantaneous
                        Point Source: Release in the Environment and in a
                        Living Cell .......................................................... 128
              Illustration 4.2: Net Rate of Global Carbon Dioxide
                        Emissions ........................................................... 129
              Illustration 4.3: Finding a Solution in a Related Discipline:
                        The Effect of Wind on the Dispersion
                        of Emissions ...................................................... 131
      4.1.2   Nonsource Problems ............................................... 133
              4.1.2.1   Diffusion into a Semi-Infinite Medium ...................... 133
              Illustration 4.4: Penetration of a Solute into a Semi-Infinite
                        Domain ................................................................ 134
              Illustration 4.5: Cumulative Uptake by Diffusion for
                        the Semi-Infinite Domain ............................... 135
              4.1.2.2   Diffusion in Finite Geometries: The Plane Sheet,
                        the Cylinder, and the Sphere ......................... 136
              Illustration 4.6: Manufacture of Transformer Steel ............... 138
              Illustration 4.7: Determination of Diffusivity in Animal
                        Tissue ................................................................. 139
              Illustration 4.8: Extraction of Oil from Vegetable Seeds......... 140
4.2   Diffusion and Reaction ...................................................... 140
      4.2.1   Reaction and Diffusion in a Catalyst Particle ......................... 141
      4.2.2   Gas–Solid Reactions Accompanied by Diffusion:
              Moving-Boundary Problems ....................................... 142
      4.2.3   Gas–Liquid Systems: Reaction and Diffusion in the Liquid
              Film ........................................................................ 143
              Illustration 4.9: Reaction and Diffusion in a Catalyst Particle.
                        The Effectiveness Factor and the Design of Catalyst
                        Pellets ................................................................ 143
              Illustration 4.10: A Moving Boundary Problem: The
                        Shrinking Core Model ..................................... 148

        Illustration 4.11: First-Order Reaction with Diffusion in a
                Liquid Film: Selection of a Reaction Solvent ........... 151
Practice Problems .................................................................................... 153

**Chapter 5  A Survey of Mass Transfer Coefficients** .................. 157
5.1  Dimensionless Groups ......................................................... 158
        Illustration 5.1: The Wall Sherwood Number ........................ 159
5.2  Mass Transfer Coefficients in Laminar Flow: Extraction from
     the PDE Model ...................................................................... 160
    5.2.1  Mass Transfer Coefficients in Laminar Tubular Flow .......... 162
    5.2.2  Mass Transfer Coefficients in Laminar Flow around Simple
            Geometries ................................................................. 163
        Illustration 5.2: Release of a Solute into Tubular Laminar
                Flow: Transport in the Entry Region ......................... 164
5.3  Mass Transfer in Turbulent Flow: Dimensional Analysis
     and the Buckingham π Theorem .......................................... 166
    5.3.1  Dimensional Analysis ................................................. 167
    5.3.2  The Buckingham π Theorem ........................................ 168
        Illustration 5.3: Derivation of a Correlation for Turbulent
                Flow Mass Transfer Coefficients Using Dimensional
                Analysis ............................................................. 169
        Illustration 5.4: Estimation of the Mass Transfer Coefficient
                $k_Y$ for the Drying of Plastic Sheets ..................... 172
5.4  Mass Transfer Coefficients for Tower Packings .................. 173
        Illustration 5.5: Prediction of the Volumetric Mass Transfer
                Coefficient of a Packing ...................................... 177
5.5  Mass Transfer Coefficients in Agitated Vessels ................. 177
        Illustration 5.6: Dissolution of Granular Solids in an
                Agitated Vessel .................................................. 179
5.6  Mass Transfer Coefficients in the Environment: Uptake and Clearance
     of Toxic Substances in Animals: The Bioconcentration Factor ......... 180
        Illustration 5.7: Uptake and Depuration of Toxins:
                Approach to Steady State and Clearance
                Half-Lives ......................................................... 183
Practice Problems .................................................................................... 185

**Chapter 6  Phase Equilibria** ............................................................. 189
6.1  Single-Component Systems: Vapor Pressure ...................... 190
        Illustration 6.1: Maximum Breathing Losses from a Storage
                Tank ................................................................. 193
6.2  Multicomponent Systems: Distribution of a Single
     Component ............................................................................ 195
    6.2.1  Gas–Liquid Equilibria ................................................. 196
        Illustration 6.2: Carbonation of a Soft Drink ..................... 197
    6.2.2  Liquid and Solid Solubilities ...................................... 199

Illustration 6.3: Discharge of Plant Effluent into a River ...... 200
6.2.3   Fluid–Solid Equilibria: The Langmuir Isotherm .................201
Illustration 6.4: Adsorption of Benzene from Water
    in a Granular Carbon Bed ...........................................205
Illustration 6.5: Adsorption of a Pollutant from
    Groundwater onto Soil ............................................. 208
6.2.4   Liquid–Liquid Equilibria: The Triangular Phase
    Diagram ....................................................................209
Illustration 6.6: The Mixture or Lever Rule in the Triangular
    Diagram.....................................................................213
6.2.5   Equilibria Involving a Supercritical Fluid .............................215
Illustration 6.7: Decaffeination in a Single-Equilibrium
    Stage .................................... .........................................218
6.2.6   Equilibria in Biology and the Environment: Partitioning
    of a Solute between Compartments .........................................220
Illustration 6.8: The Octanol–Water Partition Coefficient ......221
6.3  Multicomponent Equilibria: Distribution of Several
Components............................................................................................222
6.3.1   The Phase Rule ...................................................................222
Illustration 6.9: Application of the Phase Rule .....................222
6.3.2   Binary Vapor–Liquid Equilibria .......................................223
    6.3.2.1   Phase Diagrams .....................................................224
    6.3.2.2   Ideal Solutions and Raoult's Law: Deviation from
        Ideality ...................................................................226
    6.3.2.3   Activity Coefficients .....................................................229
6.3.3   The Separation Factor $\alpha$: Azeotropes............................ 231
Illustration 6.10: The Effect of Total Pressure on $\alpha$ .................235
Illustration 6.11: Activity Coefficients from Solubilities ........236
Practice Problems ........................................................................238

Chapter 7  Staged Operations: The Equilibrium Stage ............ 243
7.1  Equilibrium Stages ...........................................................................245
7.1.1   Single-Stage Processes .......................................................245
Illustration 7.1: Single-Stage Adsorption: The Rectangular
    Operating Diagram ..................................................248
Illustration 7.2: Single-Stage Liquid Extraction: The
    Triangular Operating Diagram .................................249
7.1.2   Single-Stage Differential Operation .......................................251
Illustration 7.3: Differential Distillation: The Rayleigh
    Equation ..................................................................252
Illustration 7.4: Rayleigh's Equation in the Environment:
    Attenuation of Mercury Pollution in a Water
    Basin .......................................................................254
7.1.3   Crosscurrent Cascades ......................................................257

          Illustration 7.5: Optimum Use of Adsorbent or Solvent in
                Crosscurrent Cascades .................................................260
          Illustration 7.6: A Crosscurrent Extraction Cascade in
                Triangular Coordinates ..............................................263
     7.1.4   Countercurrent Cascades ........................................................264
          Illustration 7.7: Comparison of Various Stage Configurations:
                The Kremser–Souders–Brown Equation ....................269
     7.1.5   Fractional Distillation: The McCabe–Thiele Diagram ..........273
          7.1.5.1   Mass and Energy Balances: Equimolar Overflow
                  and Vaporization ..............................................275
          7.1.5.2   The McCabe–Thiele Diagram ...........................278
          7.1.5.3   Minimum Reflux Ratio and Number of Plates..........282
          Illustration 7.8: Design of a Distillation Column in the
                McCabe-Thiele Diagram....................................285
          Illustration 7.9: Isotope Distillation: The Fenske
                Equation .......................................................288
          Illustration 7.10: Batch-Column Distillation: Model
                Equations and Some Simple Algebraic
                Calculations ..................................................290
     7.1.6   Percolation Processes ..............................................................296
          Illustration 7.11: Contamination and Clearance of Soils
                and River Beds ...............................................299
7.2   Stage Efficiencies......................................................................... 299
     7.2.1   Distillation and Absorption ..................................................301
     7.2.2   Extraction ...............................................................................301
     7.2.3   Adsorption and Ion-Exchange ............................................301
          Illustration 7.12: Stage Efficiencies of Liquid–Solid
                Systems ........................................................302
     7.2.4   Percolation Processes ............................................................304
          Illustration 7.13: Efficiency of an Adsorption or
                Ion-Exchange Column .....................................305
Practice Problems ...............................................................................306


**Chapter 8 Continuous-Contact Operations ............................... 313**
8.1   Packed-Column Operations .......................................................314
          Illustration 8.1: The Countercurrent Gas Scrubber
                Revisited .......................................................314
          Illustration 8.2: The Countercurrent Gas Scrubber Again:
                Analysis of the Linear Case .............................319
          Illustration 8.3: Distillation in a Packed Column: The Case
                of Constant $\alpha$ at Total Reflux .....................322
          Illustration 8.4: Coffee Decaffeination by Countercurrent
                Supercritical Fluid Extraction .........................324
8.2   Membrane Processes ...................................................................326
     8.2.1   Membrane Structure, Configuration, and Applications.........327

8.2.2   Process Considerations and Calculations ...................................333
       Illustration 8.5: Brian's Equation for Concentration
            Polarization .....................................................................335
       Illustration 8.6: A Simple Model of Reverse Osmosis ...........336
       Illustration 8.7: Modeling the Artificial Kidney: Analogy
            to the External Heat Exchanger ....................................338
       Illustration 8.8: Membrane Gas Separation: Selectivity $\alpha$
            and the Pressure Ratio $\phi$ ..............................................342
Practice Problems ......................................................................................345

**Chapter 9**  **Simultaneous Heat and Mass Transfer** .................... 349
9.1   The Air–Water System: Humidification and Dehumidification,
     Evaporative Cooling ...............................................................................350
     9.1.1   The Wet-Bulb Temperature .......................................................350
     9.1.2   The Adiabatic Saturation Temperature and the
            Psychrometric Ratio ................................................................352
           Illustration 9.1: The Humidity Chart .........................................353
           Illustration 9.2: Operation of a Water-Cooling Tower ...........357
9.2   Drying Operations ...............................................................................361
           Illustration 9.3: Debugging of a Vinyl Chloride Recovery
            Unit ..................................................................................363
9.3   Heat Effects in a Catalyst Pellet: The Nonisothermal
     Effectiveness Factor ...............................................................................365
           Illustration 9.4: Design of a Gas Scrubber:
            The Adiabatic Case .......................................................369
Practice Problems........................................................................................371

**Selected References**.................................................................. 373

**Appendix A1**  **The *D*-Operator Method** ...................................... 379

**Appendix A2**  **Hyperbolic Functions and ODEs** ...................... 381

**Subject Index** ........................................................................... 383

# 1

## *Some Basic Notions: Rates of Mass Transfer*

We begin our deliberations by introducing the reader to the basic rate laws that govern the transport of mass. In choosing this topic as our starting point, we follow the pattern established in previous treatments of the subject, but depart from it in some important ways. We start, as do other texts, with an introduction to Fick's law of diffusion, but treat it as a component of a broader class of processes, which is termed *gradient-driven transport*. This category includes the laws governing transport by molecular motion, Fourier's law of conduction and Newton's viscosity law, as well as Poiseuille's law for viscous flow through a cylindrical pipe and D'Arcy's law for viscous flow through a porous medium, both of which involve the bulk movement of fluids. In other words, we use as common ground the *form of the rate law,* rather than the underlying physics of the system. This treatment is a departure from the usual pedagogical norm and is designed to reinforce the notion that transport of different types can be drawn together and viewed as driven by a potential gradient (concentration, temperature, velocity, pressure), which diminishes in the direction of flow.

The second departure is the early introduction of the reader to the notion of a linear driving force, or potential *difference* as the agent responsible for transport. One encounters here, for the first time, the notion of a transport coefficient that is the proportionality constant of the rate law. Its inverse can be viewed as the resistance to transport and in this it resembles Ohm's law, which states that current transport $i$ is proportional to the voltage difference $\Delta V$ and varies inversely with the Ohmian resistance $R$.

Associated with the transport coefficients is the concept of an effective film thickness, which lumps the resistance to transport into a fictitious thin film adjacent to a boundary or interface. Transport takes place through this film driven by the linear driving force across it and impeded by a resistance that is the inverse of the transport coefficient. The reader will note in these discussions that a conscious effort is made to draw analogies between the transport of mass and heat and to occasionally invoke as well the analogous case of transport of electricity.

The chapter is, as are all the chapters, supplemented with worked examples, which prepare the ground for the practice problems given at the end of the chapter.

## 1.1  Gradient-Driven Transport

The physical laws that govern the transport of mass, energy, and momentum, as well as that of electricity, are based on the notion that the flow of these entities is induced by a driving potential. This driving force can be expressed in two ways. In the most general case, it is taken to be the *gradient* or *derivative of that potential* in the direction of flow. A list of some rate laws based on such gradients appears in Table 1.1. In the second, more specialized case, the gradient is taken to be constant. The driving force then becomes simply the *difference in potential* over the distance covered. This is taken up in Section 1.2, and a tabulation of some rate laws based on such potential differences is given in Table 1.2.

Let us examine how these concepts can be applied in practice by taking up a familiar example of a gradient-driven process, that of the conduction of heat.

The general reader knows that heat flows from a high temperature $T$, which is the driving potential here, to a lower temperature at some other location. The greater the difference in temperature per unit distance, $x$, the larger the transport of heat; i.e., we have a proportionality:

**TABLE 1.1**

Rate Laws Based on Gradients

| Name | Process | Flux | Gradient |
|---|---|---|---|
| 1. Fick's law | Diffusion | $N/A = -D\dfrac{dC}{dx}$ | Concentration |
| 2. Fourier's law | Conduction | $q/A = -k\dfrac{dT}{dx}$ | Temperature |
| 3. Alternative formulation | | $q/A = -\alpha\dfrac{d(\rho CpT)}{dx}$ | Energy concentration |
| 4. Newton's viscosity law | Molecular momentum transport | $F_x/A = \tau_{qx} = -\mu\dfrac{dv_x}{dy}$ | Velocity |
| 5. Alternative formulation | | $F_x/A = \tau_{yx} - \nu\dfrac{d(\rho v_x)}{dy}$ | Momentum concentration |
| 6. Poiseuille's law | Viscous flow in a circular pipe | $q/A = v_x = -\dfrac{d^2}{32\mu}\dfrac{dp}{dx}$ | Pressure |
| 7. D'Arcy's law | Viscous flow in a porous medium | $q/A = v_x = -\dfrac{K}{\mu}\dfrac{dp}{dx}$ | Pressure |

**TABLE 1.2**

Rate Laws Based on Linear Driving Forces

| Process | Flux or Flow | Driving Force | Resistance |
|---|---|---|---|
| 1. Electrical current flow (Ohm's law) | $i = \Delta V / R$ | $\Delta V$ | $R$ |
| 2. Convective mass transfer | $N/A = k_C \Delta C$ | $\Delta C$ | $1/k_C$ |
| 3. Convective heat transfer | $q/A = h\Delta T$ | $\Delta T$ | $1/h$ |
| 4. Flow of water due to osmotic pressure $\pi$ | $N_A/A = P_w \Delta \pi$ | $\Delta \pi$ | $1/P_w$ |

$$q \propto -\frac{\Delta T}{\Delta x} \tag{1.1}$$

The minus sign is introduced to convert $\Delta T/\Delta x$, which is negative quantity, to a positive value of heat flow $q$. In the limit $\Delta x \to 0$, the difference quotient converts to the derivative $dT/dx$. Noting further that heat flow will be proportional to the cross-sectional area normal to the direction of flow and introducing the proportionality constant $k$, known as the thermal conductivity, we obtain

$$\text{Heat flow } q(\text{J / s}) = -kA\frac{dT}{dx} \tag{1.2a}$$

or equivalently

$$\text{Heat flux } q / A(\text{J / sm}^2) = -k\frac{dT}{dx} \tag{1.2b}$$

These two expressions, shown graphically in Figure 1.1b, are known as Fourier's law of heat conduction. It can be expressed in yet another alternative form, which is obtained by multiplying and dividing the right side by the product of density $\rho$ (kg/m³) and specific heat $C_p$ (J/kg K). We then obtain (Item 3 of Table 1.1)

$$q / A = -\alpha\frac{d(\rho C_p T)}{dx} \tag{1.3}$$

where $\alpha = k/\rho C_p$ is termed the thermal diffusivity. We note that the term $\rho C_p T$ in the derivative has the units of J/m³ and can thus be viewed as an energy concentration.

The reason for introducing this alternative formulation is to establish a link to the transport of mass (Item 1 of Table 1.1). Here the driving potential

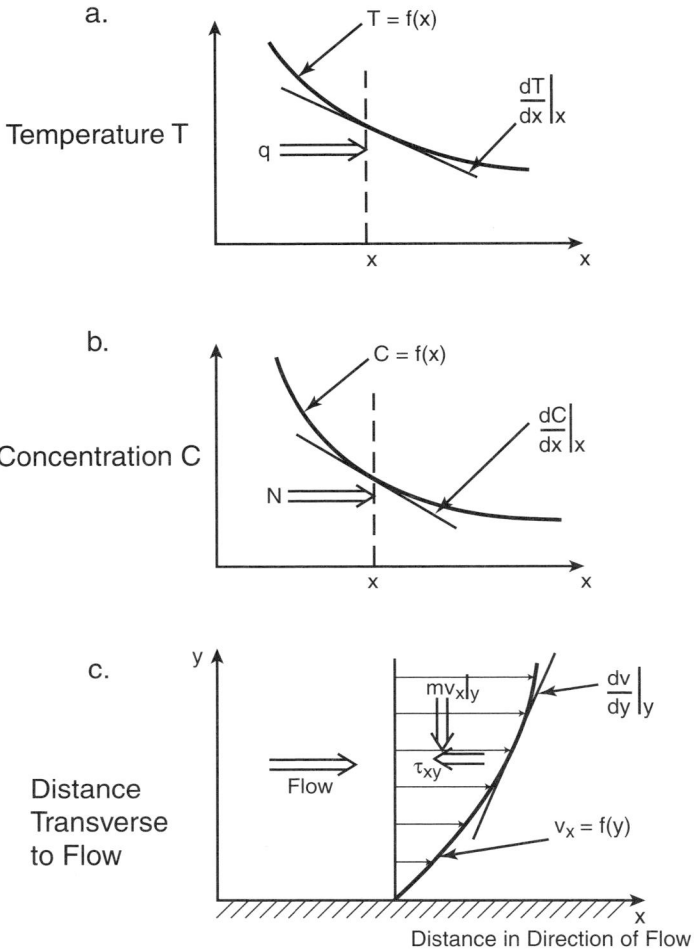

**FIGURE 1.1**
Diffusive transport: (a) heat; (b) mass; (c) momentum.

is expressed in terms of the molar concentration gradient $dC/dx$ and the proportionality constant $D$ is known as the (mass) diffusivity of the species, paralleling the thermal diffusivity $\alpha$ in Equation 1.3. Transport takes place from a point of high concentration to a location of lower concentration. Noting, as before, that the molar flow will be proportional to the cross-sectional area $A$ normal to the flow, we obtain

$$\text{Molar flow } N \text{ (mol/s)} = -DA\frac{dC}{dx} \qquad (1.4a)$$

and equivalently

$$\text{Molar flux } N/A \text{ (mol/sm}^2) = -D\frac{dC}{dx} \qquad (1.4b)$$

These two relations, depicted in Figure 1.1a, are known as Fick's law of diffusion.

There is a third mode of diffusive transport, that of momentum, that can likewise be induced by the molecular motion of the species. Momentum is the product of the mass of the molecular species and its velocity in a particular direction, for example, $v_x$. As in the case of the flow of mass and heat, the diffusive transport is driven by a gradient, here the velocity gradient $dv_x/dy$ transverse to the direction of flow (Figure 1.1c). It takes place from a location of high velocity to one of lower velocity, paralleling the transport of mass and heat. As the molecules enter a region of lower velocity, they relinquish part of their momentum to the slower particles in that region and are consequently slowed. There is, in effect, a braking force acting on them, which is expressed in terms of a shear stress $F_x/A = \tau_{yx}$ pointing in a direction opposite to that of the flow. The first subscript on the shear stress denotes the direction in which it varies, and the second subscript refers to the direction of the equivalent momentum $mv_x$. The relation between the induced shear stress and the velocity gradient is attributable to Newton and is termed Newton's viscosity law. It is, like Fick's and Fourier's law, a linear negative relation and is given by

$$F_x / A = \tau_{yx} = -\mu\frac{dv_x}{dy} \qquad (1.5)$$

Equation 1.5 can be expressed in the equivalent form:

$$\tau_{yx} = -\frac{\mu}{\rho}\frac{d(\rho v_x)}{dy} = -\nu\frac{d(\rho v_x)}{dy} \qquad (1.6)$$

where $\nu$ is termed the kinematic viscosity in units of $m^2/s$ and the product of density $\rho$ and velocity $v_x$ can be regarded as a momentum *concentration* in units of $(kg\ m/s)/m^3$. This version of Newton's viscosity law brings it in line with the concentration-driven expressions for diffusive heat and mass transport, Equation 1.3 and Equation 1.4. A summary of the relevant relations appears in Table 1.1.

Table 1.1 contains two additional rate processes, which are driven by gradients. The first is Poiseuille's law, which applies to viscous flow in a circular pipe, and a similar expression, D'Arcy's law, which describes viscous flow in a porous medium. Both processes are driven by pressure gradients and both vary inversely with viscosity, which is to be expected.

We now proceed to demonstrate the use of these rate laws with three illustrative examples. The first illustration examines several gradient-driven processes in which the gradient vanishes at a particular location of the system, yet transport still takes place. Such zero gradients are important in the solution of the differential equations of diffusion because they provide boundary conditions that can be used in the evaluation of integration constants. The second and third examples scrutinize diffusional processes that take place in different geometries. The solutions here are all effected by simple integration using the method of separation of variables. This procedure is employed extensively throughout the text. Occasional use is also made of the *D*-operator method, which is outlined in the Appendix.

### Illustration 1.1: Transport in Systems with Vanishing Gradients

It frequently happens in transport processes that the driving gradient vanishes at some position in the system, without inhibiting the flow of mass, heat, or momentum. There are two special situations that give rise to such behavior:

First, the potential exhibits a maximum or a minimum at a point or axis of symmetry. These locations can be the centerline of a slab, the axis of a cylinder, or the center of a sphere. Figure 1.2a and Figure 1.2b consider two such cases. Figure 1.2a represents a spherical catalyst pellet in which a reactant of external concentration $C_0$ diffuses into the sphere and undergoes a reaction. Its concentration diminishes and attains a minimum at the center. Figure 1.2b considers laminar flow in a cylindrical pipe. Here the state variable in question is the axial velocity $v_x$, which rises from a value of zero at the wall to a maximum at the centerline before dropping back to zero at the other end of the diameter. Here, again, symmetry considerations dictate that this maximum must be located at the centerline of the conduit.

The second case of a vanishing derivative arises when flow or flux ceases. Because the proportionality constants in the rate laws cannot themselves vanish, zero flow must perforce imply that the gradient becomes zero. This situation arises when flow or diffusional flux is brought to a halt by a physical barrier. Figure 1.2c and Figure 1.2d depict two such cases. Figure 1.2c shows a capillary that is filled with a solvent and is suddenly exposed to a solution containing a dissolved solute of concentration $C_0$. This configuration has been used in the past to determine diffusivities. As the solute diffuses into the capillary, a concentration profile develops within it, which changes with time until the concentration in the capillary equals that of the external medium. As these profiles grow, they maintain at all times a zero gradient at the sealed end of the capillary. This must be so since $N$, the diffusional flow in Equation 1.4, can only vanish if the gradient $dC/dx$ itself becomes zero. Figure 1.2d depicts a polymer extruder in which molten polymer enters one end of a pipe and exits as a thin sheet through a lateral slit. Here the barrier is the sealed end of the pipe, which prevents an axial outflow of the polymer melt and forces it instead into the lateral channel. The only way for flow to cease, $Q/A = 0$, is for the pressure gradient $dp/dx$ to vanish at this point. The resulting axial pressure profile is shown in Figure 1.2d.

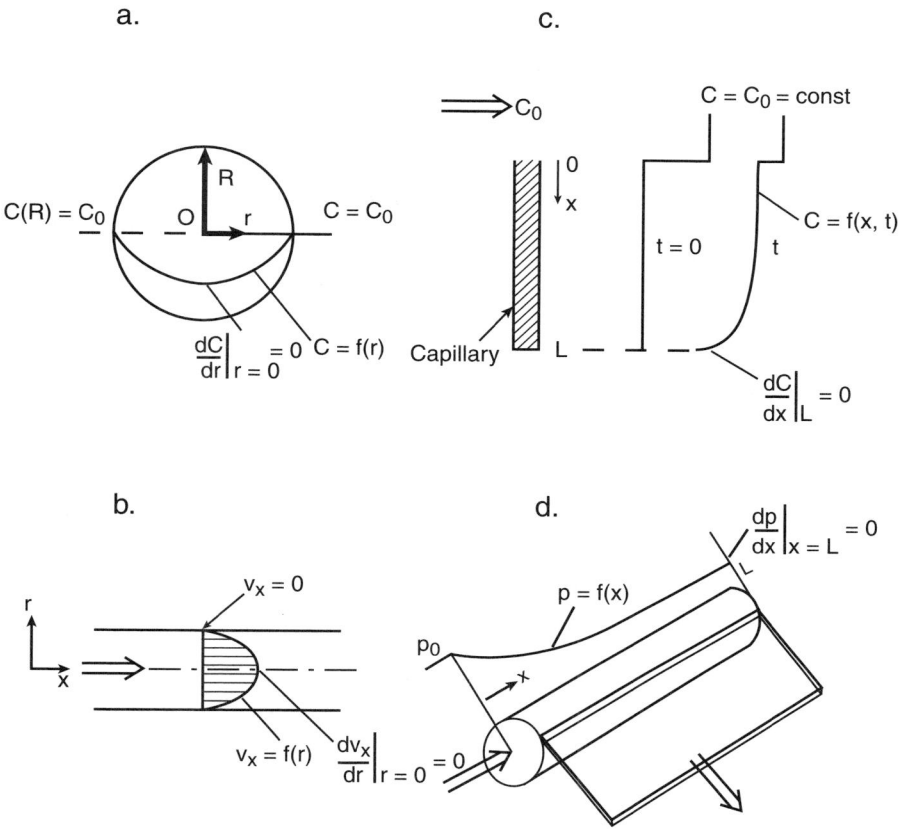

**FIGURE 1.2**
Systems with vanishing gradients: (a) catalyst pellet; (b) viscous flow in a pipe; (c) diffusion into capillary; (d) polymer extruder.

*Comments:*

These examples draw the readers' attention to the appearance of zero gradients in transport processes. Because these are confined to a specific location, they can serve, along with boundary values of the dependent variable itself, as *boundary conditions* in the solution of the model equations. Thus the catalyst pellet shown in Figure 1.2a has two such conditions, one at the center, where the flux vanishes, and a second at the surface, where the reactant concentration attains a constant value. The pellet is encountered again in Chapter 4 (Illustration 4.9) where the underlying model is found to be a second-order differential equation. Such equations require the evaluation of two integration constants, and must therefore be provided with two boundary conditions.

### *Illustration 1.2: Diffusion through a Hollow Cylinder*

The problem addressed here is the diffusional transport through a cylindrical wall of substantial thickness shown in Figure 1.3. Such processes can occur, for example, in the case of fluids contained in a cylindrical enclosure under high pressure.

We consider two problems. The first, and more important one, is the determination of the diffusional flux that results under these conditions. The second problem is the derivation of the concentration profile and is of mainly academic interest. Both problems involve the solution of a simple ordinary differential equation by the technique of separation of variables.

### 1. *Diffusional flow N*

The starting point here is Fick's law of diffusion, which is applied to a cylindrical surface of radius $r$ and length $L$ (Figure 1.3). We obtain

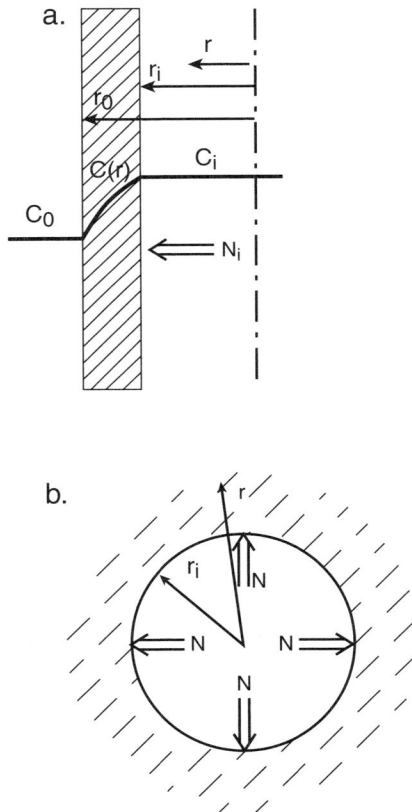

**FIGURE 1.3**
Diffusional flow from (a) a hollow cylinder; (b) a spherical cavity.

$$N = -D2\pi rL \frac{dC}{dr} \qquad (1.7a)$$

where $N$ = constant since we assume steady operation.

Separating variables and formally integrating between the limits of internal and external concentrations $C_i$ and $C_o$ we obtain

$$\int_{C_i}^{C_o} dC = -\frac{N}{D2\pi L} \int_{r_i}^{r_o} \frac{dr}{r} \qquad (1.7b)$$

and after evaluation of the integrals and rearrangement

$$N = D2\pi L \frac{(C_i - C_o)}{\ln r_o / r_i} \qquad (1.7c)$$

where $i$ and $o$ denote the inner and outer conditions. This is the desired relation, which expresses diffusion rate $N$ in terms of a driving force $C_i - C_o$ and the geometry of the system.

By multiplying numerator and denominator by $(r_o - r_i)$, this expression can be cast into the frequently used alternate form:

$$N = DA_m \frac{(C_i - C_o)}{r_o - r_i} = \frac{(C_i - C_o)}{R} \qquad (1.7d)$$

where $A_m$ is the so-called logarithmic mean of the inner and outer areas, given by

$$A_m = \frac{A_o - A_i}{\ln A_o / A_i} \qquad (1.7e)$$

and $R$ is a resistance defined by

$$R = \frac{r_o - r_i}{DA_m} \qquad (1.7f)$$

The introduction of the resistance carries the advantage that one can, in the case of a composite cylinder made up of different materials, describe the system by simply adding resistances. This principle of adding resistances in series is routinely applied in electrical circuits. For a cylinder composed of three different materials, for example, one obtains

$$N = \frac{(C_i - C_o)}{R_1 + R_2 + R_3} \tag{1.7g}$$

where the resistances can be expressed respectively as

$$R_1 = \frac{r_1 - r_i}{D_1 A_{m1}} \qquad R_2 = \frac{r_2 - r_1}{D_2 A_{m2}} \qquad R_3 = \frac{r_o - r_2}{D_3 A_{m3}} \tag{1.7h}$$

2. *Concentration profile* C = f(r)

We return to Fick's law, Equation 1.7a and integrate again, but this time only up to an arbitrary radius *r* and the concentration *C* at that point. We obtain

$$\int_{r_i}^{r} \frac{N}{D 2\pi L} \frac{dr}{r} = -\int_{C_i}^{C} dC \tag{1.8a}$$

or, since *N* is a constant given by Equation 1.7c,

$$C(r) = C_i - \frac{C_i - C_o}{\ln r_o / r_i} \ln r / r_i \tag{1.8b}$$

This equation expresses the concentration profile within the cylindrical wall.

*Comments:*

Of the various expressions presented, Equation 1.7c and its extension Equation 1.7g are the ones most frequently used in engineering applications. They allow the desired calculation of the mass flow *N*, which is the quantity of greatest practical interest. The concentration profile is not of immediate use, but reveals the surprising fact that *C(r)* is independent of diffusivity. It is these unexpected results that are the most rewarding feature of modeling. One should never set aside a solution without scrutinizing it first for unusual features of this type. We shall make frequent use of this maxim in subsequent illustrations.

### Illustration 1.3: Underground Storage of Helium: Diffusion through a Spherical Surface

The previous illustration considered the rate of diffusion through a cylindrical wall and the resulting concentration profile within that wall. A similar approach can be used to calculate these quantities for diffusion through a *spherical* wall (Figure 1.3b). This case arises much less frequently, as it requires the steady production of the diffusing species within the spherical cavity, or else assumes the diffusion rate to be sufficiently small so that the internal concentration remains essentially constant.

The case to be considered here falls in the latter category and involves the diffusional losses of helium from an underground storage facility. The background to this problem is as follows:

Helium is present in air at a concentration of about 1 ppm, which is far too small for the economic recovery of this gas. It also occurs in natural gas (methane $CH_4$), where its concentration is considerably higher, of the order of 0.1 to 5%, making economic extraction possible. Because helium is a nonrenewable resource, regulations were put in place starting in the early 1960s that required all shipped natural gas be treated for helium recovery. With supply by far outweighing the demand, ways had to be found to store the excess helium. One suggested solution was to pump the gas into abandoned and sealed salt mines where it remained stored at high pressure.

The problem here will be to estimate the losses that occur by diffusion through the surrounding salt and rock, assuming a solid-phase diffusivity $D_s$ of helium of $10^{-8}$ in.$^2$/s, i.e., more than three orders of magnitude less than the free-space diffusivity in air. The helium is assumed to be at a pressure of 10 MPa (~100 atm) and a temperature of 30°C. The cavity is taken to be spherical and of radius 100 m. Applying Fick's law, Equation 1.4a, and converting to pressure we obtain

$$N = -D_s 4\pi r^2 \frac{dC}{dr} = -\frac{D_s 4\pi r^2}{RT}\frac{dp}{dr} \tag{1.9a}$$

Separating variables and integrating yields

$$-\frac{4\pi D_s}{RT}\int_p^0 dp = N\int_{r_1}^{\infty}\frac{dr}{r^2} = N\frac{1}{r_1} \tag{1.9b}$$

and consequently

$$N = \frac{4\pi D_s r_1 p}{RT} = \frac{4\pi 10^{-8} \times 10^2 \times 10^7}{8.314 \times 303} \tag{1.9c}$$

$$N = 0.05 \text{ mol/s} \tag{1.9d}$$

*Comments:*

We have here an example of some practical importance, which nevertheless yields to a simple application and integration of Fick's law. Two features deserve some mention. The first is the formulation of the upper integration limit in Equation 1.9b. We use the argument that "far away" from the spherical cavity, i.e., as $r \to \infty$, the concentration and partial pressure of helium tends to zero, i.e., we assume the cavity to be embedded in an infinite region. The second point that needs to be examined is the assumption of a constant cavity pressure. We compute for this purpose the *yearly* loss and show that even over this lengthy period, the change in cavity pressure will be negligibly small. Thus,

$$\text{Yearly loss} = 0.05 \text{ (mol/s)} \times 60 \times 365 = 2.6 \times 10^4 \text{ mol/year}$$

i.e., about 100 kg per year. By comparison,

$$\text{Cavity contents: } n = \frac{pV}{RT} = \frac{10^7 \frac{4}{3}\pi10^6}{8.31 \times 303} = 1.17 \times 10^{10} \text{ mol}$$

and therefore

$$\% \text{ loss/year} = (2.6 \times 10^4 / 1.7 \times 10^{10})100 = 1.5 \times 10^{-4}\% / \text{ year}$$

Even if $D$ were raised to that prevailing in free air ($\sim 10^{-4}$ m$^2$/s), the losses would still amount to only 1.5% per year. This justifies the use of our assumption.

We shall have occasion to examine this problem again in Illustration 2.9.

## 1.2    Transport Driven by a Potential Difference: The Film Concept and the Mass Transfer Coefficient

In the gradient-driven procedure we considered previously, the operative gradient varied in the direction of transport, as indicated in Figure 1.1. In a number of important cases, however, the gradient either is constant or is assumed to be of constant value.

Let us first consider a case where no such assumption is made. We turn, for this purpose, to the familiar Ohm's law, which relates the flow of electrical current $i$ ($C/S$ or $A$) to the applied voltage:

$$i = \Delta V / R \tag{1.10a}$$

This is a linear law, involving a linear driving force $\Delta V$ and a proportionality constant, which is the inverse of the electrical resistance $R(\Omega)$ of the conductor. There is, at first sight, no gradient involved. Closer scrutiny of the resistance $R$ reveals, however, that it must vary directly with the length of the conductor $\Delta L$, and inversely with its cross-sectional area $A_C$, which we assume to be constant. We can then write

$$R = R_s \Delta L / A_C \qquad (1.10b)$$

where $R_s$ is termed the *specific resistance* of the conductor in units of $\Omega m$. Ohm's law can then be expressed as follows:

$$i = \left( R_s / A_C \right) \frac{\Delta V}{\Delta L} \qquad (1.10c)$$

This equivalent form of the celebrated law shows that current is in fact driven by a gradient, which does, however, turn out to be constant.

Items 2 and 3 of Table 1.2 concern what we term convective mass and heat transfer. Let us illustrate these terms by making use of Figure 1.4a. This figure depicts turbulent flow of either a gas or liquid past a liquid or solid boundary shown crosshatched on the left. That boundary can be the confining wall of a duct, or the interface separating two phases. Mass transfer is assumed to occur from a concentration $C_{A2}$ of the boundary to a lower concentration $C_{A1}$ in the bulk of the flowing fluid. This can come about if the boundary consists of a soluble substance or if a volatile liquid evaporates into a flowing gas stream.

These two operations, as well as the reverse processes of condensation and crystallization, are shown in Figure 1.5. In all four cases shown, the concentrations and partial pressures in the fluid phase are in equilibrium with the neighboring condensed phase. This condition is denoted by an asterisk. Thus, $p^*$ is the equilibrium vapor pressure of the liquid, and $C^*$ is the equilibrium solubility of the solid.

Mass transfer takes place initially through a laminar sublayer, or boundary layer, which is located immediately adjacent to the interface. Transfer through this region, also known as an "unstirred layer" in biological applications, is relatively slow and constitutes the preponderant portion of the resistance to mass transport. This layer is followed by a transition zone where the flow gradually changes to the turbulent conditions prevailing in the bulk of the fluid. In the main body of the fluid we see macroscopic packets of fluid or eddies moving rapidly from one position to another, including the direction toward and away from the boundary. Mass transfer in both the transition zone and the fully turbulent region is relatively rapid and contributes much less to the overall transport resistance than the laminar sublayer. One notes in addition that with an increase in fluid velocity there is an attendant increase in the degree of turbulence and the eddies are able to penetrate more deeply into the transition and boundary layers. The latter consequently diminish in thickness and the transport rate experiences a corresponding increase in magnitude. Thus, high flow rates mean a greater degree of turbulence and hence more rapid mass transfer.

Concentrations in the turbulent regime typically fluctuate around a mean value shown in Figure 1.4a and Figure 1.4b. These fluctuations cannot be easily quantified, nor do they lend themselves readily to the formulation of

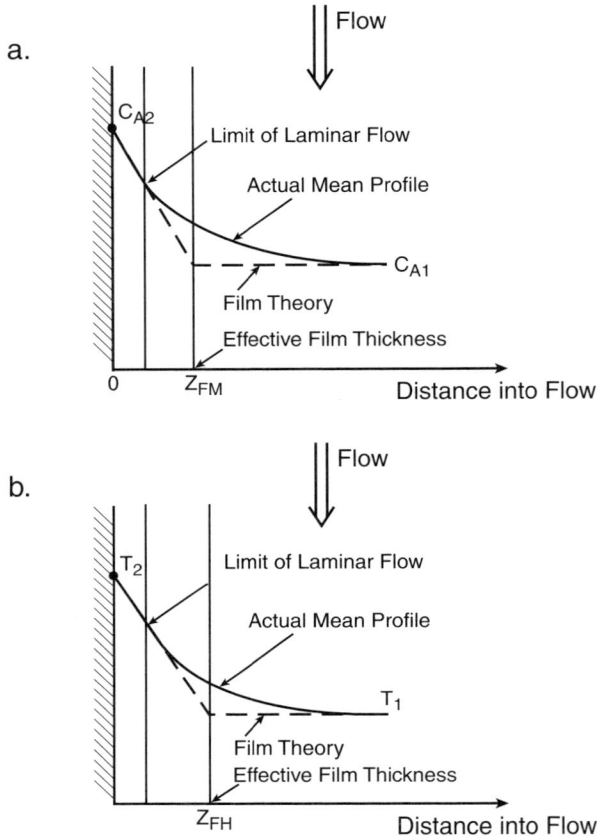

**FIGURE 1.4**
The effective film in the transport of (a) mass and (b) heat.

a rate law. This can be overcome by postulating the existence of an equivalent linear concentration profile that extends from the boundary into the bulk fluid. This postulate is enshrined in the concept known as *film theory,* and the dimension of the film in question is termed the *effective film thickness,* denoted as $z_{FM}$ in Figure 1.4a.

Let us see how this concept can be quantified into a rate law. We start with Fick's law, and applying it to the constant gradient of film theory we obtain

$$N_A / A = -D \frac{dC_A}{dz} = D \frac{(C_{A2} - C_{A1})}{z_{FM}} \tag{1.11a}$$

The ratio of diffusivity to film thickness $D/z_{FM}$ is coalesced into a single term called the mass transfer coefficient $k_C$ and we obtain

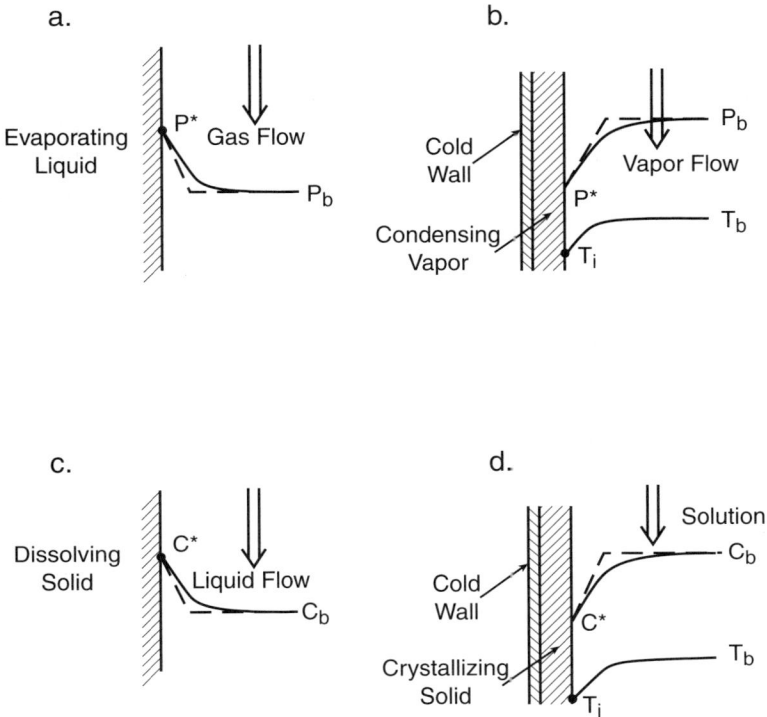

**FIGURE 1.5**
Four types of single-film mass transfer: (a) evaporation; (b) condensation; (c) dissolution; (d) crystallization.

$$N_A / A = \frac{D}{z_{FM}}(C_{A2} - C_{A1}) = k_C(\Delta C_A) \tag{1.11b}$$

as shown in Table 1.2.

A similar film theory can be postulated for the case of heat transfer, shown in Figure 1.4b. The conditions here parallel those shown for mass transfer with temperature replacing concentration as the driving potential. Starting with Fourier's law, Equation 1.2b, we then obtain

$$q / A = -k\frac{dT}{dz} = k\frac{(T_2 - T_1)}{z_{FH}} \tag{1.12a}$$

Coalescing the ratio $k/z_{FH}$ into a single term $h$ then leads to

$$q / A = \frac{k}{z_{FH}}(T_2 - T_1) = h\Delta T \tag{1.12b}$$

where $h$ is now termed the heat transfer coefficient.

The effective film thicknesses for the two cases, $z_{FM}$ and $z_{FH,}$ are not in general equal, but depend in a complex functional form on the physical properties, the geometry, and the velocity of flow of the system. That functional form will be explored in greater detail in Chapter 5. In addition, the transport rate depends *linearly* on the potential difference, a feature that is often referred to as a *linear driving force*. All three items 1 to 3 have this characteristic in common.

A special type of driving force arises in Item 4 of Table 1.2. The process here is the selective transport of water through a semipermeable membrane from a dilute solution (high water concentration) to a more concentrated solution (low water concentration). The driving force is in this case the difference of the so-called osmotic pressure $\pi$, which makes its appearance in transport through cell membranes as well as in industrial processes termed *reverse osmosis*. We have occasion to take a closer look at osmotic-pressure-driven processes in Chapter 8.

## 1.3    Units of the Potential and of the Mass Transfer Coefficient

In deriving the mass transfer rate law, Equation 1.11b, we started with Fick's law, which uses molar concentration $C$ in units of $mol/m^3$ as the driving potential. This quantity was retained to describe the driving force in the final expression (Equation 1.11b). It is a convenient quantity to use in many gas–liquid operations and carries the advantage of imparting units of m/s to the mass transfer coefficient. $k_C$ can thus be viewed as the velocity with which the rate process proceeds. It frequently happens, however, that molar concentrations are inconvenient to use in the description of certain mass transfer operations. In distillation, for example, the preferred concentration unit is the mole fraction since the associated vapor–liquid equilibrium is commonly expressed in liquid and vapor mole fractions $(x, y)$. In the evaporation of liquids, the vapor pressure is the potential of choice, and it then becomes convenient to use a pressure difference as the driving force. Yet another operation that calls for a change in concentration units is humidification, in which the preferred concentration is the absolute humidity in units of kg water/kg dry air. In each of these cases, the change in concentration units carries with it a change in the units of the mass transfer coefficient. The pertinent rate laws, driving forces, and mass transfer coefficients, together with their units, are summarized in Table 1.3. Pressure heads the list of driving potentials for gases and vapors, followed by mole fraction, moles/volume, and mass ratio in the order of preference and frequency of use. For liquid systems, molar concentration is the unit of choice, followed by mole fraction.

**TABLE 1.3**

Rate Laws and Transfer Coefficients for Diffusion through a Stagnant Film

| Flux (mol/m² s) | Driving Potential | Mass Transfer Coefficient |
|---|---|---|
| *Gases* | | |
| $N_A/A = k_G \Delta p_A$ | $p_A$ (Pa) | $k_G$ (mol/m² s Pa) |
| $N_A/A = k_y \Delta y_A$ | $y_A$ (mole fraction) | $k_y$ (mol/m² s mole fraction) |
| $N_A/A = k_C \Delta C_A$ | $C_A$ (mol/m³) | $k_C$ (m/s) |
| $W_A/A = k_Y \Delta Y_A$ (kg/m²/s) | $Y_A$ (kg A/kg B) | $k_Y$ (mol/m² s $\Delta Y_A$) |
| *Liquids* | | |
| $N_A/A = k_L \Delta C_A$ | $C_A$ (mol/m³) | $k_L$ (m/s) |
| $N_A/A = k_x \Delta x_A$ | $x_A$ (mole fraction) | $k_x$ (mol/m² s mole fraction) |
| *Conversion Factors* | | |
| Gases | $k_G = k_Y/P_T = k_C/RT = k_Y/M p_{BM}$ | |
| Liquids | $k_L C = k_x$ | |

Also listed in Table 1.3 are conversion factors for the transformation of mass transfer coefficients from one set of units to another. These are frequently required to convert literature values of $k$ given in a particular set of units, to one needed in a different application. This type of conversion is taken up in Illustration 1.4. Of note as well in Table 1.3 is the appearance of the term $p_{BM}$, the so-called logarithmic mean, or log-mean driving force, defined by

$$p_{BM} = \frac{p_{B2} - p_{B1}}{\ln \dfrac{p_{B2}}{p_{B1}}} \tag{1.13}$$

where the subscript $B$ denotes the second component in a binary system; the first is the component $A$ being transferred. Derivation of this quantity, and its appearance in the conversion factor, is addressed in Section 1.4.

### Illustration 1.4: Conversion of Mass Transfer Coefficients

In a particular application related to air flowing over a water surface, the following data were reported at $T = 317$ K:

$$p_{A1} = 2{,}487 \text{ Pa}$$
$$p_{B1} = 101{,}300 - 2487 = 98{,}813 \text{ Pa}$$
$$p_{B2} = P_T = 101{,}300 \text{ Pa}$$
$$k_{CA} = 0.0284 \text{ m/s}$$

Water evaporates into the dry airstream at a total pressure $P_T = 101.3$ kPa and is denoted by the subscript $A$; $B$ refers to the air component. $k_C$ is the mass transfer coefficient, here of water. We wish to calculate the corresponding value for $k_Y$ in units of kg $H_2O/m^2\,s\,\Delta Y$. The conversion formula given in Table 1.3 is

$$k_Y = k_C \frac{M_B}{RT} p_{BM} \tag{1.14}$$

where $M_B$ is the molar mass of air $= 29 \times 10^{-3}$.
  We obtain

$$k_Y = 0.0284 \frac{29 \times 10^{-3}}{8.31 \times 317} \frac{98,813 - 101,300}{\ln \dfrac{98,813}{101,300}}$$

and therefore

$$k_Y = 0.0312 \text{ kg } H_2O/m^2\,\Delta Y$$

## 1.4 Equimolar Counterdiffusion and Diffusion through a Stagnant Film: The Log-Mean Concentration Difference

So far our treatment has been confined to mass transfer due to diffusion only. We have considered diffusion in a stationary or unmixed medium, which has led to the use of Fick's Equation 1.4. When a stirred or turbulent medium was involved, we invoked film theory and the linear driving force concept to describe transport in such situations. This led to the formulation of the expression (Equation 1.11b).

  Mass transport can, however, also come about as a result of the bulk motion or flow of a fluid. To take this factor into account, we postulate the total flux of a component $A$ to be the sum of a diffusive flux term and a bulk flow term. Thus, for a gaseous mixture

$$N_A/A \quad = \quad -CD_{AB}\frac{dy_A}{dz} \quad + \quad y_A(N_A/A + N_B/A) \tag{1.15a}$$

$$\text{Flux of } A \qquad \text{Diffusional Flux} \qquad\qquad \text{Bulk Flow}$$

  Here we have replaced the $C_A$, which appears in Fick's law, by the equivalent term $Cy_A$ where $C =$ total molar concentration, assumed to be constant.
  For ideal gases we have $y_A = p_A/P_T$ and $C = P_T/RT$, where $P_T =$ total pressure, so that Equation 1.15a becomes

$$N_A / A = -\frac{D_{AB}}{RT}\frac{dp_A}{dz} + \frac{p_A}{P_T}(N_A / A + N_B / A) \qquad (1.15b)$$

This is the form we wish to develop and simplify.

Two special cases of Equation 1.15a are to be noted: equimolar counter-diffusion and diffusion through a stagnant film.

### 1.4.1  Equimolar Counterdiffusion

In this case we have

$$N_A = -N_B \qquad (1.16)$$

and Equation 1.15b reduces to Fick's law. This situation arises in the inter-diffusion of pure fluids of equal molar volume or in binary adiabatic distillation processes of substances with identical molar heats of vaporization. Straightforward integration of Fick's law then leads to, for a gaseous system:

$$N_A' / A = \frac{D_{AB}}{RT(z_2 - z_1)}(p_{A_1} - p_{A_2}) = \frac{D_{AB}}{RTz_{FM}}(p_{A_1} - p_{A_2}) \qquad (1.17a)$$

or in short

$$N_A'/A = k_G' \, \Delta p_A \qquad (1.17b)$$

where we use the prime symbol to denote equimolar counterdiffusion. Similarly, for a liquid system, using mole fraction as the driving potential,

$$N_A' / A = \frac{CD_{AB}}{(z_2 - z_1)}(x_{A_1} - x_{A_2}) = \frac{CD_{AB}}{z_{FM}}(x_{A_1} - x_{A_2}) \qquad (1.17c)$$

or in short

$$N_A'/A = k_x' \Delta x_A \qquad (1.17d)$$

Because $\Delta p/z_M$ and $\Delta x/z_{FM}$ are constant, the concentration profiles in both cases are linear. This is shown for a liquid system in Figure 1.6a. A summary of the pertinent rate laws, transfer coefficients, and conversion factors that apply to equimolal counterdiffusion appears in Table 1.4.

a.

b.

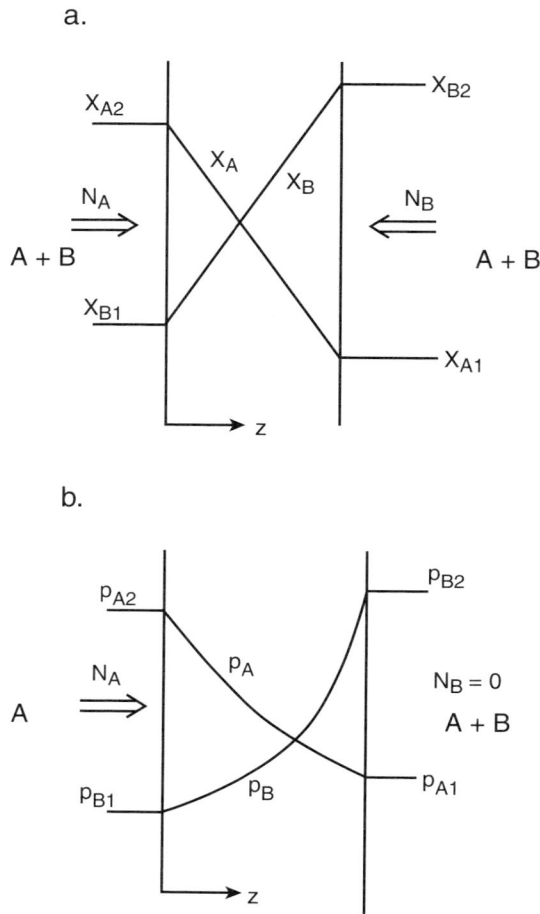

**FIGURE 1.6**
Two modes of transport: (a) equimolar counterdiffusion; (b) diffusion through a stagnant film.

### 1.4.2 Diffusion through a Stagnant Film

Here the flux of the species $B$ is zero and we have

$$N_B/A = 0 \tag{1.18a}$$

so that Equation 1.15b, after solving for $N_A/A$, is reduced to the expression

$$N_A / A = -\frac{D_{AB}P_T}{RT(P_T - p_A)}\frac{dp_A}{dz} \tag{1.18b}$$

**TABLE 1.4**

Rate Laws and Transfer Coefficients for Equimolar Diffusion

| Flux (mol/m² s) | Driving Potential | Mass Transfer Coefficient |
|---|---|---|
| *Gases* | | |
| $N_A'/A = k_G' \Delta p_A$ | $p_A$ (Pa) | $k_G'$ (mol/m² s Pa) |
| $N_A'/A = k_y' \Delta y_A$ | $y_A$ (mole fraction) | $k_y'$ (mol/m² s mole fraction) |
| $N_A'/A = k_C \Delta C_A$ | $C_A$ (mol/m³) | $k_C$ (m/s) |
| *Liquids* | | |
| $N_A'/A = k_L' \Delta C_A$ | $C_A$ (mol/m³) | $k_C'$ (m/s) |
| $N_A'/A = k_x' \Delta x_A$ | $x_A$ (mole fraction) | $k_x$ (mol/m² s mole fraction) |

*Conversion Factors*

Gases
$$k_G' = \frac{k_y y}{P_T} = \frac{k_C'}{RT}$$

Liquids
$$k_L' C = k_y'$$

*Conversion from Equimolal to Stagnant Film Coefficients*

Gases

$$k_G' = k_G \frac{p_{BM}}{P_T} \qquad k_y' = k_y \frac{p_{BM}}{P_T} = k_y y_{BM} \qquad k_C' = k_C \frac{p_{BM}}{P_T} = k_C \frac{C_{BM}}{C}$$

Liquids

$$k_C' = k_C \frac{C_{BM}}{P} = k_C x_{BM} \qquad k_x' = k_x x_{BM}$$

which can be integrated to yield

$$N_A/A = -\frac{D_{AB} P_T}{(z_2 - z_1)RT} \ln \frac{(P_T - p_{A1})}{(P_T - p_{A2})} = \frac{D_{AB} P_T}{(z_2 - z_1)RT} \ln \frac{p_{B1}}{p_{B2}} \qquad (1.18c)$$

Because of the logarithmic terms, the profiles for both component *A* and component *B* are nonlinear. This is depicted in Figure 1.6b.

We now introduce a clever device to reduce the nonlinear terms in Equation 1.18c to the product of a linear driving force in the diffusing species *A*, $p_{A1} - p_{A2}$, and a constant mass transfer coefficient $k_G$. This is done by writing:

$$1 = \frac{p_{B2} - p_{B1}}{p_{B2} - p_{B1}} = \frac{P_T - p_{A2} - P_T + p_{A1}}{p_{B2} - p_{B1}} = \frac{p_{A1} - p_{A2}}{p_{B2} - p_{B1}} \qquad (1.18d)$$

Using the definition of the log-mean pressure difference $p_{BM}$ given by Equation 1.13 and setting $z_2 - z_1 = z_{FM}$ as before, we obtain

$$N_A / A = \frac{D_{AB} P_T}{z_{FM} RT p_{BM}} (p_{A2} - p_{A1}) \quad\quad (1.18e)$$

or in short

$$N_A/A = k_G \Delta p_A \quad\quad (1.18f)$$

We have thus reduced a complex nonlinear situation to one that fits the linear driving force and film concepts and agrees with the tabulations of Table 1.3.

### Illustration 1.5: Estimation of Mass Transfer Coefficients and Film Thickness. Transport in Blood Vessels

It is of some interest to the practicing engineer to have a sense of the order of magnitude both of the mass transfer coefficient and of its associated film thickness. This would appear to be an impossible task, given the wide range of flow conditions, geometrical configurations, and physical properties encountered in practice. Surprisingly, we can arrive at some reasonable estimates of upper and lower bounds in spite of this diversity. This is due to three factors: First, it is common engineering practice to associate the upper limit of normal turbulent flow with velocities of the order 1 m/s in the case of liquids and 10 m/s for gases. This applies to industrial systems (pipe and duct flow) as well as within an environmental context (wind, river flow) and holds even in extreme cases. Hurricane-force winds, for example, may range as high as to 30 m/s but are still within the order of magnitude cited. Second, the diffusivities for a wide range of substances are, as we shall see in Chapter 3, surprisingly constant. They cluster, in the case of gases, around a value of $10^{-5}$ m²/s and for liquids around $10^{-9}$ m²/s. Third, if we confine ourselves to flow over a plane as a representative configuration, it will be found that mass transfer coefficients vary inversely with the 2/3 power of the ratio $(\mu/\rho D)$ (see Table 5.5). That ratio, termed the *Schmidt number*, Sc, is again surprisingly constant. It is of the order 1 for gases, and some three orders of magnitude higher for transport of modest-sized solutes in liquids.

Drawing on the correlation given in Table 5.5:

$$\frac{k_C}{v} = 0.036 \text{ Re}^{-0.2} \text{ Sc}^{-0.67} \quad\quad (\text{Re} > 10^6)$$

and noting the extremely weak dependence on the Reynolds number, Re, we obtain the following order-of-magnitude estimates for turbulent flow mass transfer:

For gases: $k_C \sim 10^{-2}$ m/s
For liquids: $sk_C \sim 10^{-5}$ m/s

For the effective film thickness, $z_{FM} = D/k_C$ the corresponding values are

For gases: $z_{FM} \sim 1$ mm
For liquids: $z_{FM} \sim 0.1$ mm

The $k_C$ values given here represent the *order of magnitude* of the outer limits of what can be accomplished, i.e., the maximum rate of mass transfer obtainable or the minimum time required to achieve the transfer of a given mass to or from a flat surface.

Mass transfer by molecular diffusion resides at the other end of the spectrum. It yields the *lowest* possible mass transfer rate and sets an upper limit on time requirements. The solvent spill considered in Practice Problem 4.4, for example, requires several *days* to complete evaporation by diffusion into stagnant air. The same data applied to turbulent air flow at the same temperature yield an estimate of several *minutes*, lower by three orders of magnitude. The factor of 1000 can thus be viewed as separating the two extremes of diffusive and turbulent mass transfer.

Transport in blood vessels presents another interesting subcase. Here the species involved in mass transfer, typically proteins, have much lower diffusivities than ordinary solutes, of the order $10^{-10}$ to $10^{-11}$ m²/s. Flow is generally laminar, and of a pulsatile nature. A further departure from the norm is the complex geometry of the vascular systems, which involves multiple branchings, and constrictions in flow as well as expansions. Because of these complexities, it has become customary to measure *local* mass transfer coefficients confined to a typical wall area of 1 mm². A host of such measurements has by now been reported. The surprising fact that has emerged from these studies is that irrespective of location or configuration, the vast majority of $k_C$ values clusters around a value of $10^{-5}$ m/s. Thus,

$$k_C \sim 10^{-5} \text{ m/s}$$

and for a protein with a diffusivity of $10^{-10}$ m²/s,

$$z_{FM} \sim 10^{-2} \text{ mm}$$

The studies referred to have been immensely helpful in analyzing the progress of vascular diseases, which are often associated with the migration of proteins to the vessel wall followed by an interaction with wall cells, leading, for example, to blood coagulation. Mass transfer is also important in the performance of vascular grafts and of controlled release devices (see Practice Problems 1.7 and 1.10).

## 1.5   The Two-Film Theory

Our considerations so far have been limited to transport through a single phase; i.e., in postulating the film theory it was assumed that a single film resistance was operative. This was the case for a pure liquid evaporating into a gas stream, or when a solid dissolved into a solvent stream. We now consider the extension of this process to encompass simultaneous transport in two adjacent phases. This leads to the formation of two film resistances, and brings us to the so-called *two-film theory*, which is taken up below.

Consider two phases, I and II, in turbulent flow and in contact with each other, as shown in Figure 1.7a. Transport takes place in the first place, from a high concentration $y_A$ through the effective film associated with Phase II to the interface. Here the Phase II concentration $y^*_{Ai}$ is assumed to be in equilibrium with the Phase I interfacial concentration $x_{Ai}$, so that

$$y^*_{Ai} = mx_{Ai} \tag{1.19}$$

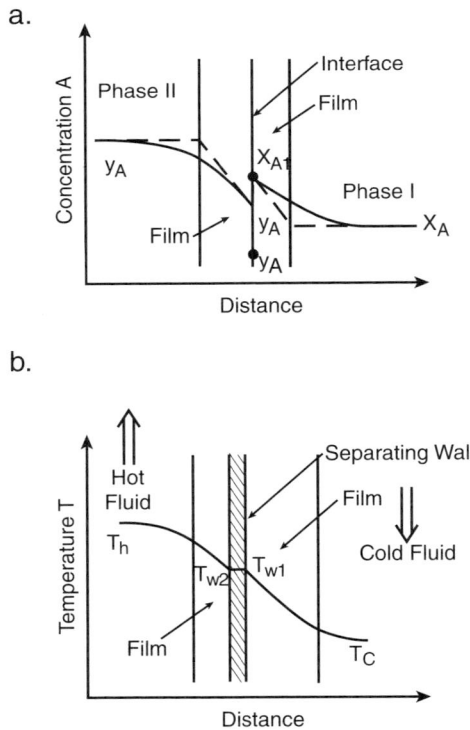

**FIGURE 1.7**
The two-film concept: (a) mass transfer; (b) heat transfer.

where $m$ is the local slope of the equilibrium curve and the asterisk serves to denote equilibrium conditions. Transport then continues from the interface, through the second film to the bulk of Phase I of concentration $x_A$. We can write for the entire process:

$$N_A/A = k_y(y_A - y_{Ai}) = k_x(x_{Ai} - x_A) \tag{1.20}$$

These expressions, while valid under the constraints of two-film theory, are nevertheless ill-suited for practical use, as neither of the interfacial concentrations $x_{Ai}$ or $y_{Ai}$ is generally known. We avoid this difficulty by postulating an equivalent rate law, given by

$$N_A/A = K_{oy}(y_A - y_A^*) \tag{1.21a}$$

$$N_A/A = K_{ox}(x_A^* - x_A) \tag{1.21b}$$

where $K_{oy}$ and $K_{ox}$ are termed overall mass transfer coefficients and the asterisked quantities are the concentrations in equilibrium with the bulk concentration of the neighboring phase. These are generally known and are displayed in Figure 1.8. It is shown in the following section that the equations are valid provided the overall coefficients are related to the film coefficients as follows:

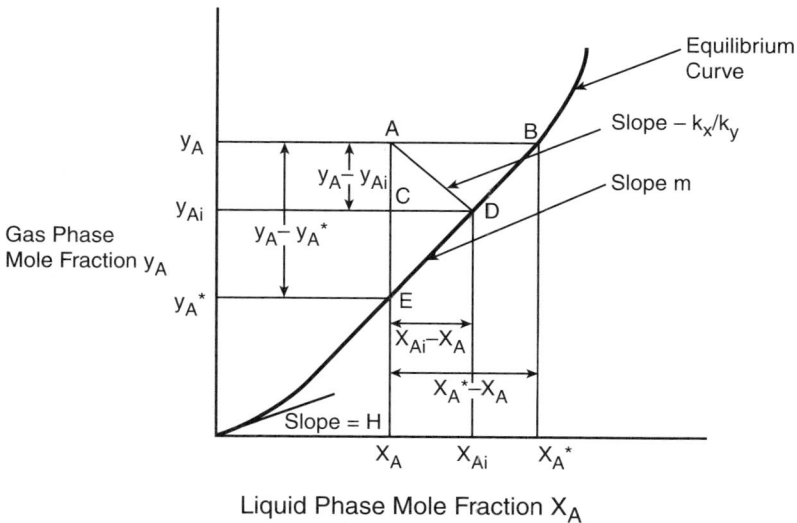

**FIGURE 1.8**
The driving forces.

$$\frac{1}{K_{oy}} = \frac{1}{k_y} + \frac{m}{k_x} \qquad\qquad (1.22a)$$

$$\frac{1}{K_{ox}} = \frac{1}{mk_y} + \frac{1}{k_x} \qquad\qquad (1.22b)$$

The reciprocal terms appearing in these expressions may be regarded as resistances to the mass transfer process. Thus, in Equation 1.22a, $1/k_y$ is the resistance due to the gas film, $m/k_x$ represents the liquid film resistance, and the sum of the two yields the overall resistance $1/K_{oy}$. This is often referred to as the law of *additivity of resistances,* which was previously encountered in Equation 1.7g. Similar arguments apply to Equation 1.22b. Note that when $m$ is small compared to $k_x$, we obtain the approximate relation

$$\frac{1}{K_{oy}} \doteq \frac{1}{k_y} \qquad m \ll 1, k_L \gg 1 \qquad\qquad (1.22c)$$

This implies that the gas is highly soluble and that most of the resistance is likely to reside in the gas phase. We speak of the process as being gas-film controlled. Conversely, if $m$ is large, i.e., the gas is sparsely soluble, we obtain

$$\frac{1}{K_{ox}} \doteq \frac{1}{k_x} \qquad m \ll 1 \qquad\qquad (1.22d)$$

and we refer to the process as being liquid-film controlled. Depending on which resistance predominates, we choose either Equation 1.22a or Equation 1.22b to express the transfer process.

Overall coefficients are widely used to describe transport between two flowing phases in contact with each other. They are usually determined experimentally and reported as *lumped averages* over the span of equilibrium constants $m$ encountered in the operation. Alternatively and much less frequently, empirical correlations of the film coefficients, if available, can be used to compute $K$ from Equation 1.25. Such correlations are discussed in Chapter 5.

The two-film concept can also be applied to heat transfer operations, as shown in Figure 1.7b. The process is similar to that of mass transport, but differs from it in two important aspects. First, the two fluids (hot and cold) are usually, but not always, separated by a solid partition. This is in contrast to mass transfer operations where direct contact of the phases is the norm. Second, no phase-equilibrium relation needs to be invoked at the interface. Instead, convergence of the two temperature profiles on either side of an interface leads to one and the same temperature at this point. No jump-discontinuities in temperature occur at any location along an interface. We can consequently express the rate of heat transfer in the following alternative ways:

$$q/A = h_h(T_h - T_{w_2}) = \frac{k}{L}(T_{w_2} - T_{w_1}) = h_C(T_{w_1 - T_c}) \qquad (1.23a)$$

|  |  |  |
|---|---|---|
| Transfer from hot fluid | Transfer through wall | Transfer to cold fluid |

where the subscripts $h$ and $c$ represent hot and cold fluids, respectively, $k$ is the thermal conductivity, and $L$ is the thickness of the partition. Note that when the partition is curved, as it is for pipes, variations in the heat transfer area have to be taken into account.

It is left to the exercises to show that the individual resistances can be added to obtain an overall heat transfer coefficient U, given by

$$\frac{1}{U} = \frac{1}{h_h} + \frac{1}{h_c} + \frac{L}{k} \qquad (1.23b)$$

which is associated with an overall heat transfer driving force, i.e., we have

$$q/A = U(T_h - T_c) \qquad (1.23c)$$

The corresponding mass transfer rate expression is given by Equation 1.25. For the convenience of the reader we have summarized the various parameters that appear in the foregoing equations in Table 1.5.

## 1.6 Overall Driving Forces and Mass Transfer Coefficients

We embark here on the proof of the validity of Equation 1.21 and Equation 1.22, i.e., that of the rate law based on an overall driving force, and the law of additivity of resistances. The procedure starts with an examination of the diagram of Figure 1.8, which represents a plot of the gas-phase concentration of the diffusing species, $y_A$, against its liquid-phase counterpart, $x_A$. It contains an equilibrium curve, which is generally nonlinear, but is assumed to contain a short segment BE, which is considered linear. Also indicated on the diagram are the various operative film and overall driving forces. Thus, for the gas phase, AC represents the film driving force $y_A - y_{Ai}$, which we have previously shown in Figure 1.7a, and the distance AE equals the overall gas phase driving force $y_A - y_A^*$.

We start by noting that the slope of the line AD is given by the ratio AC/CD and hence, by virtue of Equation 1.20,

**TABLE 1.5**

Mass and Heat Transfer Parameters for Two-Phase Transport

| | Mass Transfer[a] | Heat Transfer |
|---|---|---|
| Driving force | $\Delta y, \Delta x$ | $\Delta T$ |
| Single film coefficient | $k_x, k_y$ | $h_h, h_C$ |
| Overall coefficient | $K_{ox}, K_{oy}$ | $U$ |
| Single film resistance | $\dfrac{1}{k_x}, \dfrac{1}{k_y}$ | $\dfrac{1}{h_h}, \dfrac{1}{h_C}$ |
| Overall resistance | $\dfrac{1}{K_{oy}} = \dfrac{1}{k_y} + \dfrac{m}{k_x}$ | $\dfrac{1}{U} = \dfrac{1}{h_h} + \dfrac{1}{h_C} + \dfrac{L}{k}$ |
| Single film rate of transfer | $N_A/A = k_y(y_A - y^*_A) = k_x(x_{Ai} - x_A)$ | $q/A = h_h(T_h - T_{w2}) = h_C(T_{w1} - T_C)$ |
| Overall rate of transfer | $N_A/A = K_{oy}(y_A - y_A^*) = K_{ox}(x_A^* - x_A)$ | $q/A = U(T_h - T_C)$ |
| Equilibrium relation | $y^* = mx$ | — |

[a] Items listed are based on mole fraction concentration units. For conversion to other units, see Table 1.3.

$$\frac{AC}{CD} = \frac{(y_A - y_{Ai})}{-(x_{Ai} - x_A)} = -\frac{k_x}{k_y} \qquad (1.24a)$$

Similarly, we have for the local slope of the equilibrium curve

$$\frac{CE}{CD} = m \qquad (1.24b)$$

Now, the overall driving force $y - y^*$ is given by the sum of the two segments AC and CE so that from Equation 1.24a and Equation 1.24b

$$y_A - y_A^* = AC + CE = \frac{k_x}{k_y}(x_{Ai} - x_A) + m(x_{Ai} - x_A) \qquad (1.24c)$$

Further, since we had postulated an overall rate law of the form $N_A/A = K_{oy}(y_A - y_A^*)$, it follows from Equation 1.20 and Equation 1.24c that

$$N_A / A = K_{oy}(y_A - y_A^*) = K_{oy}\left(\frac{k_x}{k_L} + m\right)(x_{Ai} - x_A) = k_x(x_{Ai} - x_A) \qquad (1.25a)$$

Consequently,

$$K_{oy}\left(\frac{k_x}{k_y} + m\right) = k_x \tag{1.25b}$$

and therefore

$$\frac{1}{K_{oy}} = \frac{1}{k_y} + \frac{m}{k_x} \tag{1.25c}$$

This proves the validity of Equation 1.21a and Equation 1.21b. Similar arguments can be used to verify the validity of the liquid phase counterparts.

*Comments:*

It will be recalled that this entire development arose from the need to replace the interfacial concentrations in Equation 1.20, which are generally unknown, by some other known or measurable quantity. Inspection of the diagram in Figure 1.8 shows that there is only one such quantity for the gas phase, mainly $y_A^*$, the mole fraction in equilibrium with the bulk liquid concentration $x_A$. Hence, it was natural to replace the film driving force by an overall driving force $y_A - y_A^*$.

This was accomplished by making clever use of the diagram and Equation 1.20 to relate the segments AC and CE to $k_x$, $k_y$, and $m$, and to establish that the sum of the two equals the overall driving force $y_A - y_A^*$. Introduction of the rate law $N_A/A = K_{oy}(y - y^*)$ then culminated in the derivation of the law of additivity of resistances, Equation 1.22a and Equation 1.22b. Thus, it was possible to resolve, by a series of simple moves, a seemingly intractable problem.

We pause at this point to take stock of the principles established so far and to test our grasp of those principles with the following example.

### Illustration 1.6: Qualitative Analysis of Concentration Profiles and Mass Transfer

The reader is here confronted with a series of hypothetical concentration profiles near a gas–liquid interface, which are sketched in Figure 1.9. The task to be addressed is twofold: We wish to establish whether in each of these instances mass transfer does in fact take place and, if so, in which direction it will proceed. The main principle we have to apply is that mass transfer can only occur along a *negative concentration gradient*. With this fact firmly in mind, we can proceed as follows:

*Case 1:* Here the concentrations in both the gas and liquid phase diminish in the positive direction, causing solute to transfer from the gas to the liquid phase. The fact that the interfacial liquid mole fraction is higher than the gas concentration is no impediment. It is merely an indication of high gas solubility, a perfectly normal and acceptable phenomenon.

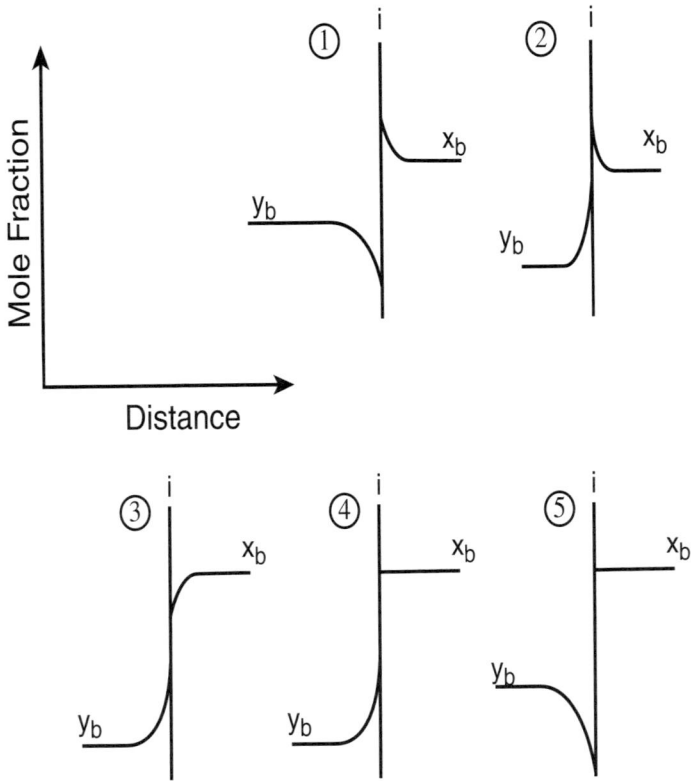

**FIGURE 1.9**
Hypothetical concentration profiles near a gas–liquid interface.

*Case 2:* The gas-phase concentration here increases in the positive direction so that no transfer of solute from gas to liquid can take place. Neither can there be any transfer in the opposite sense, because the liquid concentration rises in the negative direction. Such profiles arise only in cases when solute is generated by chemical reaction at the gas–liquid interface. The product solute then diffuses from the interface into the bulk fluids.

*Case 3:* This case involves decreasing concentrations in both phases, but the decrease is in the negative direction. Solute will therefore desorb from the liquid into the gas phase. Gas solubility is low because the interfacial concentrations are nearly identical.

*Case 4:* The flat liquid-phase profile indicates that the liquid phase is well stirred and shows no mass transfer resistance. Because the gas phase concentration diminishes in the negative direction, the transfer will be from liquid to gas.

*Case 5:* What was stated for Case 4 applies here as well, but the transfer this time is from the gas. This follows from the fact that gas-phase concentration decreases in the positive direction.

The question is now asked whether the results would still be the same if the gas-phase profiles had in each case been located *above* the liquid-phase counterparts. The answer is yes; transfer would still take place as indicated before. The only change would be in the equilibrium solubility of the gas, which would now be lower than before.

*Comments:*

The answers being sought here require a firm understanding of the principles involved. There were several pitfalls to be avoided on the way. We may be led to think, for example, that transfer from gas to liquid is not possible because of the higher level of $x_B$. That level, as we have seen, merely indicates a high equilibrium solubility of the gas but does not preclude movement from gas to liquid. A second pitfall lies in the tendency to focus on transfer in the positive direction only. Evidently, movement in the opposite direction is equally possible and should be kept in mind as an alternative.

We now make an abrupt departure from our theoretical deliberations to introduce the reader to the practical task of making rough estimates from a minimum of information. The need to do this arises quite frequently in a real-world context when time constraints do not allow elaborate calculations or a lengthy search for appropriate data to be made. This is often referred to as a "back-of-the-envelope calculation" and the associated problem as a "Fermi problem." In contrast to Illustration 1.5, the estimates here are based on personal observations and are therefore more subjective.

### Illustration 1.7: Drying with an Air Blower: A Fermi Problem

Readers are familiar with the use of warm-air blowers for the purpose of drying their hands. What we wish to do here is to estimate the associated mass transfer coefficient $k_C$ in units of m/s. To do this, the following approximate estimates will be needed: 1. the amount of moisture to be evaporated, 2. the drying time, and. 3. temperature and corresponding vapor pressure of the adhering moisture.

*1. Amount of Moisture.* Let us assume the thickness $L$ of the moisture to be 0.1 mm. This leads to a rate of evaporation of

$$N_A / A = \frac{A \times L \times \rho / M}{A \times t} \tag{1.26a}$$

where $A$ = surface area of both hands, $\rho$ = density of water, $M$ = molar mass of water, and $t$ = time. Note that the actual amount of water need not be calculated, only the average thickness of the water film.

2. *Drying Time.* A survey of various individuals who had made use of the device led to an agreed average value of $t = 100$ s. We then obtain from Equation 1.26a

$$N_A / A = \frac{(10^{-4}\,\text{m})(10^3\,\text{kg}/\text{m}^3)}{(18 \times 10^{-3}\,\text{kg}/\text{mol})(100\,\text{s})} = 0.056\,\text{mol}/\text{m}^2\text{s} \qquad (1.26b)$$

3. *Temperature and Vapor Pressure.* The moisture temperature may vary anywhere from 10 to 40°C, depending on whether hot or cold water was used in washing the hands, and the effect of heat transfer from the hot air as well as evaporative cooling. Let us assume that the temperature stabilizes at 25°C and make use of a rule that states that in the range 5 to 40°C, the value of the vapor pressure of water in millimeters of mercury equals that of the corresponding temperature in degrees centigrade. Thus, at 25°C, the vapor pressure, which we denote by $p^*$, is approximately 25 mmHg. If we assume water content in the hot air negligible, we obtain

$$N_A / A = k_C C^* = k_C \frac{p^*}{RT} \qquad (1.26c)$$

and hence

$$k_C = \frac{N_A / A}{p^* / RT} = \frac{0.056}{(20/760)10^5 / 8.314 \times 293} = 0.041\,\text{m}/\text{s} \qquad (1.26d)$$

This is in line with the order-of-magnitude estimates given in Illustration 1.5.

*Comments:*

How valid is the result? To answer this question, let us examine in turn the effect of a change in each of the variables.

1. Raising the temperature of the moisture to 40°C or lowering it to 10°C will merely change the result by a factor of two; i.e., the order of magnitude of the answer remains the same.

2. Water content of the air has a relatively minor effect. It would have to rise to a value of $\frac{1}{2}\,p^*$, i.e., to a relative humidity of 50% for the result to increase by a factor of two. More about the concept of humidity appears in Chapter 9, and it is seen there that for a given water content, the relative humidity drops rapidly with an increase in temperature. It is therefore reasonable to assume that at the temperature of the drying air, typically of the order of 50°C, this relative humidity level will be quite low.

3. The last variable to be examined is the average thickness $L$ of the moisture film, which we set at 0.1 mm. To see if this is reasonable,

let us calculate the corresponding mass of water held by each hand. Assuming a hand length of 15 cm and width of 10 cm, both sides will hold a water mass of $2 \times 15 \times 10 \times 0.01 \times 1 \text{ g/cm}^3 = 3 \text{ g H}_2\text{O}$ or 3 ml per hand. This is clearly a reasonable amount.

In summary, if all three values erred by a factor of two in the *same* direction, the result would change by a factor of $2^3 = 8$, not quite an order of magnitude. This is highly unlikely, and it is more reasonable to assume that the answer is correct to within a factor of two to three. We feel content with this.

The type of problem addressed here is often referred to as a Fermi problem, after the physicist and Nobel Laureate Enrico Fermi. Fermi used to regale his student audiences by showing them how to estimate the number of pianotuners in the city of Chicago. By using the same techniques that are applied above, the estimated answer usually came within a factor of two of the number of piano tuners listed in the local telephone book. Fermi was also responsible for estimating the yield of the first atomic bomb immediately after the explosion and long before the pertinent instrument recordings had been analyzed. He did this by dropping small pieces of paper into the path of the oncoming shock wave and measuring the distance over which they were entrained. This allowed him to calculate the velocity of the shock, from which he was able to deduce the energy produced by the explosion. His estimate came remarkably close to the actual value of 10,000 tons of TNT.

## 1.7 Conclusion

We note in closing that this chapter has presented the reader with a first look at the basic rate laws that govern mass transport. Chapters 3 and 4 address these notions in greater detail and use is made of the relations presented here in numerous subsequent situations.

## Practice Problems

1.1. *Gradient-Driven Processes*

   a. Do the flux relations listed in Table 1.1 apply to time-varying processes?

   b. Under what conditions is the concentration gradient in Fick's law a constant, and when does it become a variable?

   c. Under what conditions does the diffusivity become a variable?

d. The gradients that appear in Fick's law and Fourier's law, $dC/dx$ and $dT/dx$, are normally negative, because the potentials of $C$ and $T$ diminish in the direction of increasing values of $x$. Do these gradients ever become positive?

e. Under steady flow conditions in a cylindrical pipe, flow rate $Q$ and velocity $v_x$ in Poiseuille's law are constants, and hence so is the gradient $dp/dx$, which becomes $\Delta p/\Delta x$. This is the form commonly encountered in pipeline calculations. Can you envisage conditions that would lead to a *variable* gradient?

1.2. *More about Driving Forces and Transport Coefficients*

a. Would you expect the effective film thickness to increase or diminish with an increase in velocity of the flowing fluid?

b. Liquid-phase diffusivities are some four orders of magnitude smaller than the corresponding gas-phase diffusivities. Would you therefore expect all gas–liquid operations to be liquid-film controlled?

c. Gases with high solubility have a low slope $m$ of the equilibrium curve $p_A^* = f(C_A)$. Does this imply that the liquid phase driving force is small? (*Hint:* Consult Figure 1.8.)

d. Consider evaporation from a falling water droplet. Would you expect the local mass transfer coefficient to vary with angular position $\varphi$? If so, where would it be highest, and where would it be the lowest?

1.3. *Diffusion through a Stagnant Film*

a. In Figure 1.6b, the stagnant component $B$ exhibits a considerable concentration gradient. Why, then, is there no Fickian diffusion along it?

b. Show that the bulk-flow component $x_A N_A/A$ of the diffusing species is given by

$$x_A N_A / A = D_{AB} \frac{dC_B}{dz}$$

(*Hint:* Use the answer to Part a.)

c. Show that for dilute gas mixtures, i.e., low concentrations of the diffusing species $A$, the log-mean pressure difference $p_{BM}$ tends to the total pressure $P_T$.

1.4. *Conversion Factors for Mass Transfer Coefficients*
   Prove the relation:

   $$k_G{}' = k_G \frac{p_{BM}}{p_T}$$

   given in Table 1.4 and show that for dilute gases, $k_G{}' \doteq k_G$.

1.5. *Diffusional Concentration Profiles in a Spherical Geometry*
   Derive the concentration profile $C = f(r)$ in a spherical shell, which arises when a solute with uniform internal concentration $C_i$ diffuses through the shell to an external medium held at a fixed concentration $C_o$.

   $$\text{Answer: } C(r) = C_i - (C_i - C_o)\frac{(1 - r_i / r_o)}{(1 - r_i / r)}$$

1.6. *Ohm's Law Again*
   State under what conditions the Ohmian resistance, usually taken to be a constant, might become a variable for a conductor of fixed length.

1.7. *Effective Film Thickness Near a Controlled-Release Drug Delivery Device*
   In the conventional method of drug intake, the drug is administered either orally or by injection, with the expectation that the blood circulation will convey it to the site where it is required. This procedure is not the most efficient method of delivery because the drug is diluted by the blood of the entire body. This carries with it the risk of undesirable side effects if the drug concentration at the delivery site is to be high enough for optimal effectiveness. One method of overcoming these drawbacks is by implantation near the desired site of a wafer loaded with the medication. The drug is released at a controlled rate into the bloodstream and conveyed by it to the affected organ. *Consider the following case:* A particular implant has been designed to release the drug at a constant rate of 0.5 μg/cm² s. The drug has a solubility in blood of 10 g/l and a diffusivity of $10^{-6}$ cm²/s. The effective therapeutic concentration is one tenth of the solubility. What is the effective film thickness in the blood?

   *Answer:* 0.18 mm

1.8. *Mass Transfer through a Membrane*
   In the most general case of mass transfer through a permeable membrane, transport proceeds from an internal bulk concentration $C_{bi}$, through an internal film resistance, to an internal membrane concentration $C_{mi}$, then passes through the membrane of thickness $L$ to

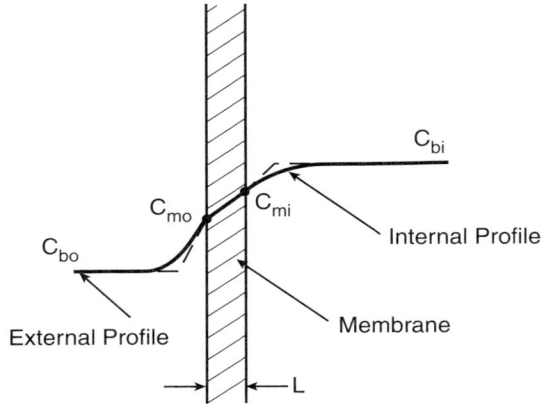

**FIGURE 1.10**
Mass transport through a membrane (Practice Problem 1.8).

an external membrane concentration $C_{mo}$ and from there through an external film resistance to its final destination, $C_{bo}$ (Figure 1.10). Note that concentrations are continuous; i.e., no interfacial equilibrium needs to be invoked. Using the principle of additivity of resistances, derive the following relation:

$$\frac{1}{K_{OC}} = \frac{1}{k_i} + \frac{1}{k_o} + \frac{L}{D_m}$$                                    (1.27)

where $K_O$ = overall mass transfer coefficient and $k_i$ and $k_o$ are the internal and external film coefficients.

1.9. *Mass Transfer between Ocean Waters and the Atmosphere*
    The following have been determined for the transport of carbon monoxide (CO) between ocean waters and air:

    Overall Mass Transfer Coefficient: $K_{OL}$ = 20 cm/h

    Equilibrium Constant $H$: = 62,000 atm/mol fraction

    Global Ocean Surface Area: $A$ = 3.6 × 10$^{18}$ cm$^2$

    Mean Concentration of CO in Air: 0.13 ppm by volume

    Mean Concentration of CO in Water: 6 × 10$^{-8}$ cc STP/cc H$_2$O

    a. In what direction is the transfer of carbon monoxide?
    b. What is the transfer rate in g/year? (*Hint:* Transform the carbon monoxide concentrations to units of mol/cm$^3$, and derive the equilibrium concentration $C^*$.)

    *Answer:* 4.5 × 10$^{13}$ g/year

1.10. *The Blood Coagulation Trigger*

Blood coagulation, which takes place at the site of an injury or in response to exposure to a foreign surface, is triggered by a series of enzymatic reactions, which culminate in the production of the enzyme thrombin. Thrombin is responsible for the formation of fibrin, which together with the platelets present in blood is a key ingredient of a blood clot. Most of these events take place at the contact site. Assume that in response to an event requiring blood coagulation, thrombin is produced in accordance with an overall first-order rate $k_r C$. Its concentration in the flowing blood can be taken as constant. Show that the likelihood of coagulation increases dramatically as $k_r$ approaches a value of $10^{-5}$ m/s. (*Hint:* Consult Illustration 1.5.) *Note:* To prevent coagulation, grafts are suitably modified chemically or coated to render them "inert."

# 2

# Modeling Mass Transport: The Mass Balances

Problems involving mass transport, and the solution and analysis of such problems, almost always require the formulation of a mathematical model of the process. The term model, as used here, refers to the equation, or set of equations, that describe the physical system or process under consideration. Such models were already encountered, at a modest level, in the solution of Illustration 1.2, Illustration 1.3, and Illustration 1.5. In all three of these examples, a single expression was applied to model the process, which consisted of the fundamental law of diffusion represented by Fick's law. No other basic equations were required, and we were able to proceed to a solution of the problem without invoking any additional principles.

This simple procedure, requiring only the application of a single and established expression, is the exception rather than the rule in problems involving mass transport. In the vast majority of cases we have to draw on additional tools to complete the mathematical formulation of the process. The tools required comprise various forms of the law of conservation of mass, supplemented by what we term *auxiliary relations*. These latter relations are largely empirical in nature and include the equations of transport seen in the previous chapter, as well as expressions describing chemical reaction rates and phase equilibria.

Not infrequently, we have to make use of more than one conservation law. Thus, if mass transport is accompanied by heat effects, we may have to invoke both the law of conservation of mass and the law of conservation of energy. We will see some examples of this dual case in Chapter 9. The law of conservation of momentum, on the other hand, is much more sparingly used in mass transport problems and can, for the purposes of this book, be set aside.

## 2.1   The Compartment and the One-Dimensional Pipe

The question that has to be addressed now is how to cast these laws and auxiliary relations into mathematical expressions, i.e., into mathematical models. We consider, for this purpose, the two physical entities shown in Figure 2.1, which are widely used to model transport of mass and energy, and to a lesser degree that of momentum. Figure 2.1a represents a well-stirred tank, also often referred to as a compartment. In the most general case, mass or energy flow by bulk movement into and out of the tank is generated or consumed by chemical reactions, or is exchanged with the surroundings. Within the tank or compartments, concentrations, temperature, and the physical properties in general are uniform; i.e., they vary at most with time, and not at all with distance. To apply the law of conservation of mass to the system, we take an inventory of the mass of a particular species, or make what is termed a mass or material balance. To do this, we argue that, for mass to be conserved, the difference between input and output must equal the change undergone by the tank contents. We include under *input* the mass generated by a chemical reaction within the tank, or that received from the surroundings, and under *output* the mass consumed by

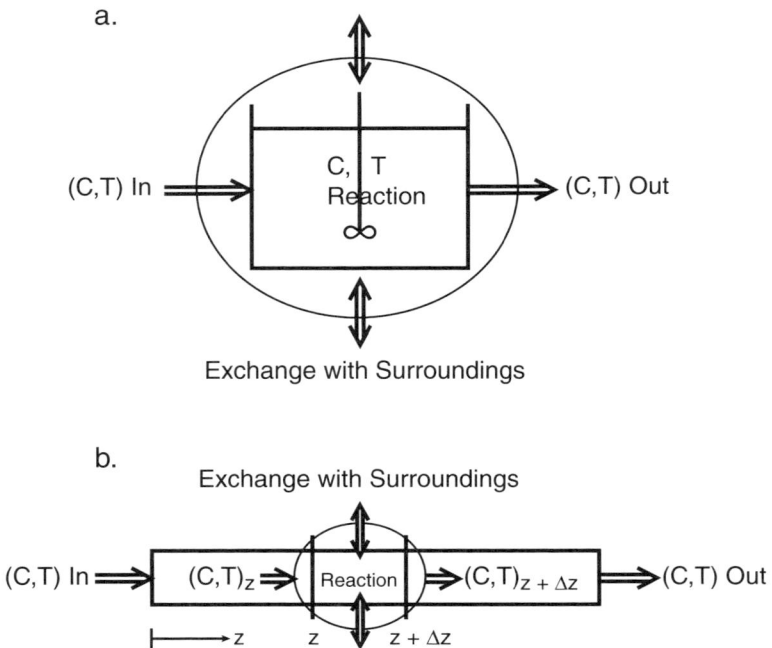

**FIGURE 2.1**
Two basic physical models: (a) the stirred tank or compartment; (b) the 1-D pipe.

reaction or that transferred to the surroundings. This leads to the general scheme:

$$\text{Rate of mass in} - \text{Rate of mass out} = \begin{matrix} \text{Rate of change} \\ \text{of mass in} \\ \text{the tank} \end{matrix} = \frac{d}{dt}(\text{mass}) \quad (2.1)$$

$$\begin{bmatrix} \text{by bulk flow,} \\ \text{by reaction,} \\ \text{by transfer} \\ \text{from the surroundings} \end{bmatrix} - \begin{bmatrix} \text{by bulk flow,} \\ \text{by reaction,} \\ \text{by transfer} \\ \text{to the surroundings} \end{bmatrix}$$

A similar procedure can be applied to the device we call a one-dimensional (1-D) pipe, shown in Figure 2.1b. In this configuration, the properties of the system are time invariant, but vary with distance in the direction of flow. The system is said to be at steady state and distributed in one spatial coordinate. This is the exact reverse of the conditions that prevailed in a compartment. The physical phenomena that take place in the two cases, however, are similar. Mass is again transported by bulk flow, enters and leaves the device by exchange with the surroundings, and is generated or consumed by chemical reaction. The only difference here is that mass can also enter and leave by diffusion, which was not the case in the compartmental model.

To obtain an expression for the steady-state distribution of the system variables, the mass balance must now be taken over an incremental element extending from $z$ to $z + \Delta z$. This is necessary to bring the distance variable into the model. The increment $\Delta z$ need not worry us here because we will ultimately allow it to go to zero, thereby transforming the original difference equation into the more familiar form of a differential equation. This leads to the following representation of the mass balance:

$$\text{Rate of mass in at } z - \text{Rate of mass out at } z + \Delta z = 0 \quad (2.2)$$

$$\text{and over } \Delta z \qquad \text{and over } \Delta z:$$

$$\begin{bmatrix} \text{by bulk flow and diffusion,} \\ \text{by reaction,} \\ \text{by transfer} \\ \text{from the surroundings} \end{bmatrix} - \begin{bmatrix} \text{by bulk flow and diffusion,} \\ \text{by reaction,} \\ \text{by transfer} \\ \text{to the surroundings} \end{bmatrix}$$

Expression 2.1 and Expression 2.2 are our principal starting tools for modeling mass transport. We make extensive use of them in the sections and chapters that follow. To convey to the reader a flavor of how these tools are applied in practice, we consider in the following two cases drawn from the field of environmental engineering. The first involves a compartmental model in which inflow and outflow of a substance take place, as well as its transfer to the surroundings (Figure 2.2a). The second illustration considers a 1-D distributed system, in which inflow and outflow are accompanied by the uptake of substance from its surroundings (Figure 2.2b). Both of these cases are frequently encountered as components of more elaborate environmental models.

### Illustration 2.1: Evaporation of a Solute to the Atmosphere

Consider a body of water such as a lake, which receives an inflow of $Q$ m³/s of water and discharges it at the same rate (Figure 2.2a). The volume of water in the lake is consequently constant. Dissolved in the intake is a pollutant, such as a pesticide, at a concentration level of $C_f$(kg/m³). The lake is initially devoid of any contaminant.

During its passage through the basin, which is assumed to be uniform in concentration, the pollutant is partially transferred to the atmosphere by evaporation and leaves at the concentration $C$ (kg/m³), which prevails in

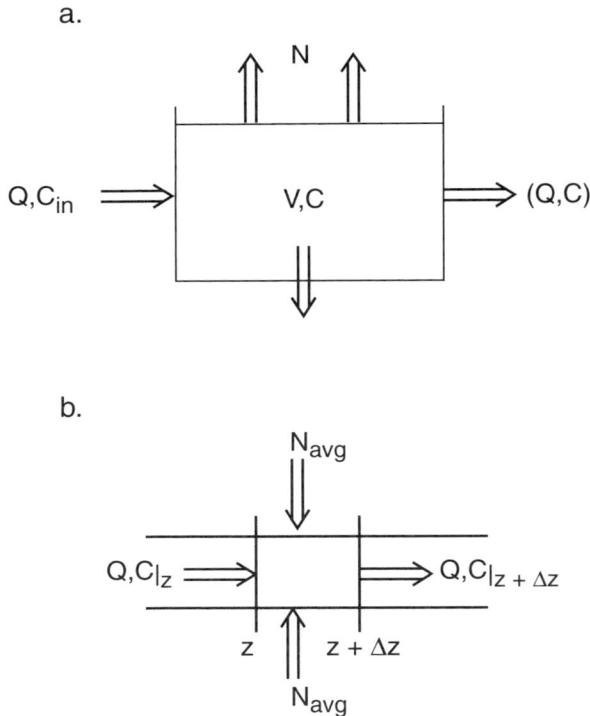

**FIGURE 2.2**
Two environmental models: (a) evaporation from a basin; (b) reaeration of a river.

the basin. The task here is to determine the time course of the pollutant concentration.

Drawing on the scheme represented by Equation 2.1 and assuming the basin contact to be well mixed, we write

$$
\text{Rate of pollutant in} - \text{Rate of pollutant out} =
\begin{array}{c}
\text{Rate of} \\
\text{Change of} \\
\text{Pollutant Content}
\end{array}
$$

$$
QC_f \quad - \quad [QC + N] \quad = \quad \frac{d}{dt}VC \tag{2.3a}
$$

If one assumes mass transfer to be controlled by the water phase, we can write

$$
N = K_{OL}A(C - C^*) \tag{2.3b}
$$

where $A$ = surface area (m$^2$), $K_{OL}$ = overall mass transfer coefficient (m/s), and $C^*$ is the pollutant concentration in the water, which is assumed to be in equilibrium with the concentration in the atmosphere (kg/m$^3$). Equation 2.3a now becomes

$$
(QC_f + K_{OL}AC^*) - (QC + K_{OL}AC) = V\frac{dC}{dt} \tag{2.3c}
$$

which yields after separating variables and formally integrating the result:

$$
\int_0^C \frac{VdC}{(QC_f + K_{OL}AC^*) - (Q + K_{OL}A)C} = \int_0^t dt \tag{2.3d}
$$

and consequently

$$
\frac{V}{(Q + K_{OL}A)}\ln\frac{QC_f + K_{OL}AC^*}{(QC_f + K_{OL}AC^*) - (Q + K_{OL}A)C} \tag{2.3e}
$$

This expression can be written in the equivalent exponential form

$$
C = \frac{QC_f + K_{OL}AC^*}{Q + K_{OL}A}\left[1 - \exp\left(-\frac{Q + K_{OL}A}{V}t\right)\right] \tag{2.3f}
$$

which gives us the desired final result, the time dependence of the pollutant concentration in the water basin.

*Comments:*

Before setting aside the solution of a model, it is always useful to verify its validity and to examine it for unusual features. This is done by first setting time $t = 0$. This yields $C = 0$ as required. Next we allow $t$ to go to infinity. This reduces the exponential term to zero and we obtain

$$C_{ss} = \frac{QC_f + K_{OL}AC^*}{Q + K_{OL}A} \qquad (2.3g)$$

Equation 2.3g represents the ultimate steady-state value of the pollutant concentration $C_{ss}$, which is attained after a long period of time. Note that this result is also obtained by setting the time derivative in Equation 2.3c equal to zero. The entire solution curve is shown in Figure 2.3a and demonstrates the asymptotic approach to steady-state conditions as time goes to infinity.

A third way of verifying the validity of the solution is to eliminate the evaporative term $N$ in Equation 2.3a or setting $K_{OL} = 0$. Equation 2.3f and Equation 2.3g then become

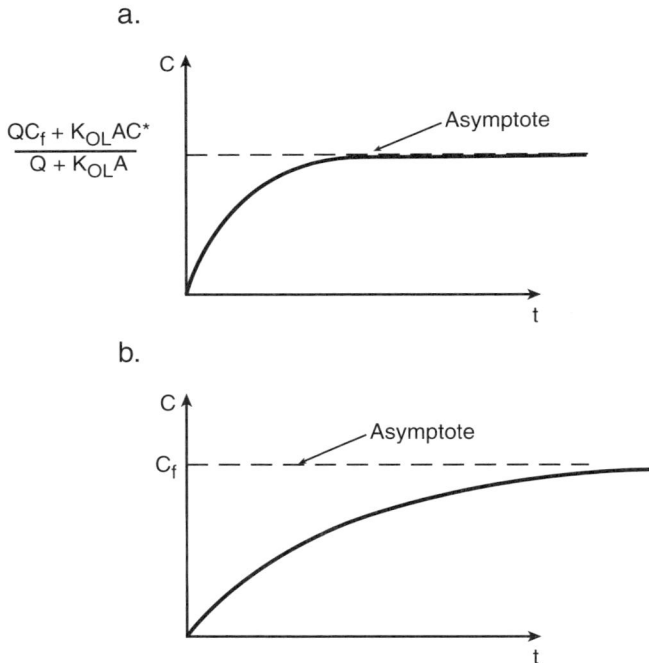

**FIGURE 2.3**
Time course of a pollutant concentration in a water basin: (a) with evaporative loss; (b) without evaporative loss.

$$C / C_f = 1 - \exp\left(-\frac{Q}{V}t\right) \tag{2.4a}$$

and for the steady state, $t \to \infty$,

$$C_{ss} = C_f \tag{2.4b}$$

These last expressions represent the response of pollutant concentration in the water basin to a jump increase in the inflow from $C = 0$ to $C = C_f$. Here, again, there is an initial rapid increase in pollutant levels, which diminishes with time and ultimately approaches the asymptotic steady-state value of $C = C_f$. This is shown in Figure 2.3b. The difference in the two responses lies principally in the time of approach to steady state. This can be shown by computing the time required to attain 95% of the respective steady-state values. For Figure 2.3a with evaporative loss we obtain by division of Equation 2.3f and Equation 2.3g

$$\frac{C}{C_{ss}} = 0.95 = 1 - \exp\left(-\frac{Q + K_{OL}A}{V}t_{0.95}\right) \tag{2.5a}$$

and consequently:

$$t_{0.95} = -\frac{V}{Q + K_{OL}A}\ln 0.05 = 3.0\frac{V}{Q + K_{OL}A} \tag{2.5b}$$

For Figure 2.3b (no evaporative loss) we obtain by setting $K_{OL} = 0$:

$$t_{0.95} = 3.0V/Q \tag{2.5c}$$

The result shows that the approach to steady state is faster in Figure 2.3a with evaporation loss than when there is none. This was not immediately anticipated on intuitive grounds and demonstrates the power of models to reveal the unexpected, or at least resolve uncertainties.

Equation 2.5b and Equation 2.5c reveal another interesting feature of system behavior. To show this, we recast the equations in the form:

Figure 2.3a

$$\frac{Qt_{0.95}}{V} = \frac{3.0}{1 + K_{OL}A/Q} \tag{2.5d}$$

Figure 2.3b

$$\frac{Qt_{0.95}}{V} = 3.0 \qquad\qquad (2.5e)$$

Because $Qt_{0.95}$ equals the total volume that has passed through, it is apparent that the ratio $Qt_{0.95}/V$ represents the number of volume changes, or turnover, undergone by the basin in the course of attaining 95% of the steady-state pollutant level. For Figure 2.3b, three volume changes are required whereas for Figure 2.3a the turnover number is less than 3. Depending on the size of the basin, this may take from days to weeks or even years.

A final question to be considered is the following: How realistic is the model we have just presented in an environmental or "real-world" context and what are its limitations, if any? We start by noting that the model assumes uniform distribution of the pollutant within the basin. No account is taken of the existence of dead-water zones where the rise in pollutant level will be slow, or of portions of the flow that may proceed rapidly and with little loss to the atmosphere from inlet to outlet, in effect short-circuiting the mixing process. The best we can expect from the model, therefore, is that it will give us the course in time of the *average* concentration in the basin, ignoring local highs and lows. This is nevertheless useful information to have, as it provides us with a semiquantitative time frame for the contamination process.

A second limitation resides in the fact that we have confined pollutant losses to those that occur by evaporation to the atmosphere. This ignores the role played by bottom sediments as well as solids suspended in the water of the basin in removing solute by *adsorption*, and of possible biodegradation of the solute by bacterial action. These are important mechanisms that add to the loss incurred by evaporation and have to be taken into account in comprehensive models of pollutant fate. By ignoring these processes, we have in effect set an upper limit to the pollutant concentration in the water. In other words, things will not be as bad as our model predicts, at least as far as the aqueous phase is concerned, because a good deal of the pollutant may disappear as a result of adsorption and biodegradation.

We turn in the next illustration to an examination of the reverse process, that of uptake of a substance of a body of water. The system considered differs from the basin examined previously in two important ways: First, the substance in question is a benign one, namely, oxygen, which is taken up by the water from the atmosphere. This process is termed *reaeration*, and is a highly desirable one as it helps maintain aquatic life and aids in the biodegradation of objectionable substances. Second, the body of water is assumed to be in steady flow, such as a river. The system is no longer considered to be well mixed except in the vertical direction and we can therefore expect a steady increase in oxygen concentration in the direction of flow. This calls for the use of what we termed a 1-D pipe model, i.e., one in which the mass balance is performed over an increment $\Delta z$ in the direction of flow (see Figure 2.1b). There are no variations with time, and distance $z$ now becomes the independent variable. The increment over which the mass balance is to be performed is shown in Figure 2.2b, and the expression we use is Equation 2.2. Let us see how this works out in practice.

### *Illustration 2.2: Reaeration of a River*

Consider a river that has, at a point we refer to as $z = 0$, an oxygen content $C_o$, which is below the saturation solubility of oxygen. This means that oxygen concentration will steadily increase in the direction of flow due to uptake from the atmosphere and will ultimately, as $z \to \infty$, attain the equilibrium solubility of oxygen, $C^*$. Application of Equation 2.2 yields, in the first instance

$$\begin{matrix} \text{Rate of O}_2 \text{ in at } z \\ \text{and over } \Delta z \end{matrix} \ - \ \begin{matrix} \text{Rate of O}_2 \text{ out at } z + \Delta z \\ \text{and over } \Delta z \end{matrix} = 0$$

$$(QC_z + N_{avg}) \qquad - \qquad (QC_{z+\Delta z}) = 0 \qquad (2.6a)$$

If we assume the liquid-phase resistance to be controlling, $N_{avg}$ will be of the form

$$N_{avg} = K_{OL}A(C^* - C)_{avg} \qquad (2.6b)$$

where $C^*$ is the aqueous oxygen concentration that would be in equilibrium with the atmosphere. Note that we have subscripted the driving force $(C^* - C)$ with the term "avg" (average) to denote that $C$ varies over the increment $\Delta z$, and that one must consequently apply an average value of $C^* - C$ over that distance. We need not concern ourselves with the exact magnitude of this term since it will ultimately, when we go to the limit $\Delta z \to 0$, shrink to a point quantity $(C^* - C)$, which is everywhere well defined.

A second question concerns the definition of the interfacial area $A$. It is relatively rare to find that this quantity coincides with the area of a flat surface. More commonly, the river has surface ripples and waves of an unknown interfacial area. The value becomes particularly uncertain in the case of rapids in which air is entrained into the water phase in the form of bubbles or foam, again, of an unknown interfacial area. It has become common practice in these cases to lump this unknown factor into the mass transfer coefficient in the form of an average specific area "$a$" with units of m² interfacial area/m³ river volume. We write

$$N = K_{OL}a(C^* - C)_{avg} \text{ (incremental river volume)} \qquad (2.6c)$$

or

$$N = K_{OL}a(C^* - C)_{avg}A_C\Delta z \qquad (2.6d)$$

where $A_C$ is the cross-sectional area of flow of the river, and $K_{OL}a$ is the so-called *volumetric mass transfer coefficient* with units of reciprocal time (s$^{-1}$). The mass balance (Equation 2.6a) then becomes

$$QC_z - QC_{z+\Delta z} - K_{OL}a(C^* - C)_{avg}A_C\Delta z = 0 \qquad (2.6e)$$

Dividing by $A_C\Delta z$, and letting $\Delta z$ go to zero, the difference quotient $\Delta C/\Delta z$ is converted into a derivative and we obtain

$$Q/A_C\frac{dC}{dz} + K_{OL}a(C^* - C) = 0 \qquad (2.6f)$$

Equivalently, since $Q/A_C$ equals the river flow velocity $v$, we can write

$$v\frac{dC}{dz} + K_{OL}a(C^* - C) = 0 \qquad (2.6g)$$

Integrating by separation of variables yields

$$\int_{Co}^{C}\frac{dC}{C^* - C} = \frac{K_{OL}a}{v}\int_{0}^{z}dz \qquad (2.6h)$$

and hence

$$\ln\frac{C^* - C_o}{C^* - C} = \frac{K_{OL}a}{v}z \qquad (2.6i)$$

Solving for $C$ leads to the final exponential oxygen distribution

$$C(z) = C^* - (C^* - C_o)\exp\left(-\frac{K_{OL}a}{v}z\right) \qquad (2.6j)$$

*Comments:*
Let us start by verifying the validity of this expression. For $z \to 0$, the equation reduces to $C = C_o$, in agreement with the inlet condition we had specified. For $z \to \infty$, the exponential term vanishes and we obtain $C = C^*$;

i.e., the river is fully saturated with oxygen and is in equilibrium with the atmospheric air. This is again as it should be.

Let us next consider the case where the river is initially devoid of all oxygen, i.e., $C_o = 0$. Equation 2.6j then reduces to the equation

$$C/C^* = 1 - \exp\left(-\frac{K_{OL}a}{v}z\right)$$ (2.6k)

This expression is identical in form to Equation 2.4a and one can therefore expect the oxygen distribution to be of the form shown in Figure 2.3b, with distance $z$ replacing time $t$, and the equilibrium solubility $C^*$ taking the place of the inlet concentration to the basin, $C_f$.

A number of underlying assumptions need to be noted. First, it was assumed that the river cross section, and hence its flow velocity, was constant. In general, there will be some variation in these factors, which can, however, be easily incorporated into the model provided their functional dependence on distance is known or can be estimated. One merely has to recast Equation 2.6h into the form

$$\int_{C_o}^{C} \frac{dC}{C^* - C} = K_{OL}a \int_0^1 v(z)dz$$ (2.6l)

Second, we neglected any consumption of oxygen in the river, i.e., the use of oxygen by aquatic life and its consumption in the biodegradation of dissolved substances. These two processes will act to delay the reaeration process so that the result (Equation 2.6j) is to be regarded as a lower threshold value. The consumption of oxygen by the above processes can be incorporated without undue difficulty into the reaeration model presented above.

## 2.2 The Classification of Mass Balances

The reader will have noted from the foregoing illustrations that there exist a number of different types of mass balances, which are dictated by the type of process under consideration. Thus, in Illustration 2.1, the mass balance was taken over a finite entity, the water basin, and the balance space in Illustration 2.2 was an incremental quantity $\Delta z$, which ultimately shrunk to a point in space. A further distinction is the appearance of time in Illustration 2.1 and an absence of spatial coordinates. The reverse was the case in Illustration 2.2. Clearly, both time and spatial geometry determine the type of mass balance that has to be made.

### 2.2.1   The Role of Balance Space

The space over which a mass balance is taken generally falls into two categories.

First, the space is finite in size. This occurs when the balance is taken over a finite entity such as a tank or compartment, a finite length of pipe, a column, or a sphere. We speak of the balance as being an "integral" or a "macroscopic" balance. Balances involving compartments are invariably of this type and lead to either algebraic equations (AE) or ordinary differential equations (ODE).

Second, the balance is taken over an incremental space element, $\Delta x$, $\Delta r$, or $\Delta V$. The mass balance equation is then divided by these quantities and the increments allowed to go to zero. This reduces the difference quotients to derivatives and the mass balance now applies to an infinitesimal point in space. We speak in this case of a "difference" or "differential" balance, or alternatively of a "microscopic" or "shell" balance. Such balances arise whenever a variable such as concentration undergoes changes in space. They occur in all systems that fall in the category of the device we termed a 1-D pipe (Figure 2.1b). When the system does not vary with time, i.e., is at steady state, we obtain an ODE. When variations with time do occur, the result is a partial differential equation (PDE) because we are now dealing with two independent variables. Finally, if we discard the simple 1-D pipe for a multidimensional model, the result is again a PDE.

### 2.2.2   The Role of Time

When a process is time dependent, we speak of "unsteady," "unsteadystate," or "dynamic" systems and balances. If, on the other hand, there are no variations with time, the process is said to be at steady state. There are several categories of unsteady balances depending on the balance space and time framework we use. They are summarized below.

#### 2.2.2.1   *Unsteady Integral Balance*

Here the balance is taken over a well-stirred tank or compartment such as the water basin we considered in Illustration 2.1. To describe the process, we use the scheme Rate in – Rate out = $d/dt$ contents, which was given by Equation 2.1. It follows from this expression that all unsteady integral balances lead to first-order ODEs. When the time derivative in Equation 2.1 is zero, the system reverts to a steady state and the result is an algebraic equation (AE).

#### 2.2.2.2   *Cumulative (Integral) Balance*

This is a special type of balance involving time as a variable, which is rarely, if ever, singled out for discussion in textbooks. It consists of considering a finite time interval $(0, t)$ and making an inventory using the mass of a species present originally at time $t = 0$, the mass consumed, or lost to the surroundings over the time interval $(0, t)$, and that left over at the end. We put this formally as follows:

$$\text{Mass initially present } (t = 0) = \text{Mass consumed or lost over } (0, t)$$
$$+ \text{ Mass left over at } t \qquad (2.7)$$

Cumulative balances invariably lead to algebraic equations. Paradoxically and in spite of the occurrence of time in the statement (Equation 2.7), they may arise in both steady- and unsteady-state processes. We shall have occasion to demonstrate this in a number of illustrations and practice problems.

### 2.2.2.3 Unsteady Differential Balances

It has previously been noted that, when the 1-D pipe model is applied to an unsteady process, the mass balance will lead to a PDE. We generalize this into the following statements: Whenever a variable is distributed in both time and distance, i.e., varies with $t$ and $x$ ($y, z$), the resulting mass balance will be a PDE.

We can summarize the results of the foregoing discussion as follows. A mass balance can be time-dependent or time-independent, and it can be applied over a finite entity or a differential increment of either time or distance. It can further depend on a single space variable, or several such variables, or none at all. Depending on which combination of factors applies to a particular problem, this will result in an AE, or an ODE, or a PDE. These features are summarized for the convenience of the reader in Table 2.1.

### 2.2.2.4 Dependent and Independent Variables

Before concluding our discussion, we remind the reader of the distinction that has to be made between dependent and independent variables. This

**TABLE 2.1**

Categories of Balances and the Resulting Equations

| Type of Balance | Equation |
|---|---|
| *A. Integral Balances* | |
| 1. Steady-state balance | AE |
| 2. Unsteady balance | ODE |
| 3. Cumulative balance | AE |
| *B. Differential Balances* | |
| 1. Steady-state 1-D balance | ODE |
| 2. Unsteady 1-D balance | PDE |
| 3. Steady-state multidimensional balance | PDE |
| 4. Unsteady multidimensional balance | PDE |

necessity arises in the context of differential equations, which for ODEs generally lead to solutions of the form:

$$u = f(x) \tag{2.8a}$$

or

$$u = f(t) \tag{2.8b}$$

In the case of PDEs and rectangular coordinates,

$$u = f(x, y, z) \tag{2.8c}$$

$$u = f(x, t)$$

and

$$u = f(x, y, z, t)$$

Here $u$ is the dependent variable and $x$, $y$, $z$, and $t$ are the independent variables.

The dependent variable — also termed a *state variable* — is, for mass transfer operations, usually represented by the concentration of a system, or its total mass. Concentration can be expressed in a variety of ways, the most common of which is $kg/m^3$ or $mol/m^3$, or in terms of mole and mass fractions and ratios, whereas total mass is represented in terms of kg or mol. The reader is reminded that the number of dependent variables equals the number of unknowns. For a system to be fully specified, the number of equations must therefore equal the number of unknowns, in other words, the number of dependent variables. The model is then said to be complete.

The independent variables are usually represented by time $t$ and distance $x$, $y$, $z$, or, in the case of radial coordinates, by the radial distance variable $r$. Occasionally and paradoxically, distance may depend on time and then becomes the dependent variable. This is the case, for example, with spherical particles, which undergo a change in size due to reaction, dissolution, or deposition of material. The attendant change in mass is then expressed by the derivative

$$\text{Rate of change of mass} = \frac{dm}{dt} = \rho_p \frac{d}{dt} 4 / 3 \pi r^3 = \rho_p 4 \pi r^2 \frac{dr}{dt} \tag{2.9}$$

where distance $r$ is now the dependent variable.

Another departure from the normal definition of variables occurs when two first-order differential equations are combined by division into a single ODE.

Consider, for example, the system

$$\frac{du}{dt} = f(u,v) \qquad\qquad (2.10a)$$

$$\frac{dv}{dt} = g(u,v) \qquad\qquad (2.10b)$$

which may be the result of two unsteady integral balances.

Division of the two equations leads to the result

$$\frac{du}{dv} = \frac{f(u,v)}{g(u,v)} \qquad\qquad (2.10c)$$

which, given suitable forms of $f$ and $g$, can be integrated by separation of variables. In the process of dividing the two equations (Equation 2.10a and Equation 2.10b) we have eliminated the independent time variable $t$ and replaced it by the former dependent variable $v$. Such transformations are frequently used to obtain partial solutions of systems of simultaneous ODEs.

### Illustration 2.3: The Countercurrent Gas Scrubber: Genesis of Steady Integral and Differential Mass Balances

Gas scrubbers are widely used devices designed to remove impurities or recover valuable substances from gases by contacting them with a suitable solvent, such as water. A gas scrubber typically consists of a cylindrical shell filled with plastic or ceramic particles designed to enhance the contact area between the two phases (Figure 2.4a and Figure 2.4b). Solvent enters the column at the top and trickles down through the packing where it contacts the gas phase, which enters the scrubber at the bottom and flows upward countercurrent to the solvent stream. The purified gas stream leaves the column at the top while used solvent containing the impurity exits at the bottom.

In the present illustration we use the gas scrubber as a vehicle to demonstrate the genesis of steady-state integral and differential balances. Four such balances are shown in Figure 2.4, with the balance space indicated by an envelope drawn around it. Solvent enters the envelope with a flow rate $L_s$ (kg solvent/s) and a solute concentration $X_2$ (kg solute/kg solvent) and leaves with the same solvent flow rate but an increased concentration $X_1$. For the gas stream, the corresponding quantities are $G_s$ (kg carrier gas/s) and $Y_2$, $Y_1$ in units of kg solute/kg carrier gas. The term *carrier* denotes the gaseous component, which is not absorbed by the solvent. Typically, that component is air or some other inert gas, such as hydrogen or nitrogen.

We start by considering the two integral balances shown in Figure 2.4a and Figure 2.4b.

The balance space in these two cases is a finite one, consisting either of a part of the column (Figure 2.4a) or the entire column itself (Figure 2.4b). A solute balance for the two cases then leads to the following expression:

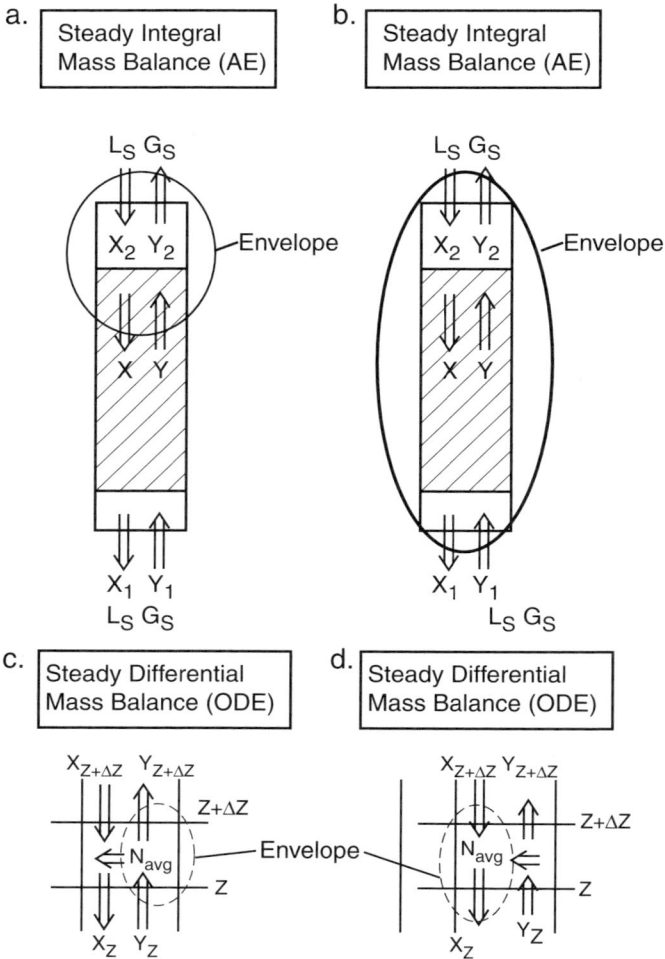

**FIGURE 2.4**
The packed gas absorber: (a), (b) types of mass balances leading to AE; (c), (d) types of mass
balances leading to ODE.

<u>Figure 2.3a</u>

$$\text{Rate of solute in} - \text{Rate of solute out} = 0$$

$$(L_s X_2 + G_s Y) - (L_s X + G_s Y_2) = 0 \tag{2.11a}$$

<u>Figure 2.3b</u>

$$\text{Rate of solute in} - \text{Rate of solute out} = 0$$

$$(L_s X_2 + G_s Y_1) - (L_s X_1 + G_s Y_2) = 0 \tag{2.11b}$$

These two expressions are processed further by casting them in the following difference form

$$\frac{Y - Y_2}{X - X_2} = \frac{L_s}{G_s} \tag{2.11c}$$

and

$$\frac{Y_1 - Y_2}{X_1 - X_2} = \frac{L_s}{G_s} \tag{2.11d}$$

We return to these expressions, and those that follow, for a more detailed examination in Chapter 8, Illustration 8.1. It is shown there that Equation 2.11c and Equation 2.11d represent the so-called operating lines, which are used in the graphical representation of scrubber performance. For our present purposes we limit ourselves to the observation that these balances resulted in algebraic equations, as predicted and stipulated in Table 2.1. We note further that neither Equation 2.11c nor Equation 2.11d contains the distance variable $z$. They can therefore tell us nothing about the concentration variations as a function of column height nor help us establish the size of column required to effect a reduction of solute content from $Y_1$ to $Y_2$.

We must, for those purposes, turn to differential balances which upon integration will yield the desired functional dependence of solute concentration as column height, i.e., $Y = f(z)$. Two such balances are sketched in Figure 2.4c and Figure 2.4d. To set up the corresponding equations, we write for the case of the gas-phase differential balance (Figure 2.4c):

Rate of solute in − Rate of solute out = 0

$$(G_s Y_z) - (G_x Y_{z+\Delta z} + N_{avg}) = 0 \tag{2.12a}$$

For the liquid-phase balance (Figure 2.4d):

$$(L_s X_{z+\Delta z} + N_{avg}) - (L_s X_z) = 0 \tag{2.12b}$$

We must next formulate an expression for the mass transfer rate $N_{avg}$ and here we encounter the same difficulty we had seen in Illustration 2.2 dealing with the reaeration of rivers. In both cases the interfacial area is unknown and we must therefore resort again to the use of a volumetric mass transfer coefficient $K_o a$ where the unknown interfacial area $a$ ($m^2/m^3$ column volume) is lumped together with $K_o$. If the gas phase is assumed to be controlling, we can write

$$N_{avg} = - K_{oY} a (Y - Y^*)_{avg} A_C \Delta z \tag{2.12c}$$

where $A_C$ is the column cross-sectional area and $Y^*$ is the gas-phase solute content in equilibrium with the liquid-phase concentration $X$.

Introducing this expression into Equation 2.12a and Equation 2.12b, dividing by $A_C \Delta z$, and letting $\Delta z$ go to zero, we obtain the twin result

$$G_s \frac{dY}{dz} + K_{oY}a(Y - Y^*) = 0 \qquad (2.12d)$$

and

$$L_s \frac{dX}{dz} + K_{oY}a(Y - Y^*) = 0 \qquad (2.12e)$$

where $G_s$ and $L_s$ are now the mass *velocities* of carrier gas and solvent, respectively, with units of kg/s m² column cross section.

Since the two equations, 2.12d and 2.12e, contain three dependent variables, $X$, $Y$, and $Y^*$, we require a third relation to complete the model. This will be given by the equilibrium relation, which can be written in the general form

$$Y^* = f(X) \qquad (2.12f)$$

The model, consisting of the three equations (Equation 2.12d, Equation 2.12e, and Equation 2.12f), can now be said to be complete. We return to it in Illustration 8.1 where the solutions to this model are taken up. We also examine the role of the integral balances (Equation 2.11c and Equation 2.11d), which, at the moment at least, are seemingly adrift with no apparent use in modeling the system.

*Comments:*

This example was intended to draw the reader's attention to the multitude of mass balances that can be applied even in cases of only modest complexity. In fact, there are three additional balances that can be performed on the scrubber:

<u>Differential mass balance over both phases</u>

$$\text{Rate of solute in} - \text{Rate of solute out} = 0$$

$$(L_s X_{z+\Delta z} + G_s Y_z) - (L_s X_z + G_s Y_{z+\Delta z}) = 0 \qquad (2.12g)$$

<u>Integral mass balance over the gas phase</u>

$$\text{Rate of solute in} - \text{Rate of solute out} = 0$$

$$(G_s Y_1) - (G_s Y_2 + N_{\text{Tot}}) = 0 \qquad (2.12h)$$

Integral mass balance over the liquid phase

Rate of solute in – Rate of solute out = 0

$$(L_s X_2 + N_{Tot}) - (L_s X_1) = 0 \tag{2.12i}$$

This profusion of balances will have made it clear that the choice of the proper balances and balance space is often not a straightforward one and calls for good judgment or leads to some trial-and-error work.

### Illustration 2.4: Two Examples from Biology: The Quasi-Steady-State Assumption

In simple biological models the human or animal body is assumed to be composed of a number of well-stirred compartments of uniform concentration. These compartments encompass body fluids (plasma, intercellular fluids) as well as body tissues such as fat, muscle, and bones, and are described by one or more first-order ODEs in time. The first example taken up below is of this type, and is depicted in Figure 2.5a.

A second simple model assumes that the concentrations in one or more of the compartments are distributed in one direction but vary only slowly with time so that the time derivative of the concentration is very small and can be neglected. This is referred to as the *quasi-steady-state assumption* and leads to an ODE in distance. The second example taken up below belongs to this category and is shown in Figure 2.5b.

### 1. Two Well-Mixed Compartments in Instantaneous Equilibrium

The first example considered involves a "well-mixed" tissue region surrounding a blood vessel whose contents are likewise taken to be well mixed and uniform in concentration. Blood entering and leaving the heart muscle and conveying to it a drug dissolved in the blood can be considered representative of this type of situation. If passage of the solute into the muscle tissue is rapid enough, the two phases may be taken to be at equilibrium at all times. It is common at the low concentrations involved to assume a linear equilibrium relation of the form

$$C_B = K C_T \tag{2.13a}$$

where $K$ is termed the partition coefficient. The physical configuration and pertinent variables for the present case are displayed in Figure 2.5a. We proceed to write an integral mass balance over the entire system, as shown in the figure, and obtain

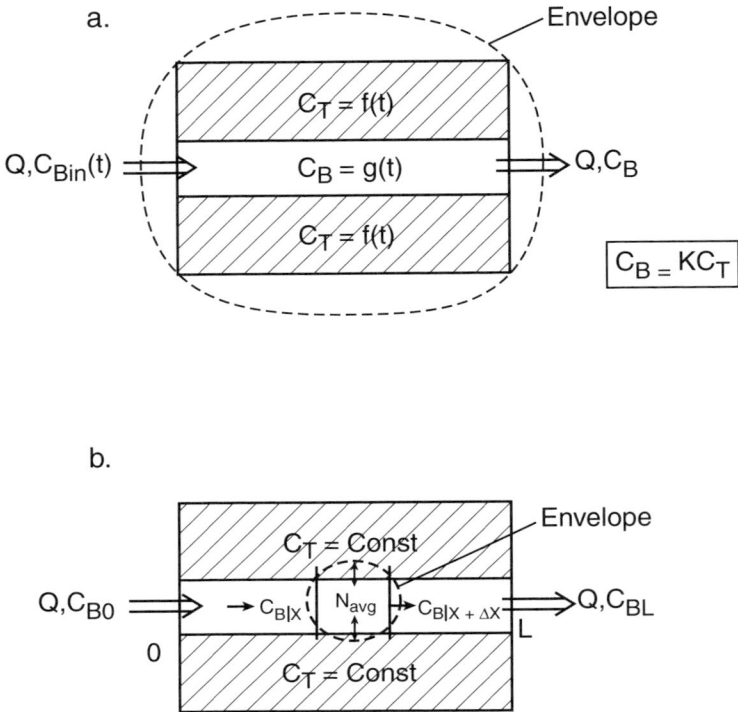

a.

b.

**FIGURE 2.5**
A well-mixed tissue region surrounding a blood vessel: (a) uniform concentration in blood; (b) concentration in blood distributed and in quasi-steady state.

$$
\text{Rate of solute in} - \text{Rate of solute out} = \begin{array}{c} \text{Rate of} \\ \text{change of} \\ \text{solute contents} \end{array}
$$

$$
Q\,C_{Bin}(t) - Q\,C_B = V_T\frac{dC_T}{dt} + V_B\frac{dC_B}{dt} \tag{2.13b}
$$

where $C_{B\,in}(t)$ is in general an arbitrary function of time, $Q$ = volumetric flow rate in cm$^3$/s, and $V_B$ and $V_T$ are the blood and tissue volumes, respectively (cm$^3$). There are two equations (2.13a and 2.13b) in the two dependent variables or unknowns $C_B$ and $C_T$. The model is consequently complete.

Substitution of Equation 2.13a into Equation 2.13b yields a single expression in either blood or tissue concentrations, which can be integrated in a straightforward fashion by separation of variables. We obtain for $C_B$ in the first instance

$$QC_{Bin} - QC_B = \frac{(V_T + V_B)}{K}\frac{dC_B}{dt} \tag{2.13c}$$

For the case where a drug is suddenly introduced at a constant rate into the blood, i.e., $C_{B\,in}$ goes from zero to a constant value $C_{Bo}$ at $t = 0$, integration results in

$$\int_0^{C_B} \frac{dC_B}{C_{Bo} - C_B} = \frac{V_T + V_B}{KQ}\int_0^t dt \tag{2.13d}$$

and consequently

$$-\ln\frac{C_{Bo} - C_B}{C_{Bo}} = \frac{V_T + V_B}{KQ}t \tag{2.13e}$$

or equivalently in exponential form

$$C_B = C_{Bo}\left[1 - \exp\left(-\frac{QKt}{V_B + V_T}\right)\right] \tag{2.13f}$$

*Comments:*
The validity of the solution can be quickly established by letting $t$ go to zero and infinity, respectively. For the former case we obtain $C_B = 0$, as required, and for the latter $C_B = C_{Bo}$; i.e., the concentration in the blood asymptotically approaches the constant inlet value $C_{Bo}$. This behavior is identical to that which was seen in Illustration 2.1: Evaporation of a Solute to the Atmosphere, which shows the same asymptotic approach to a constant value, as displayed in Figure 2.3. When evaporative loss is omitted, leading to Equation 2.4a, the resemblance is even more pronounced, with only a slight change of variables, from $V$ and $Q$ in Equation 2.4a to $V_B + V_T$ and $KQ$ in Equation 2.13f bringing about complete coincidence. Such a coincidence of solution forms for different physical situations is a frequent occurrence in modeling but does not come as a complete surprise because the original ODEs are in these cases also identical in form. The lesson for the reader here is that if the solution of a model appears to be intractable, we should look in other disciplines that yield identical model equations for the possible existence of a ready-made solution. This is both convenient and relieves us of the onus of having reinvented the wheel.

A second comment concerns the use of the model solutions, Equation 2.13e and Equation 2.13f. We start by noting that a semilog plot of Equation 2.13e, $\ln(C_{Bo} - C_B)$ vs. $t$, using measured values of the drug concentration $C$, yields a slope with the value $(V_T + V_B)/KQ$. In many cases of interest, the blood volume is much smaller than that of the surrounding tissue, $V_T \gg V_B$, so that the slope becomes $V_T/KQ$. The flow rate $Q$ is usually known from independent measurements (see Practice Problem 2.6), while the partition coefficient can often be determined *in vitro*, i.e., in the laboratory using extracted tissue. The value of the slope can then be used to determine the unknown tissue volume $V_T$. Conversely, if tissue volume is known from independent measurements, the slope will yield a value for the partition coefficient $K\alpha$. Note that the tissue concentration is then given by

$$C_T = C_B / K = (C_{Bo} / K)\left[1 - \exp\left(-\frac{QKt}{V_B + V_T}\right)\right] \qquad (2.13g)$$

Once these parameters have been established from appropriate experiments, we can use Equation 2.13g to establish the time required to attain a desired therapeutic concentration.

2. *A Well-Mixed Tissue Compartment in Contact with Flowing Blood with a Varying Concentration*

The case in question is sketched in Figure 2.5b, and involves the exchange of a substance between a tissue region of uniform concentration and blood flowing in a capillary along which concentration varies. The membrane separating the two has a transport resistance represented by the mass transfer coefficient $k_m$, which equals diffusivity divided by the thickness of the membrane. Let us consider first the case of substance uptake by the tissue. If that substance enters the capillary at a flow rate $Q$, we can write for the difference element shown in Figure 2.5b

Rate of solute in − Rate of solute out = 0

$$QC_B\big|_z - (QC_B\big|_{z+\Delta z} + N_{avg}) = 0 \qquad (2.14a)$$

or, using an explicit expression for $N_{avg}$ and setting $C_B^* = KC_T$, as given by Equation 2.13a,

$$QC_B\big|_z - QC_B\big|_{z+\Delta z} - k_m P\Delta x(C_B - KC_T) = 0 \qquad (2.14b)$$

where $P$ = perimeter of capillary. Here, use has been made of the quasi-steady-state assumption; i.e., the process was taken to be slow enough that over a short finite period of observation (say, a few minutes) neither the tissue nor the blood concentrations vary significantly.

Dividing by $\Delta x$ and going to the limit $\Delta x \rightarrow 0$ we obtain the ODE

$$Q\frac{dC_B}{dz} + k_m P(C_B - \alpha C_T) = 0 \tag{2.14c}$$

Integration by separation of variables then leads to

$$\int_{C_{Bo}}^{C_B} \frac{dC_B}{C_B - KC_T} = -\frac{k_m P}{Q}\int_0^L dz \tag{2.14d}$$

and consequently

$$\ln\frac{C_B - KC_T}{C_{Bo} - KC_T} = -\frac{k_m A}{Q} \tag{2.14e}$$

or alternatively

Uptake: $\qquad \dfrac{C_B - KC_T}{C_{Bo} - KC_T{}^0} = \exp\left(-\dfrac{k_m A}{Q}\right) \tag{2.14f}$

where $A$ = interfacial area.

For the reverse process, i.e., when the substance is removed by blood that enters the capillary devoid of it, we have

Clearance: $\qquad C_B / KC_T = 1 - \exp\left(-\dfrac{k_m A}{Q}\right) \tag{2.14g}$

This process is referred to as clearance of the tissue.

*Comments:*

A quick check for validity of solution gives $C_B = C_{Bo}$ for $A \to 0$, as it should, and $C_B = KC_T$ for $A \to \infty$. The latter is the result of equilibration of blood and tissue that occurs as $x$, and consequently area $A$, goes to infinity.

Expression 2.14f and Expression 2.14g are used particularly for the determination of the mass transfer coefficient, assuming that values of $K$ and $Q$ have been determined independently. This can be done in the usual fashion by making a semilog plot of the concentration fraction in Equation 2.14e against the reciprocal $1/Q$, i.e., by running a series of experiments at different volumetric flow rates $Q$. For clearance experiments, $C_{Bo}$ is set equal to zero. The slope of the plot then yields the product $k_m A$. Because detailed anatomical information for the determination of the interfacial area $A$ is rarely available, we must be content to deal with the product $k_m A$ itself. That quantity, however, is still highly useful because it provides a measure of the overall permeability of the tissue–capillary interface.

   Experiments of the type just described are critical in evaluating the effectiveness of drugs designed to increase the passage of metabolically important materials into tissue. Such drugs are referred to as *vasoactive* and they act by increasing the area available for transfer, or by increasing the permeability of the membrane itself. This is reflected in an increase in the slope of the plot of Equation 2.14e.

### Illustration 2.5: Batch Distillation: An Example of a Cumulative Balance

Batch distillation is practiced with considerable frequency both on a laboratory and an industrial scale for the purpose of separating and purifying liquid mixtures.

   A simple version of the process that comes close to that used in laboratory practice is sketched in Figure 2.6. The still, shown on the left, is loaded with a liquid mixture of composition $x_W^0$ and total mass $W^0$ and subsequently brought to a boil by internal or external heating. The vapor produced at a rate $D$ mol/s is passed into a water-cooled condenser and the resulting liquid condensate is collected in a receiver shown on the right of Figure 2.6. The composition and total mass in the still at any instant are denoted by $x_W$ and $W$ and that in the receiver by $x_D'$ and $D'$. We note that in industrial practice a cylindrical column containing various vapor–liquid contacting devices such as packing of the type used in gas absorbers (see Illustration 2.3) is mounted on the still to promote additional fractionation of the vapor. The

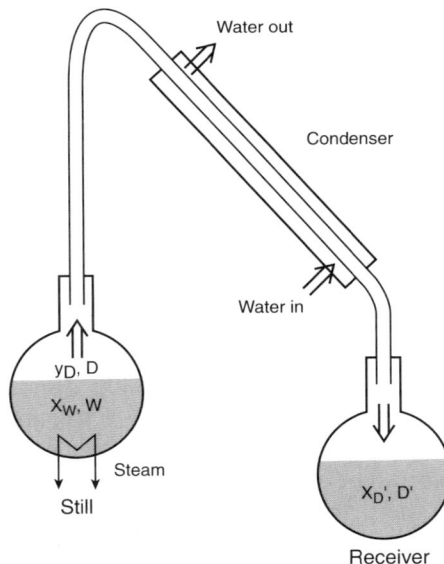

**FIGURE 2.6**
A batch-distillation apparatus.

liquid phase for this section is provided by diverting part of the liquid condensate, termed *reflux,* to the top of the column. This type of operation is discussed in greater detail in Chapter 7 (Illustration 7.10).

As is usual in systems of some complexity, a number of different balances can be made depending on the choice of balance space. We may choose, for example, to make an unsteady integral balance about the still, or to make a similar balance about the receiver. They can be instantaneous or cumulative in time and can involve total or component mass balances.

The balance considered here is the cumulative balance up to some point in time *t*. We apply the scheme previously given in Equation 2.7 and write:

Mass initially present = Mass left in still + Mass in receiver

$$W^0 = W + D'$$ (2.15a)

and for the component mass, assuming a binary system:

$$x_W{}^0 W^0 = x_W W + x_D' D'$$ (2.15b)

where the total mass is expressed in kg or in mol, and the compositions represent mass or mole fractions. The receiver contents $D'$ have been primed to distinguish them from the *rate* of distillation (mol/s), which is commonly given the symbol $D$.

*Comments:*

Cumulative balances such as the two simple expressions (Equations 2.15a and Equation 2.15b) are often overlooked in modeling, or else written out without much thought to their origin. It is important to note that they are quite independent of the unsteady integral balances mentioned previously and consequently serve as additional tools that can be used to supplement the model equations. Typically, they are used as adjuncts to instantaneous balances in batch distillation. We can, for example, solve for either still or distillate composition and obtain

$$x_d' = \frac{x_W{}^0 - x_W(1-f)}{f}$$ (2.15c)

$$x_W = \frac{x_W{}^0 - x_D'(1-f)}{f}$$ (2.15d)

where $f = 1 - W/W^0$ = fraction distilled.

Use of these equations is demonstrated in Illustration 7.10.

## 2.3   The Information Obtained from Model Solutions

The reader will have noted that in several of the illustrations so far, the model and its solution were used to extract a particular piece of information about the underlying physical system or process. Thus, in the two biological examples given in Illustration 2.4, the suggested use of the solution was the determination of tissue volume and permeability. We now wish to generalize this aspect of modeling and ask the question: What types of information can we expect to find in the solution of a model? This is evidently a question of some importance because, without *a priori* knowledge of the information contained in the solution, modeling would presumably not be undertaken at all.

The question can be answered in a general way by stating that *any of the quantities appearing in the solution, including the dependent and independent variables, can be the unknown or the information being sought.* These solutions generally take one of the following forms:

$$u = f(t, \text{parameters}) \qquad\qquad (2.16a)$$

$$u = g(z, \text{parameters}) \qquad\qquad (2.16b)$$

or

$$u = F(v, \text{parameters}) \qquad\qquad (2.16c)$$

Here, the dependent variable $u$, which is typically a concentration or a mass related to the system, is seen to be a function of time $t$, distance $z$, or another mass or concentration $v$ plus a set of specific parameters pertinent to the system.

Two additional forms make their appearance:

$$z = G(t, \text{parameters}) \qquad\qquad (2.16d)$$

and

$$u = H(x, y, z, t, \text{parameters}) \qquad\qquad (2.16e)$$

The first of these represents the special case where distance $z$ becomes the dependent variable and time the independent variable. Solutions of this form arise, for example, in systems involving reacting particles, where particle radius is now the dependent variable, which varies with time. The mathematical expression that results has been given by Equation 2.9. The second, Equation 2.16e, is a multidimensional one and arises from a PDE. Some solutions to such PDE models appear in Chapter 3, and the models themselves are taken up in Illustration 2.6 and Illustration 2.7.

We first note that in addition to $u$ either $t$ or $z$ may be the unknowns being sought. This is in fact the most common occurrence. We may wish, for example, to determine the time or distance necessary to affect a prescribed concentration change. Designing a scrubber, i.e., calculating its height $z$, falls in the latter categories.

Next in importance as an unknown are the parameters, which we can accommodate in the following broad categories:

1. Transport coefficients, such as diffusivities, film coefficients, overall mass transfer coefficients, and permeabilities
2. Rate constants pertaining to chemical or biological reactions taking place in the system
3. Flow rates, including those in and out of compartments, pipes and columns; carrier and solvent flow rates
4. Volumes, in particular those in addition to concentration or mass pertaining to a stirred tank or compartment
5. Parameters describing phase equilibria such as partition coefficients and Henry's law constants, relative volatilities, equilibrium solubilities, and vapor pressures
6. Inlet concentrations to compartments, pipes, and columns

There is yet another type of information contained in the model that is obtained by manipulation of the primary results. We call this *derived information*. The manipulations involved typically consist of differentiation or integration of the results. If these come in the form of concentration distributions, we obtain the following derived quantities:

By differentiation

$$N = -DA \frac{dC}{dz} \quad \text{Diffusional flow} \qquad (2.16f)$$

By integration

$$\bar{C} = \frac{\int C dA}{A} \qquad (2.16g)$$

or                                        Average concentrations

$$\bar{C} = \frac{\int C dV}{V}$$

Use is made of these expressions in Illustration 4.9 and Practice Problem 4.11 and again in Section 5.2. They can be regarded as part of the information package provided by the model solution. We have summarized both the model solutions and the information contained in them for convenient reference in Table 2.2.

**TABLE 2.2**

Models, Model Solutions, and the Information Contained in Them

| Model | Model Solutions |
|---|---|
| *A. Integral Balances* | |
| 1. Steady-state balance | $u = f(v, \text{parameters})$ |
| 2. Unsteady balance | $u = F(t, \text{parameters})$ |
| 3. Cumulative balance | $u = g(v, \text{parameters})$ |
| *B. Differential balance* | |
| 1. Steady-state 1-D balance | $u = G(z, \text{parameters})$ |
| 2. Unsteady-state 1-D balance | $u = g(z, t, \text{parameters})$ |
| 3. Steady-state multidimensional balance | $u = H(z, t, \text{parameters})$ |
| 4. Unsteady multidimensional balance | $u = k(x, y, z, t, \text{parameters})$ |
| *C. Parameters* | |
| 1. Transport coefficients | 2. Rate constants |
| 3. Flow rates | 4. Volumes |
| 5. Phase equilibrium parameters | 6. Inlet concentrations |
| *D. Derived Information* | |
| 1. Diffusional flow | |
| 2. Average concentration | |

## 2.4  Setting Up Partial Differential Equations

The mass balances that have been considered up to this point are confined to cases involving a single independent variable, time or distance. We now examine situations in which more than one such variable needs to be taken into account. Suppose, for example, that diffusion takes place from an external medium into a porous cylindrical or spherical particle, which is initially devoid of the diffusing species. Concentrations will then vary both with time and radial distance and, in the case of the cylindrical particle, with axial distance as well. This is a system that is distributed in both time and distance and that consequently leads to a PDE (see Table 2.1).

Let us assume that the variations are with respect to one distance variable $x$ and with time $t$. We have previously considered, at the ODE level, distributions in time only, or in distance only. This has led to the schemes represented by Equation 2.1 and Equation 2.2. To deal with simultaneous variations in both time and distance, we superpose the two expressions; i.e., we write

$$
\begin{array}{c}
\text{Rate of mass in} \\
\text{at } x \text{ and over } \Delta x
\end{array}
\quad - \quad
\begin{array}{c}
\text{Rate of mass out} \\
\text{at } x + \Delta x \text{ and over } \Delta x
\end{array}
$$

$$
= \quad
\begin{array}{c}
\text{Rate of change} \\
\text{of mass in } \Delta x
\end{array}
\quad = \quad
\left[ \frac{\partial}{\partial t}(\text{mass}) \right]_{\Delta x}
\tag{2.17}
$$

On occasion, mass will enter at $x + \Delta x$ and leave at $x$, in which case the scheme is adjusted accordingly.

This is the formulation that must be used when the system is distributed over distance $x$ and time $t$. When variations occur in more than one direction, we merely add appropriate terms to the left side of Equation 2.17, for example, "Rate of mass in at $y$ and over $\Delta y$" and so on. For variations in the radial direction, $y$ and $\Delta y$ are replaced by $r$ and $\Delta r$. Note that the time derivative in Equation 2.17 is now a *partial* derivative because we are dealing with more than one independent variable. The following illustration provides an example of the application of Equation 2.17.

### Illustration 2.6: Unsteady Diffusion in One Direction: Fick's Equation

Consider diffusion to be taking place into a rectangular slab, which is infinitely wide in the $y$ and $z$ directions and of finite width $L$ in the $x$ direction. The slab can be a stagnant gas or liquid, or a porous solid, and initially contains the diffusing species at a concentration level $C = C_o$. At time $t = 0$, the two sides of the slab at $x = L/2$ and $x = -L/2$ are suddenly exposed to a higher external concentration $C = C_e$. Diffusion into the slab commences, with the diffusing species entering each face and moving simultaneously toward the centerline. The resulting profiles and their development with time are shown in Figure 2.7a. Initially, at $t = t_o$ the profile is flat and uniform at the level $C = C_o$. As time progresses and solute penetrates into the interior, the profile assumes a parabolic shape, which becomes increasingly flatter until at time $t = t_\infty$ it has reached the level of the external concentration $C = C_e$. Diffusion then comes to a halt.

To model this process, we choose a difference element to the right of $x = 0$ (see Figure 2.7b). Solute movement will then be from right to left, entering at the position $x + \Delta x$ and exiting at $x$. Applying the scheme of Equation 2.18, we obtain the following result:

$$
\begin{array}{c}
\text{Rate of mass in} \\
\text{at } x + \Delta x
\end{array}
\quad - \quad
\begin{array}{c}
\text{Rate of mass out} \\
\text{at } x
\end{array}
\quad = \quad
\left( \frac{\partial}{\partial t} \text{mass} \right)_{\Delta x}
\tag{2.18a}
$$

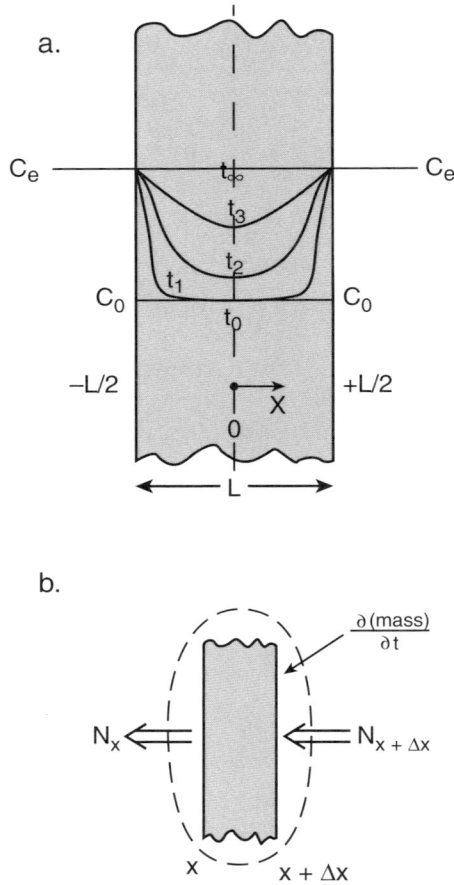

**FIGURE 2.7**
Diffusion into a slab: (a) development of concentration profiles; (b) difference element for the mass balance.

or, introducing Fick's law of diffusion, Equation 1.4a

$$DA\frac{\partial C}{\partial x}\bigg|_{x+\Delta x} - DA\frac{\partial C}{\partial x}\bigg|_{x} = \frac{\partial}{\partial t}A\Delta x C_{avg} \qquad (2.18b)$$

The reader is reminded that in the original formulation of Fick's law, concentration diminished in the direction of increasing $x$ (see Figure 1.1a) resulting in a negative gradient. The reverse is true here: Flow is in the direction of diminishing values of $x$ and $C$, yielding a positive gradient $\partial C/\partial s$. The need for a negative sign in Fick's law is thus removed, and we simply write $N = DA\partial C/\partial x$. This ensures that the diffusional flow will be a positive quantity.

To obtain the final result, Equation 2.18b is divided by $A\Delta x$ and $\Delta x$ is allowed to go to zero. This yields

$$D\frac{\partial^2 C}{\partial x^2} = \frac{\partial C}{\partial t} \tag{2.18c}$$

This is Fick's equation in one dimension. Its solution yields the time-dependent concentration profiles shown in Figure 2.7a. Illustrations dealing with Fick's law will appear in Chapter 3.

*Comments:*

Although we have confined ourselves to Fick's law in one dimension, its extension to three dimensions is straightforward. We merely have to extend Equation 2.18a to three dimensions in the increments $\Delta x$, $\Delta y$, $\Delta z$, which now represent the sides of a cube. The result is

$$D\Delta y\Delta z\left[\left(\frac{\partial C}{\partial x}\right)_{x+\Delta x} - \left(\frac{\partial C}{\partial x}\right)_{x}\right] + D\Delta x\Delta z\left[\left(\frac{\partial C}{\partial y}\right)_{y+\Delta y} - \left(\frac{\partial C}{\partial y}\right)_{y}\right] +$$

$$D\Delta x\Delta y\left[\left(\frac{\partial C}{\partial z}\right)_{z+\Delta z} - \left(\frac{\partial C}{\partial z}\right)_{z}\right] = \frac{\partial}{\partial t}\Delta x\Delta y\Delta z C \tag{2.18d}$$

On dividing by $\Delta x$, $\Delta y$, $\Delta z$ and allowing the increments to go to zero, we obtain the three-dimensional version of Fick's law in rectangular coordinates:

$$D\left(\frac{\partial^2 C}{\partial x^2} + \frac{\partial^2 C}{\partial y^2} + \frac{\partial^2 C}{\partial z^2}\right) = \frac{\partial C}{\partial t} \tag{2.18e}$$

For radial, spherical, and cylindrical coordinates, the distance derivatives are somewhat more complicated.

Although we do not, in this text, take up the actual solution of PDEs, it is important to examine one of the tools needed for this purpose, the boundary and initial conditions. The reader may recall that at the ODE level, the number of boundary conditions required equals the *order* of the equation. This concept can be extended to PDEs as follows: Each set of partial derivatives requires a number of conditions equal to its highest order. Thus, the 1-D Fick's equation (Equation 2.18c) requires two boundary conditions for the distance derivative $\partial^2 C/\partial x^2$, and one condition, also called an initial condition, for the time derivative $\partial C/\partial t$. These conditions are obtained from an examination of the physical system. Thus, from Figure 2.7a, we have the following conditions:

For the distance derivative
1. $C = C_e$ at $x = +L/2$ and any time
2. $C = C_e$ at $x = -L/2$ and any time
For the time derivative
3. $C = C_o$ at $t = 0$ and any position

These conditions can be placed in the following terse forms:

1.                         $C(t, L/2) = C_e$                         (2.18f)

2.                         $C(t, -L/2) = C_e$                        (2.18g)

3.                         $C(0, x) = C_o$                           (2.18h)

These, and other boundary conditions, are encountered again in Chapter 4 where they are used for a number of different processes and geometries. As noted, we do not undertake the actual solutions of the PDEs, but rather present them in graphical or tabular form and use them to address a number of practical problems. At this, the reader will wish to breathe a sigh of relief.

Let us next consider a process that is at steady state, but one in which the dependent variable varies in two directions. This, too, leads to a PDE.

### Illustration 2.7: Laminar Flow and Diffusion in a Pipe: The Graetz Problem for Mass Transfer

When a solute is released from a soluble tubular wall into a flowing fluid, two cases need to be distinguished:

1.  The fluid is in turbulent flow. This implies that the core is well mixed and has a uniform concentration $C_b(x)$, which varies in the direction of flow. That bulk concentration is initially zero at the inlet if we assume the feed to be pure solvent and gradually increases in level as material dissolves into the flowing fluid. Ultimately, at long distances from the inlet, the fluid becomes fully saturated with solute and the bulk concentration equals the equilibrium solubility, $C_b = C^*$. Mass transfer then comes to a halt. This situation is depicted in Figure 2.8a and leads to an ODE when a mass balance is applied to an increment $\Delta x$ in the direction of flow. A situation similar to this case has been encountered in Illustration 2.2, where oxygen from the atmosphere entered and dissolved in a river that was in turbulent flow.

2.  When the fluid is in laminar flow, the core is no longer well mixed and we see instead a gradual variation of concentration in the radial direction as solute from the wall enters and dissolves in the flowing fluid. The concentration profile assumes the shape of a parabola

a.

b.

c.

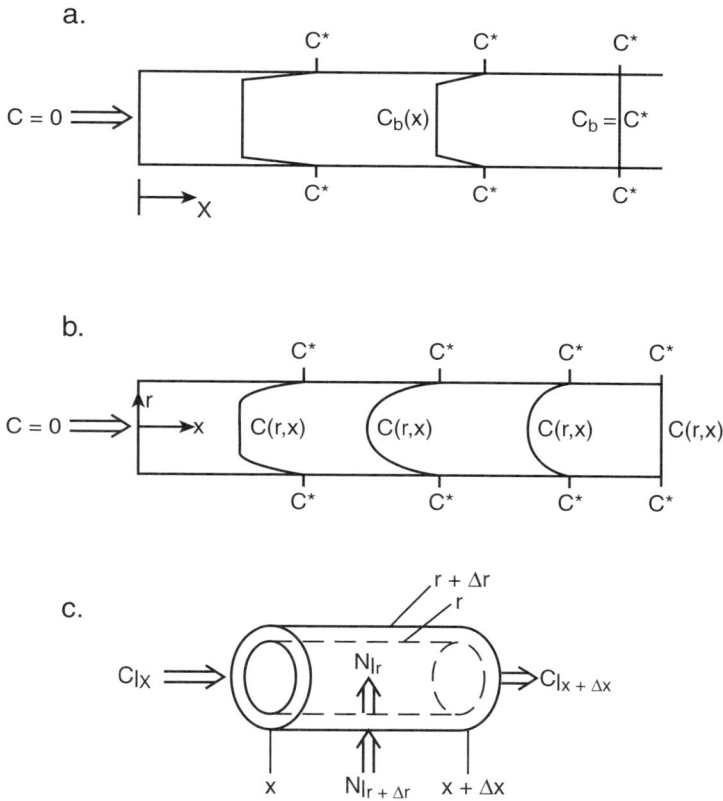

**FIGURE 2.8**
Solute dissolution into a flowing fluid: (a) turbulent flow; (b) laminar flow; (c) difference element for laminar flow.

whose height gradually diminishes in the direction of flow until the fluid is fully saturated. At this point the concentration has become uniform ($C(r, x) = C^*$), and mass transfer again ceases. This situation is depicted in Figure 2.8b.

Because we are here dealing with concentration variations in two directions, $r$ and $x$, the mass balance must be taken over an element with increments $\Delta x$ and $\Delta r$. This leads to the doughnut shape shown in Figure 2.8c, and ultimately yields a PDE.

Let us proceed to write a mass balance over this element. The volumetric flow rate into and out of the element is given by the product of velocity and area; i.e., $[v(r)]_{avg} 2\pi r \Delta r$, where the bracketed term represents the local average velocity. The mass of solute entering or leaving the element is then obtained by multiplication by the solute concentration $C$ at the two locations. Radial transport is by diffusion and here we note that, once again, as in the previous illustration, flow is in the direction of diminishing values of the distance variable and of C. The minus sign in Fick's law is consequently dropped and we use instead the form $N = DA(\partial C/\partial r)$. With these expressions in place, the mass balance becomes

$$\frac{\text{Rate of solute in}}{\text{at } x \text{ and } r + \Delta r} - \frac{\text{Rate of solute out}}{\text{at } x + \Delta x \text{ and } r} = 0$$

$$\left[\begin{array}{c} [v(r)]_{\text{avg}} 2\pi r \Delta r C_{1x} \\ +D\Delta x\left(2\pi r \dfrac{\partial C}{\partial r}\right)_{r+\Delta r} \end{array}\right] - \left[\begin{array}{c} [v(r)]_{\text{avg}} 2\pi r \Delta r C_{1x+\Delta x} \\ +D\Delta x\left(2\pi r \dfrac{\partial C}{\partial r}\right)_{r} \end{array}\right] = 0 \qquad (2.19a)$$

Dividing by $2\pi r \Delta x \Delta r$ and letting both increments go to zero, we obtain the PDE

$$v(r)\frac{\partial C}{\partial x} = D\frac{1}{r}\frac{\partial}{\partial r}\left(r\frac{\partial C}{\partial r}\right) \qquad (2.19b)$$

or in expanded form

$$v(r)\frac{\partial C}{\partial x} = D\left[\frac{\partial^2 C}{\partial r^2} + \frac{1}{r}\frac{\partial C}{\partial r}\right] \qquad (2.19c)$$

where $[v(r)_{\text{avg}}]$ has now become a point quantity. The boundary conditions are again three in number, the radial derivatives requiring two such conditions (highest order 2) and the axial derivative one. They are the following:

Condition at the wall:        $C(x, R) = C^*$                                    (2.19d)

Condition at the axis:        $\left.\dfrac{\partial C}{\partial r}\right|_{r=0} = 0$                                    (2.19e)

Condition at the inlet:        $C(0, r) = 0$                                    (2.19f)

The condition at the axis is of a type encountered before in Illustration 1.1 (see Figure 1.2a and Figure 1.2b) and reflects the fact that the concentration profile must be symmetrical about the central axis and its derivative must consequently vanish there. This system of equations (Equation 2.19c through Equation 2.19f) yields, on solution, the radial and axial concentration gradients shown in Figure 2.8b.

*Comments:*

The Graetz problem, also known as the Graetz–Nusselt problem, was originally formulated for the corresponding heat transfer case, which is represented by the PDE

$$v(r)\frac{\partial T}{\partial x} = \alpha\left[\frac{\partial^2 T}{\partial r^2} + \frac{1}{r}\frac{\partial T}{\partial r}\right] \tag{2.19g}$$

where $\alpha$ = thermal diffusivity.

In the original version, first put forward in the 1880s, two boundary conditions were considered: constant wall temperature $T(x, R) = T_w$ and constant flux

$$k\frac{\partial T}{\partial r}(z, R) = q_w$$

Some initial results were given by Graetz, but it was not until 1956 that the complete analytical solution became available.

Since its inception, the Graetz problem has been applied to a host of related problems in both heat and mass transfer with a variety of boundary conditions encompassing both Newtonian and non-Newtonian flow. In Chapter 5 we show how the solution profiles of the Graetz problem can be cast into equivalent mass transfer coefficients, which can then be used to model the process at the ODE level.

## Illustration 2.8: A Metallurgical Problem: Microsegregation in the Casting of Alloys and How to Avoid PDEs

One method of casting alloys is to pour the molten charge into a mold and allow it to cool in contact with the ambient air. The process of solidification that results is a complex one involving the transport of heat to the external medium and a simultaneous transfer of mass from the liquid to the solid phase. The progress of the proceedings is best visualized by means of a plot of temperature vs. liquid and solid compositions termed a *phase* or *melting-point diagram*. A simple version of this diagram for a binary (i.e., two-component) system appears in Figure 2.9a, and we shall see in Chapter 6 that similar diagrams can be constructed for vapor–liquid systems (see Figure 6.17a).

The upper curve, referred to as the *liquidus,* represents a plot of liquid composition vs. temperature, and the lower curve, termed the *solidus,* shows the corresponding solid compositions. Horizontal lines drawn through the diagram intersect the two curves at points representing liquid and solid compositions in equilibrium with each other. The extremities of the diagram denote the melting point of the metal components of the alloy, $T_{mp1} > T_{mp2}$.

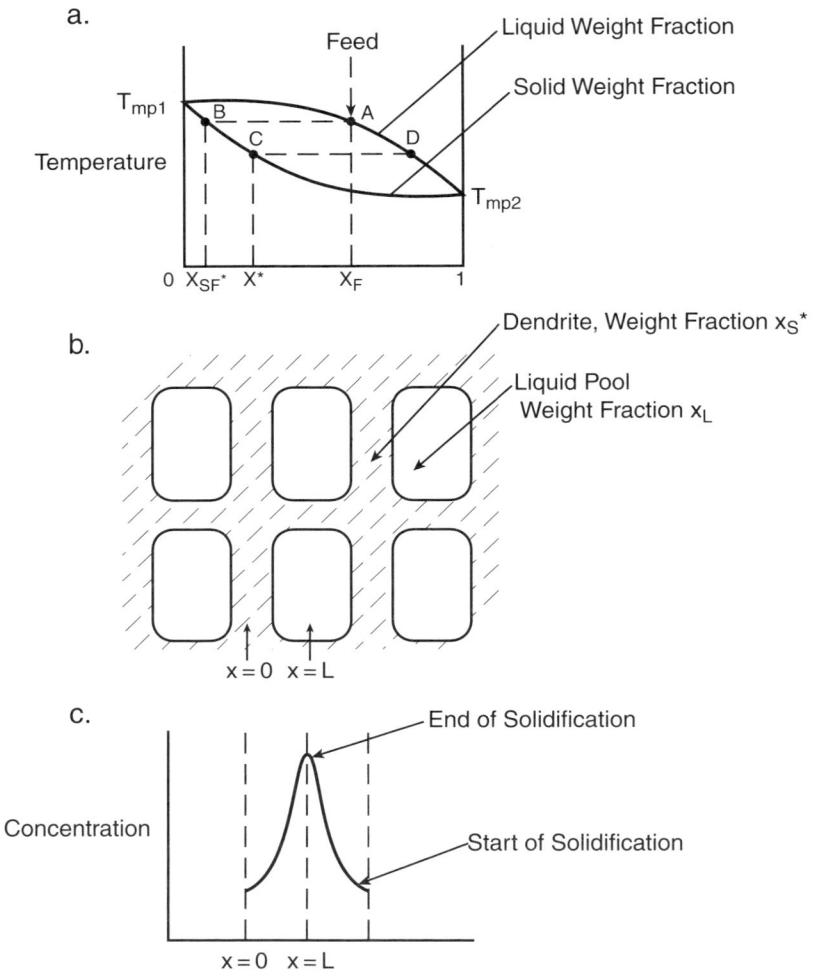

**FIGURE 2.9**
Microsegregation in the casting of alloy: (a) phase diagram; (b) dendrite formation; (c) concentration profile.

On cooling, the liquid charge proceeds along the vertical line of constant composition $x_F$ until it reaches the liquidus at point $A$. Here the first solid crystals of composition $x_{sF}^*$ are formed, which are in equilibrium with the feed composition $x_F$. On further cooling and solidification, a gradual shift of these compositions along the liquidus and solidus curves takes place, resulting in ever-increasing concentrations of component 2 in the liquid and solid phases. These two phases are at all times in equilibrium with each other, with the two compositions located at the end points $x_S^*$ and $x_L$ of a horizontal line drawn through the diagram. These two concentrations gradually approach each other and with further cooling ultimately converge to a single point representing the pure metal component 2 with the lower melting point $T_{mp2}$. Solidification is then complete.

The physical structure of the charge during this solidification process is depicted in Figure 2.9b. Typically, the solid phase is initially confined to narrow, fingerlike protrusions termed *dendrites*. These regions are surrounded by small pools of liquid, which gradually diminish in size as solidification progresses. Simultaneously, the concentration of component 2 in the thickening dendrite increases until it peaks as pure metal 2. The resulting concentration profile is a single dendrite shown in Figure 2.9c. These dendrites form a repetitive pattern of microsegregation throughout the cast and are an undesirable feature because they lead to non-uniform properties of the material. To remove these nonhomogeneities, the cast is subjected to a subsequent thermal treatment termed *homogenization* in which the concentrations are smoothed by a slow process of solid-phase interdiffusion. This process is an unsteady one and is described by Fick's equation (Equation 2.19c). We do not wish to address its solution, which is a fairly complex one, here, but note that it requires as an initial condition the concentration profile shown in Figure 2.9c. It is this distribution that initiates the homogenization process and ultimately leads to a cast of uniform properties.

Although the process of solidification is a highly complex one requiring in principle a set of PDEs, the derivation of the concentration distribution in the dendrite is apparently amenable to a simple treatment. The literature on the subject proposes the following differential equation for the description of the process:

$$(x_L - x_S^*)df_S = (1 - f_S)dx_L \tag{2.20}$$

where $f_S$ is the local weight fraction of the solid, $x$ is the weight fraction of the component that crystallizes out, and $x_S^*$ is the equilibrium weight fraction at the solid–liquid interface.

Although superficially this expression resembles a mass balance, it is not clear how it is arrived at. It does not contain time or distance as an independent variable, which invariably appears when we model a compartment or what we termed a 1-D pipe. We must also rule out a cumulative balance, which is always algebraic in form. The question then arises whether a new type of mass balance formulation is required to cover this case. Fortunately, as is shown below, this does not turn out to be the case. The three mainstay formulations — compartmental, 1-D distributed, and cumulative — are able to cover this case as well.

Let us proceed to model the system using one of our formulations, the compartmental model. To do this we assume the liquid to be well mixed with a uniform concentration of $x_L$. This is not an unreasonable assumption as the size of the pool is quite small, typically 50 to 200 μm in width. The solid phase, on the other hand, cannot be considered uniform because solid-phase diffusivities are several orders of magnitude smaller than those prevailing in the liquid (see Section 3.1 in the next chapter). Concentrations in the dendrite will consequently vary in the lateral direction, with the local interfacial concentration $x_S{}^*$ in instantaneous equilibrium with the uniform liquid pool at all times (Figure 2.9c). Drawing an envelope around the liquid pool we obtain the following mass balance:

Total mass balance

$$\text{Rate of total mass in} - \text{Rate of total mass out} = \frac{d}{dt}(\text{total contents})$$

$$0 \quad\quad - \quad\quad R \quad\quad = \frac{d}{dt}L \quad\quad\quad (2.21a)$$

Component mass balance

$$\text{Rate of solute in} - \text{Rate of solute out} = \frac{d}{dt}(\text{solute content})$$

$$0 \quad\quad - \quad\quad x_S{}^* R \quad\quad = \frac{d}{dt}x_L L \quad\quad\quad (2.21b)$$

where $R$ = rate of solidification (kg/s) and $L$ is the total mass of liquid in the pool (kg).

As expected, the time variable $t$ makes its appearance, as does the unknown variable $R$, the rate of solidification. Neither of these appears in Equation 2.20, thus bringing the proceedings to a seeming impasse.

Some thought will reveal that both of these undesirable variables may be eliminated by the simple device of dividing the two equations. We obtain

$$x_S{}^* = \frac{d(x_L L)}{dL} \quad\quad\quad (2.21c)$$

or, in expanded form,

$$x_S{}^* = x_L + L\frac{dx_L}{dL} \qu\quad\quad (2.21d)$$

This is beginning to look much more like the desired expression (Equation 2.20), and one final step will bring us to that goal. We invoke a cumulative mass balance, which serves to convert liquid mass $L$ to the solid weight fraction $f_S$. We have up to any time $t$

$$L + S = M_{\text{Tot}} \qu\quad\quad (2.21e)$$

where $S$ = mass of solid and $M_{\text{Tot}}$ = total mass (liquid pool + half the dendrite width). Consequently,

$$\frac{S}{M_{Tot}} = f_S = 1 - \frac{L}{M_{Tot}} \qquad (2.21f)$$

and Equation 2.21d becomes, after some rearrangement,

$$(x_L - x_S{}^*)df_S = (1 - f_S)dx_L \qquad (2.20)$$

which is the desired result. Because the fraction solidified $f_S$, can be written in the form

$$f_S = \frac{Ax \, \rho_S}{A\ell \, \rho_S} = \frac{x}{\ell} \qquad (2.21g)$$

the solution of Equation 2.20 can be used to derive the concentration profile in the dendrite, and thus establishes the initial condition required for the homogenization process.

*Comments:*

- Two features stand out in the treatment of this problem. First and foremost, we have reinforced our confidence in the three basic mass balance formulations at the algebraic and ODE level: the compartmental, 1-D distributed, and cumulative balances. They are vindicated as a comprehensive tool kit at this level of modeling.

- Second, we see here again the near-miraculous reduction of a highly complex process to manageable proportions. This was accomplished by assuming the liquid pool to be well mixed and in equilibrium with the solid interface, both reasonable assumptions. We followed this up with the neat "trick" of dividing the two mass balances, thereby eliminating both the independent variable $t$ and the unknown $R$. This gave us the desired relation between the solid-phase concentration and the fraction solidified.

- The analysis used here is not confined to liquid–solid systems. It can be applied to any process in which a transfer of mass takes place between two phases that are in constant equilibrium with each other. It was first applied by Rayleigh to analyze equilibrium batch distillation and to derive the attendant concentration changes as a function of the fraction distilled. This case as well as other batch processes of the same type are taken up in Chapter 7 dealing with equilibrium stage processes.

- Equation 2.20 as it stands cannot be integrated directly because it contains more than two variables, i.e., $f$, $x_L$, and $x_S{}^*$. A second equation will be required, which is given by the following equilibrium relation:

$$x_S{}^* = f(x_L) \tag{2.22a}$$

Many binary alloys have phase diagrams in which the ratio $x_S{}^*/x_L$ can be approximated as a constant $K$, termed the *partition coefficient*. Combining Equation 2.20 and Equation 2.22a, we obtain

$$[-(K-1)x_L]df = (1 - f_S)dx_L \tag{2.22b}$$

which, after separating variables and formal integration, yields the result

$$-(K-1)\int_{x_L0}^{x_L} \frac{dx_L}{x_L} = \int_0^r \frac{df_S}{1-f_S} \tag{2.22c}$$

Evaluating the integrals, we obtain

$$(K-1)\ln \frac{x_L}{x_L{}^0} = \ln(1-f_S) \tag{2.22d}$$

or alternatively

$$\frac{x_L}{x_L{}^0} = (1-f_S)^{K-1} \tag{2.22e}$$

The corresponding solid-phase concentration is given by

$$\frac{x_S{}^*}{x_L{}^0} = K(1-f_S)^{K-1} \tag{2.22f}$$

where $K < 1$.

This form of expression, known as the Rayleigh equation, is encountered again in Chapter 7 (Illustration 7.3 and Illustration 7.4).

The equation applies only over a limited concentration range, as it predicts that $x_S{}^*$ goes to infinity as $f_S$ attains unity. A more realistic representation of the equilibrium is given by the expression

$$x_S{}^* = \frac{x_L}{(\alpha-1)x_L + \alpha} \tag{2.22g}$$

where $\alpha$ is the so-called separation factor, defined as

$$\alpha = \frac{x_L(1 - x_S{}^*)}{x_S{}^* (1 - x_L)}$$ (2.22h)

with $\alpha > 1$. At low concentrations $\alpha \to 1/K$.

Application of this expression is left to the exercises (Practice Problem 2.10).

## 2.5   The General Conservation Equations

The method we have described for setting up PDEs has a cumbersome feature attached to it. It must be repeated each time there is a change in geometrical configuration or in the process conditions. A switch from rectangular to cylindrical coordinates, for example, requires a new balance to be made. So does the inclusion of reaction terms.

This drawback can be overcome by formulating the mass balances in a generalized vectorial form, using the symbolism of vector calculus. These symbols, or *operators* as they are termed, arise in a natural way in the formulation of generalized transport equations. They are at first sight forbidding, and the beginner will probably be best served by regarding them as a convenient shorthand, without delving into their deeper origins. The symbol $\nabla^2 u$ for example, which is termed the *Laplacian* of $u$ (and pronounced "del square u") is shorthand for a collection of *second-order* partial derivatives. The symbols $\nabla u$ and $\nabla \bullet v$, ("del u" and "del dot v") serve the same purpose for combinations of *first-order* partial derivatives. For example, in rectangular coordinates, del dot of the velocity vector $v$ is synonymous with the sum of the first-order derivatives of the velocity components. Thus,

$$\nabla \bullet v = \frac{\partial v_x}{\partial x} + \frac{\partial v_y}{\partial y} + \frac{\partial v_z}{\partial z}$$ (2.23a)

Similarly, we have for the Laplacian

$$\nabla^2 u = \frac{\partial^2 u}{\partial x^2} + \frac{\partial^2 u}{\partial y^2} + \frac{\partial^2 u}{\partial z^2}$$ (2.23b)

Thus, both del dot and del square tend to be scalar expressions. Del u, on the other hand, is a vector.

Some thought will lead us to the conclusion that del dot terms will likely arise in flowing systems, whereas the Laplacian will most probably appear in the description of diffusion processes. This is indeed the case.

The use of these operators in the formulation of mass balances leads to the following generalized conservation equations. We have, for the component mass balance

$$\mathbf{v} \bullet \nabla C_A + D\nabla^2 C_A \; \pm \; r_A \; = \; \frac{\partial C_A}{\partial t} \tag{2.24a}$$

Flow     Diffusion     Reaction     Transient

and for the total mass balance, also known as the continuity equation,

$$\nabla \bullet \mathbf{v} = 0 \tag{2.24b}$$

Two restrictions apply to these expressions. First, they are confined to *incompressible flow*, i.e., systems in which density changes can be neglected, such as liquid flow or gas flow involving low pressure drops. Second, the formulation requires continuity of the concentration within the flow field. Systems in turbulent flow in which $C_A$ undergoes an abrupt transition from linear gradient in the film to a constant value in the fluid core cannot be accommodated by these expressions. We must, in these cases, revert to the use of the classical shell balance.

To aid the reader in the use of these equations, we have compiled a "dictionary" of the operator symbols, which provides a translation into scalar form for the three principal geometries (rectangular, cylindrical, and spherical). We can use this dictionary, Table 2.3, to extract several important subsidiary relations. For example, in the absence of flow and reaction, the general conservation equation becomes

$$D\nabla^2 C_A = \frac{\partial C_A}{\partial t} \tag{2.24c}$$

which is Fick's equation in three dimensions, with the Cartesian representation (see Table 2.2)

$$D\left[\frac{\partial^2 C_A}{\partial x^2} + \frac{\partial^2 C_A}{\partial y^2} + \frac{\partial^2 C_A}{\partial z^2}\right] = \frac{\partial C_A}{\partial t} \tag{2.24d}$$

This expression is identical to Equation 2.18d, which had been derived by means of a shell balance. Its counterpart for heat conduction, known as Fourier's equation, is given by

$$\alpha\left[\frac{\partial^2 T}{\partial x^2} + \frac{\partial^2 T}{\partial y^2} + \frac{\partial^2 T}{\partial z^2}\right] = \frac{\partial T}{\partial t} \tag{2.24e}$$

The Graetz problem can be accommodated in similar fashion. Here the transient and reaction terms are dropped and we obtain

$$\mathbf{v} \bullet \nabla C_A = D \nabla^2 C_A \tag{2.24f}$$

The dot product on the left is composed by the rules of vector algebra. In other words, it equals the sum of the vector component products. Setting $v_r = \partial C_A / \partial \theta = \partial^2 C_A / \partial z^2 = 0$ and using the tabulations of Table 2.3, we obtain

$$v_z \frac{\partial C_A}{\partial z} = D \frac{1}{r} \frac{\partial}{\partial r}\left(r \frac{\partial C}{\partial r}\right) \tag{2.24g}$$

which is in agreement with Equation 2.19b.

Finally, when all but the reaction and transient forms are omitted, we are led to the result

$$\pm V r_A = V \frac{dC_A}{dt} \tag{2.24h}$$

This will be recognized as a mass balance for a batch reactor of volume $V$.

### Illustration 2.9: Laplace's Equation, Steady-State Diffusion in Three-Dimensional Space: Emissions from Embedded Sources

Steady-state diffusion has been considered in some detail in Chapter 1 at the elementary level of 1-D transport. In Cartesian space the operative expression was Fick's law (Equation 1.4), which can also be written in the equivalent form:

$$\frac{d^2 C}{dx^2} = 0 \tag{2.25a}$$

The extension to three Cartesian dimensions is given by

$$\frac{\partial^2 C}{\partial x^2} + \frac{\partial^2 C}{\partial y^2} + \frac{\partial^2 C}{\partial z^2} = 0 \tag{2.25b}$$

or in a generalized vectorial form by

$$\nabla^2 C = 0 \tag{2.25c}$$

This is the classical and much-studied expression known as Laplace's equation. It can be obtained from Fick's equation by omission of the transient

**TABLE 2.3**

Dictionary of Vector Operators

$\nabla u$

1. Cartesian $\quad (\nabla u)_x = \dfrac{\partial u}{\partial x}$ $\qquad (\nabla u)_y = \dfrac{\partial u}{\partial y}$ $\qquad (\nabla u)_z = \dfrac{\partial u}{\partial z}$

2. Cylindrical $\quad (\nabla u)_r = \dfrac{\partial u}{\partial r}$ $\qquad (\nabla u)_\theta = \dfrac{1}{r}\dfrac{\partial u}{\partial \theta}$ $\qquad (\nabla u)_z = \dfrac{\partial u}{\partial z}$

3. Spherical $\quad (\nabla u)_r = \dfrac{\partial u}{\partial r}$ $\qquad (\nabla u)_\theta = \dfrac{1}{r}\dfrac{\partial u}{\partial \theta}$ $\qquad (\nabla u)_\varphi = \dfrac{1}{r\sin\theta}\dfrac{\partial u}{\partial \varphi}$

$\nabla \bullet v$

1. Cartesian $\quad \nabla \bullet v = \dfrac{\partial v_x}{\partial x} + \dfrac{\partial v_y}{\partial y} + \dfrac{\partial v_z}{\partial z}$

2. Cylindrical $\quad \nabla \bullet v = \dfrac{1}{r}\dfrac{\partial}{\partial r}(r v_r) + \dfrac{1}{r}\dfrac{\partial v_\theta}{\partial \theta} + \dfrac{\partial v_z}{\partial z}$

3. Spherical $\quad \nabla \bullet v = \dfrac{1}{r^2}\dfrac{\partial}{\partial r}(r^2 v_r) + \dfrac{1}{r\sin\theta}\dfrac{\partial}{\partial \theta}(v_\theta \sin\theta) + \dfrac{1}{r\sin\theta}\dfrac{\partial v_\varphi}{\partial \varphi}$

$\nabla^2 u$

1. Cartesian $\quad \nabla^2 u = \dfrac{\partial^2 u}{\partial x^2} + \dfrac{\partial^2 u}{\partial y^2} + \dfrac{\partial^2 u}{\partial z^2}$

2. Cylindrical $\quad \nabla^2 u = \dfrac{1}{r}\dfrac{\partial}{\partial r}\left(r\dfrac{\partial u}{\partial r}\right) + \dfrac{1}{r^2}\dfrac{\partial^2 u}{\partial \theta^2} + \dfrac{\partial^2 u}{\partial^2 z}$

3. Spherical $\quad \nabla^2 u = \dfrac{1}{r^2}\dfrac{\partial}{\partial r}\left(r^2\dfrac{\partial u}{\partial r}\right) + \dfrac{1}{r^2 \sin\theta}\dfrac{\partial}{\partial \theta}\left(\sin\theta\dfrac{\partial u}{\partial \theta}\right) + \dfrac{1}{r^2 \sin\theta}\dfrac{\partial^2 u}{\partial \varphi^2}$

term, or from the general conservation equation (Equation 2.24a) by omitting transient, reaction, and flow terms.

A host of solutions of this problem are known from the analogous case of heat conduction and are easily adapted by substituting concentration and diffusivity for temperature and thermal conductivity. Many of these solutions, particularly those dealing with finite geometries, are forbidding in form and difficult to apply in practice. The intent here is to draw the reader's attention to some simple solutions, which are particularly useful in an environmental and biological context.

The geometry considered is a semi-infinite medium bounded by a plane surface. An object of finite dimensions (sphere, disk, cylinder) is embedded at a distance $x$ from the surface and is assumed to have a constant surface concentration $C_2$. Solute diffuses into the surrounding space and ultimately reaches the bounding surface, which is held at a constant concentration $C_1$ (Figure 2.10). $C_1$ is often near zero because of dispersion into a flowing fluid or the atmosphere. The release rate when $C_1 = 0$ exceeds that of any other practical configuration and sets an upper limit to diffusion from a source of constant concentration. It can therefore be used to estimate the maximum performance to be expected in finite spaces.

Examples of such embedded objects are underground deposits of a toxic or benign nature and artificial implants (see Practice Problem 2.13). The situation described in Illustration 1.3 — helium storage in an abandoned salt mine — is another example.

The primary information obtained from Laplace's equation concerns the three-dimensional concentration distributions, which are of no direct practical use. It is common practice to convert these results into an equivalent rate equation based on a linear driving force. For conduction, it takes the form

$$q(J/s) = kS(T_2 - T_1) \tag{2.26}$$

and for diffusion

$$N(mol/s) = DS(C_2 - C_1) \tag{2.27}$$

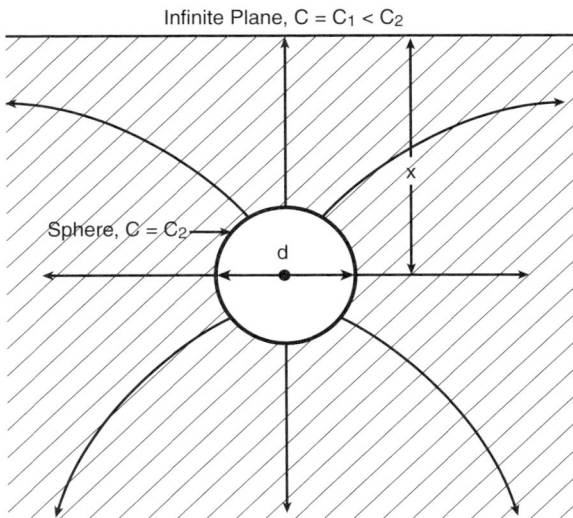

**FIGURE 2.10**
Diffusion from a sphere embedded in a semi-infinite medium.

These are, of course, precisely the type of simple expressions we wish to have on hand to calculate the flux that results in these complex geometries. Equation 2.27 requires knowledge only of the bounding concentrations and the diffusivity. The geometry of the system is accounted for through the so-called shape factor $S$, which has the dimensions of length and is extensively tabulated in standard handbooks of heat transfer. A short version is given in Table 2.4.

Let us use these tabulations to reexamine the helium storage problem of Illustration 1.3. For distances from the surface $x$ much larger than the cavity dimension, we have, with $C_1 \approx 0$

$$N = DS(C_2 - C_1) \approx D\, 2\pi d\, C_2 \tag{2.28a}$$

or equivalently

$$N = D\, 4\pi r\, p/RT \tag{2.28b}$$

This expression is identical to Equation 1.9c, obtained by a spherical shell balance, and represents diffusion into an infinite medium.

When distance $x$ is reduced, the result begins to be affected by the proximity of the surface plane, but not by much. Suppose, for example, that the center of the storage cavity is at a distance of only 100 m from the surface. We then have

$$N = D[2\pi d/(1 - 100/400)]C_2 \tag{2.28c}$$

i.e., the losses increase by 33%.

The use of a spherical shell balance within an infinite medium is thus justified at large depths but is only an approximation in the vicinity of the planar surface. This is in agreement with physical reasoning.

### Illustration 2.10: Lifetime of Volatile Underground Deposits

A shallow buried dump 100 m in diameter ($d$) contains dispersed in it an estimated 100 mol of a toxic substance with a low vapor pressure of $10^{-2}$

**TABLE 2.4**

Shape Factors for Various Geometries Embedded in a Semi-Infinite Medium

| Embedded Object | Shape Factor $S$ | Conditions |
|---|---|---|
| 1. Sphere, diameter $d$ | $2\pi d\,/(1 - d/4x)$ | $x > d/2$ |
|  | $2\pi d$ | $x > d$ |
| 2. Thin circular disk, diameter $d$ | $2d$ | $x = 0$ |
|  | $4d$ | $x > d$ |
| 3. Thin rectangular plate $a < b$ | $\pi a\,\ln(4a/b)$ | $x = 0$ |
|  | $2\pi a\,\ln(4a/b)$ | $x > a$ |
| 4. Horizontal cylinder, length $L$, diameter $d$ | $2\pi L/\ln(2L/d)$ | $x > d$ |

mmHg. It is desired to estimate the minimum time it takes for the charge to evaporate and disperse into the surrounding space.

We assume the geometry to be that of a shallow disk with a maximum shape factor $S = 4d$. Diffusivity is set at $D = 10^{-8}$ m$^2$/s. We then obtain from Equation 2.27b

$$N = DSC_2 = DSp/RT \qquad (2.29a)$$

$$= 10^{-8} \, 4 \times 100 \, (10^{-2}/760)10^5/8.314 \times 298$$

$$N = 2.12 \times 10^{-9} \text{ mol/s}$$

For a charge of 100 mol, the lifetime of the deposit works out to

$$t = \frac{100}{2.12 \times 10^{-9} \times 3600 \times 24 \times 365} \text{ years}$$

$$t = 6.69 \times 10^4 \text{ years}$$

On the other hand, for the same amount of the carcinogen benzene, which has a vapor pressure of 95 mmHg, the lifetime drops to a value of

$$t = 6.69 \times 10^4 \, (10^{-2}/95) = 7.4 \text{ years}$$

*Comments:*
In either case, the evaporative disposal time is unacceptably high. The charge to be disposed of is relatively small, 7.8 kg in the case of benzene or 100 kg if we assume a molar mass of 1000 for the less-volatile substance.

Some relief may be found by adjusting diffusivity upward. Chapter 3 provides an expression for the rough estimate of diffusivity in porous media, given by $D = D_o\varepsilon/4$. $D_o$ represents the free-space diffusivity, of the order $10^{-5}$ m$^2$/s for gases, and $\varepsilon$ is the porosity of the medium. If we assume a highly porous soil with $\varepsilon = 0.4$, $D$ is reduced by two orders of magnitude and the lifetime of the benzene deposit drops to 0.074 years = 27 days. For the less volatile substances, it remains unacceptably high, at 669 years.

---

## Practice Problems

2.1. *Compartmental Modeling in Pharmacokinetics*
   The subject of pharmacokinetics comprises the study of the fate of an injected drug in the human or animal body. A typical exercise involves the monitoring of the drug concentration in the blood after

injection and extracting from the resulting time history relevant kinetic or transport parameters. A favorite device for the interpretation of the experimental findings is the one-compartment model shown in Figure 2.11a. After the one-shot injection of the drug, the drug concentration declines exponentially in the manner shown in the accompanying diagram. That curve, determined experimentally, is usually described by a first-order rate law, which is incorporated in a mass balance around the compartment. We have

$$\text{Rate of drug in} - \text{Rate of drug out} = \frac{d}{dt}\text{contents}$$

$$0 - k_e CV = \frac{d}{dt}VC \tag{2.29a}$$

with the solution

$$\ln C/C_o = -k_e t \tag{2.29b}$$

where $k_e$ is the so-called elimination-rate constant, which can be obtained from a semilog plot of the experimental concentration-time data and which is often reported in terms of the half-life of the drug $t_{1/2} = \ln 2/k_e$, i.e., the time required to reduce its concentration to

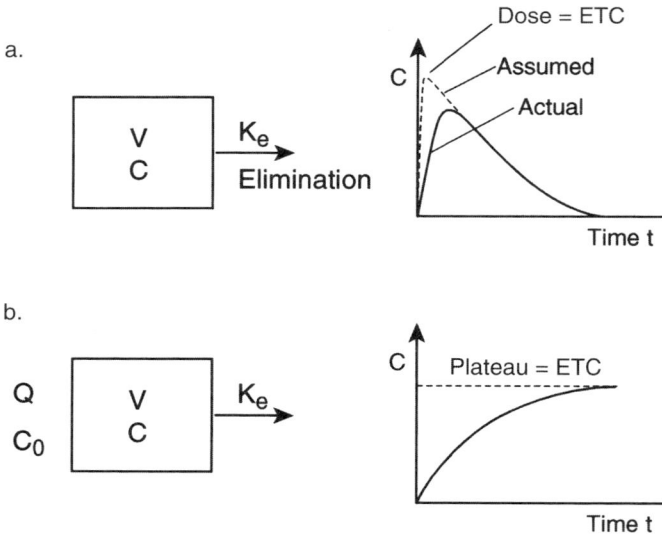

**FIGURE 2.11**
One-compartment models and the time course of drug concentration: (a) clearance following injection; (b) steady infusion.

one-half the initial level. Note that the injected dose does not enter the mass balance directly but appears instead as the initial concentration $C_o$, which is obtained by extrapolation of the data to time $t = 0$ (Figure 2.11). $k_e$ can then be used to determine the time course of drug concentration during its steady infusion, or administration, to a patient. Infusion follows the time course shown in Figure 2.11b. It reaches a plateau value termed the *effective therapeutic concentration* (ETC) when infusion and elimination rates are exactly in balance. Consider the case of a patient who has undergone major surgery and requires a slow, intravenous infusion of aminocaproic acid to control hemorrhage. It is planned to infuse at the rate of 1 g/h, and the drug is known to have a half-life of 3.9 h. If the drug is only effective above 0.048 g/l, how long does it take to reach this level after the start of infusion? Assume a fluid volume of 50 l.

*Answer*: 3.25 h

2.2. *Solute Release from a Tubular Wall into a Stream in Turbulent Flow*
A solute is released from a tubular wall into a fluid in turbulent flow. Two cases are to be considered:

a. The substance is released at a constant rate $N$ (mol/m²/s) all along the length of a tube. This case arises in some medical applications.

b. The solute concentration at the wall is constant and equal to its equilibrium solubility $C^*$. This leads to the type of profile development we had seen in Figure 2.8a.

Derive for both these cases the variation in the axial direction of the bulk concentration $C_b$ of the solute.

*Answer*:

a.
$$C_b = \frac{4N}{vd} x$$

b.
$$(C^* - C_b)/C^* = \exp(-4k_c x/vd)$$

2.3. *Clearance of a Contaminated River Bottom Sediment*
A toxic substance is released into a river at a constant concentration and over a time interval $\Delta t$. The contaminant is adsorbed onto the bottom sediment and, after contamination has ceased, is gradually cleared into the river. It is desired to calculate the time necessary for the sediment to be cleared of 95% of its contaminant. Assume the concentration in the sediment and the water to be uniform in the vertical direction but not in the direction of flow.

a.  Is the process an unsteady one or can it be considered to be at a steady state?

b.  What are the dependent and independent variables for the system?

c.  Describe the space over which the mass balance has to be applied.

d.  Describe the resulting model equations without actually deriving them.

2.4.  *Performance of a Dryer*
A batch of wet solids holding a total of 10 kg of water is to be dried by passing hot air over it. The air enters with a humidity of 0.03 kg $H_2O$/kg air and leaves with a humidity of 0.10 kg $H_2O$/kg air. The (dry) airflow rate is constant at 0.05 kg/s. Calculate the time necessary to evaporate 95% of the moisture. (*Hint:* Make a cumulative balance.)

2.5.  *The Countercurrent Heat Exchanger*
The countercurrent heat exchanger is the heat transfer analogue to the countercurrent gas scrubber described in Illustration 2.3. Cold fluid enters the central tube at a flow rate of $F_c$ kg/s countercurrent to a hot fluid, which enters the concentric shell on the right and with a flow rate $F_h$ kg/s. Heat is transferred to the cold fluid at a rate $q = UA\Delta T$ causing a change in the enthalpy $H_c$, and consequently the temperature $T_C$ of the cold fluid. Enthalpy is related to temperature through the expression $H = FC_p(T - T_{ref})$.

a.  Make an integral heat balance over the entire exchanger. The resulting expression corresponds to Equation 2.11b obtained for the countercurrent gas scrubber.

b.  Derive the thermal counterparts to the differential mass balances (Equations 2.12d and 2.12e) but do not solve.

2.6.  *The Use of the Dye Dilution Method in Determining Flow Rates*
Blood flow rates $Q$ through an organ may be determined by adding a dye or other tracer to the ingoing arterial bloodstream and monitoring the concentration of the (venous) outlet. Type and size of the chosen tracer are such that we can assume it stays strictly in the blood phase and does not permeate into the surrounding tissue. Derive an expression for $Q$ in terms of measurable quantities.

$$Answer: Q = \frac{m}{\int_0^\infty C_v dt}$$

2.7.  *Unsteady Diffusion with a Chemical Reaction*
Consider the case of a substance diffusing into a porous solid slab (see Illustration 2.7) in which it is irreversibly bound to the solid

substrate in accordance with a first-order rate law. Modify the result obtained in Illustration 2.7 to describe this process.

2.8. *Unsteady Diffusion from a Sphere into a Solution of Finite Constant Volume*

Soluble material is leached from a slurry of porous spherical particles by suspending them in a well-stirred liquid solvent of constant volume. This type of operation finds frequent use in the leaching of ores, and in the extraction of edible oils from seeds. Show that the model for this system is made up of a single PDE for the spheres and an ODE for the stirred tank. What are the relevant boundary and initial conditions?

Solutions to this problem and an illustrative example are given in Chapter 4.

2.9. *A Total Mass Balance: The Continuity Equation in Three Dimensions*

Show that the total mass balance for a flowing system with and without chemical reactions in a rectangular coordinate system is given by

$$-\left[\frac{\partial}{\partial x}\rho v_x + \frac{\partial}{\partial y}\rho v_y + \frac{\partial}{\partial z}\rho v_z\right] = \frac{\partial \rho}{\partial t} \tag{2.30}$$

where $\rho$ = density and $v_x$, $v_y$, $v_z$ are the velocity components in the three directions. (*Hint:* Consider a cube with sides $\Delta x$, $\Delta y$, $\Delta z$.)

2.10. *Use of the Separation Factor in Solidification Processes*

Solve the differential Equation 2.20, using the equilibrium relation represented by Equation 2.22h. *Hint:*

$$\int \frac{dx}{x(a+bx)} = -\frac{1}{a}\ln\frac{a+bx}{x}$$

2.11. *Eutectic Compositions and Solidification*

The phase diagram shown in Figure 2.9a is an ideal case that occurs with some frequency but not exclusively. A fairly common extension of this case arises when the lenslike region shown in Figure 2.10a extends only partway from zero weight fraction to some intermediate composition $x_{SE}$ termed a *eutectic*. It is joined there at its tip by a second lens, which covers the remaining weight fraction range from $x_{SE}$ to $x_S = 1$. Similar diagrams arise in vapor–liquid equilibria, where the eutectic composition is termed an *azeotrope* (see Figure 6.20a). Indicate how the separation factor of the preceding problem should be modified to cover this case.

2.12. *Diffusion and Reaction in a Spherical Catalyst Pellet*

Use the general conservation Equation 2.24a and Table 2.2 to derive the differential equation for uniform diffusion and reaction in a spherical catalyst pellet.

2.13. *Controlled-Release Implants*

Controlled-release devices consist in essence of encapsulated saturated suspensions of a medication. The release proceeds through a confining porous membrane into the surrounding tissue, ultimately reaching the target area or organ where the medication is assimilated. Estimate the maximum possible release rate, given the following data:

Implant: $0.5 \times 0.5$ cm cylinder

Saturation solubility: $C_s = 1$ mmol/l

Diffusivity: $D = 10^{-5}$ cm$^2$/s

# 3

## Diffusion through Gases, Liquids, and Solids

Chapter 1 introduced the reader to the notion of diffusional processes in which mass transfer takes place by molecular motion only and is proportional to the concentration gradient of the diffusing species. This proportionality is enshrined in Fick's law of diffusion and this introductory chapter was used to acquaint the reader with some simple applications of that law (see Illustration 1.2 and Illustration 1.3). The intent of the present chapter, and the one that follows, is to amplify and expand the material on diffusion presented in Chapter 1.

We examine in some detail the important diffusion coefficient, which, when viewed mathematically, is the proportionality constant in Fick's law. It is also a material property that depends on the nature of the diffusion species, the matrix through which diffusion takes place, as well as on temperature, and, in the case of gases, on pressure. We consider in some detail the diffusivities of gases within gases, within liquids, and within solids, and the diffusivities associated with the interdiffusion of liquids and solids. These coefficients are of considerable practical importance in various engineering disciplines, in materials processing, and in the biological and environmental sciences.

## 3.1 Diffusion Coefficients

### 3.1.1 Diffusion in Gases

The mechanisms by which diffusion in gases takes place are depicted schematically in Figure 3.1a. Gas molecules move in space in random motion with an average velocity $u$, repeatedly undergoing collision with other moving gas molecules, which causes them to be deflected into a new direction. The average distance traveled by a molecule is referred to as the *mean free path* $\lambda$, where $\lambda$ is of the order $10^{-7}$ m at atmospheric pressure and is a direct measure of the diffusivity of a substance. A selected list of diffusivities of gases and vapors in air appears in Table 3.1.

a.

Free Path

Diffusing
Molecule

b.

Vacancies
or Defects

Diffusing
Molecule

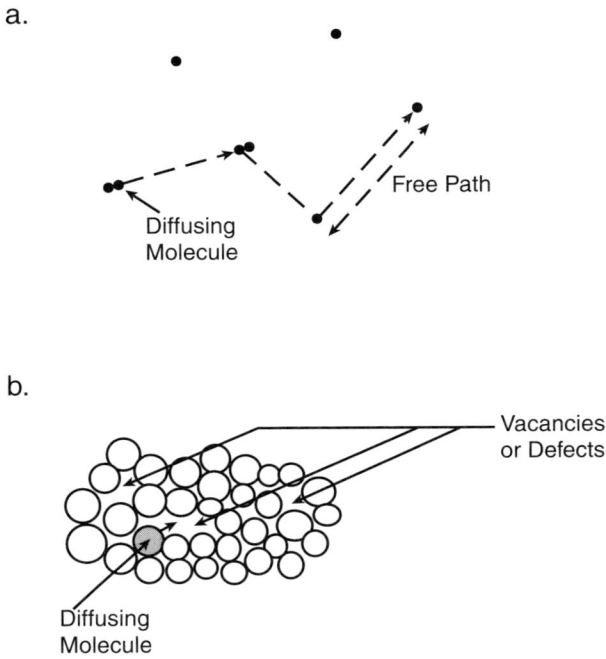

**FIGURE 3.1**
Diffusional mechanisms in (a) gases and (b) liquids and solids.

**TABLE 3.1**

Diffusivities in Air ($P$ = 1 atm, $T$ = 25°C)

| Diffusing Species | Diffusivity, cm²/s |
|---|---|
| Hydrogen | 0.78 |
| Helium | 0.70 |
| Ammonia | 0.22 |
| Water | 0.26 |
| Oxygen | 0.20 |
| Ethanol | 0.14 |
| Acetic acid | 0.12 |
| Benzene | 0.090 |
| Toluene | 0.086 |
| n-Hexane | 0.080 |
| Carbon tetrachloride | 0.083 |
| Chlorotoluene | 0.065 |
| DDT | 0.047 |
| 2,4,2',4''-Tetrachlorobiphenyl (a PCB) | 0.052 |
| Mercury | 0.13 |

A first simple expression for the diffusivity of gases was derived some 150 years ago based on the kinetic theory of gases. In this theory, the molecules are regarded as point entities, which undergo elastic collisions with each other without the intrusion of intermolecular attractive or repulsive forces. This simple model led to the expression

$$D = \frac{1}{3} u\lambda \propto \sqrt{\frac{T}{M}} \Big/ \frac{n}{V} \tag{3.1}$$

where the average velocity $u$ varies directly with the square root of absolute temperature and the mean free path is inversely proportional to the molar density $n/V$. This expression correctly predicts the order of magnitude of $D$ (see Practice Problem 3.1) and the dependence on gas pressure and molar mass $M$, but gives a much less accurate representation of the variation with temperature.

This first attempt at a prediction of $D$ was followed by a series of more elaborate theories, which took account of the finite size of the gas molecules, as well as the effect of intermolecular forces. Probably the most popular among current prediction methods is that due to Fuller, Schettler, and Giddings, who proposed the following expression for the calculation of gas diffusivities:

$$D_{AB} = \frac{1.00 \times 10^{-3} T^{1.75} (1/M_A + 1/M_B)}{P_T \left[ (\Sigma V_A)^{1/3} + (\Sigma V_B)^{1/3} \right]^2} \tag{3.2}$$

Here $D_{AB}$ is in units of cm²/s, $T$ is the absolute temperature (K), $P_T$ the total pressure in atmospheres, and $V$ are the atomic and molecular volume contributions. These are empirical constants that correspond very approximately to the molar volume of the substances in cm³/mol. They have been tabulated and a partial list for use with organic molecules appears in Table 3.2.

### Illustration 3.1: Diffusivity of Cadmium Vapor in Air

Cadmium vapor is a toxic substance whose diffusivity in air is not readily available in the literature. It is desired to calculate its diffusivity at its boiling point of 1038 K and a pressure of 1 atm. Because an empirical atomic volume is not available, we use the reported value for its *liquid* volume of 14 cm³/mol for an atomic weight of 112.4. We obtain, using Equation 3.2:

$$D_{AB} = \frac{10^{-3}(1038)^{1.75} \left( \dfrac{1}{112.4} + \dfrac{1}{29} \right)^{1/2}}{1.0 \left[ 14^{1/3} + 20.1^{1/3} \right]^2}$$

$$D_{AB} = 1.6 \text{ cm}^2/\text{s}$$

**TABLE 3.2**

Atomic and Molecular Volume Contributions for
Diffusivity Calculations

| Species | Volume $V$ (cm³/mol) |
|---|---|
| *A. Gases (Fuller, Schettler, and Giddings Method)* | |
| C | 16.5 |
| H | 1.98 |
| O | 5.48 |
| N | 5.69 |
| Cl | 19.5 |
| Aromatic ring | −20.2 |
| Air | 20.1 |
| *B. Liquids (Wilke–Chang Method)* | |
| C | 14.8 |
| H | 3.7 |
| O | 7.4 |
| O in high esters and others | 11.0 |
| O in acids (–OH) | 12.0 |
| Cl (terminal) | 21.6 |
| 6-numbered ring | −15 |

A data point for the diffusivity of Cd in $N_2$ at 273 K is available for comparison. Its value is 0.15 cm²/s and we obtain, by applying a temperature correction in line with Equation 3.2,

$$D_{AB} = 0.15 \times \left( \frac{1038}{273} \right)^{1.75} = 1.55 \text{ cm}^2/\text{s}$$

This is in good agreement with the calculated value of 1.6 cm²/s.

*Comments:*

Some remarks are in order regarding the magnitude of gas phase diffusivities. We note from Table 3.1 that at 25°C and a pressure of 1 atm, most diffusivities cluster around a value of 0.1 cm²/s. This includes metal vapors, as well as medium-sized organic molecules such as DDT and the PCBs. The reason for this lies in the relatively weak dependence of $D_{AB}$ on molar volume and mass and the limited number of gaseous or volatile substances available. The larger organic molecules such as polymers, proteins, and carbohydrates that would lead to low diffusivity values do not exist in the vapor phase. Thus, gas diffusivities lower than 0.01 cm²/s are unlikely to be encountered. An upper ceiling is provided by the lightest molecules, hydrogen and helium, which have a mutual diffusion coefficient of 1.35 cm²/s at 25°C and 1 atm. A reasonable order-of-magnitude estimate can therefore be arrived at in most cases by starting with a value of $D_{AB}$ = 0.1 cm²/s and applying temperature or pressure correction factors in accordance with Equation 3.2.

Applying this procedure to the cadmium vapor of Illustration 3.1, we obtain a value of

$$D_{Cd-\text{Air}} = 0.1 \left( \frac{1038}{273} \right)^{1.75} = 1.04 \text{ cm}^2 / \text{s}$$

which is of the correct order of magnitude.

## 3.1.2  Diffusion in Liquids

Liquid densities exceed those of gases at normal atmospheric pressures by a factor of about 1000. These differences are reflected in the intermolecular distances that exist in the two phases. In gases under standard conditions these distances are some three orders of magnitude greater than the molecular dimensions. Liquid molecules are by contrast closely packed, with intermolecular distances of the same order as the molecular size.

Gas molecules spend most of their time in transit between collisions and are only modestly affected by intermolecular forces. In liquids these forces are the dominant factor that determines the mobility of the molecules. They are notoriously difficult to quantify, and as a consequence, the prediction of liquid diffusivities has lagged behind theories describing the motion of gas molecules.

The Stokes–Einstein equation, one of the earliest theoretical expressions for liquid diffusivities, viewed the diffusion process as a hydrodynamic phenomenon in which the thermal motion of the molecules is resisted by a Stokesian drag force. This theory, along with subsequent modification by Sutherland and Eyring, established the following proportionality for the diffusion coefficient:

$$D_{AB}(\text{cm}^2 / \text{s}) = \alpha \frac{T}{\mu_B V_A^{1/3}}$$

This relation, which is most successful for large molecules ($V_A > 500$ cm³/ mol), states that diffusivity varies inversely with the viscosity of the solvent and the molecular dimension of the diffusion molecule. This agrees with our intuitive grasp of the process.

Modern theories consider the diffusing particle to be contained in a cage whose dimensions are constantly fluctuating. Local fluctuations in density periodically open holes or vacancies large enough to allow the particle to diffuse out of the cage (Figure 3.1b). Although this view has led to some progress in the quantification of the diffusion process, we are at present still constrained to using semiempirical expressions for the prediction of diffusivities. One such relation, proposed by Wilke and Chang, has been reasonably successful in predicting diffusion coefficients of small molecules in aqueous and organic systems at normal temperatures. It has the form

$$D_{AB}(\text{cm}^2/\text{s}) = 7.4 \times 10^{-8}(\varphi M_B)^{1/2} \frac{T(K)}{\mu_B(cp)V_A^{0.6}} \qquad (3.3)$$

where $M_B$ and $V_A$ (cm³/mol) are the molar mass of solvent and volume of the diffusing species and $\varphi$ is an empirical coefficient with a value of 2.6 for water and 1.0 for unassociated solvents. For organic solutes $V_A$ is composed of atomic and ring contributions, a partial list of which is given in Table 3.2B. An example of the application of Equation 3.3 is given in Practice Problem 3.3.

Diffusivities in molten salts and metals are even more difficult to predict and here we often resort to an Arrhenius-type relation to express the strong temperature dependence of the diffusion coefficient, which is concealed in the viscosity of Equation 3.2 and Equation 3.3:

$$D_{AB} = D_0 \exp(-E_a/RT) \qquad (3.4)$$

This equation also finds use as a correlation for diffusion coefficients in solids. Tabulations of $D_0$ and the activation energy $E_a$ for various species can be found in the pertinent literature. An example of their application appears in Practice Problem 3.10.

Given the uncertainties of current prediction methods for liquid diffusivities, it is clearly preferable to use measured values of $D_{AB}$. Table 3.3 lists diffusion coefficients in water at 25°C of a variety of solutes, including gases (Part A) and ions (Part C), and solutes of a biological or toxic nature (Parts D and E). In Table 3.4 we have reproduced a small selection of diffusivities in liquid metals and molten salts at different temperatures.

The remarkable feature that emerges from these tabulations is the relatively small numerical range of the diffusivities for a wide variety of different substances. Most coefficients cluster around a value of $10^{-5}$ cm²/s, with the lighter and smaller solutes ($H_2$ and He) exceeding this benchmark by a factor of 5 to 7, while very large molecules of a biological origin fall below this value by factors of 2 to 20. Diffusivities in molten metals and salts, which are quite different in nature from normal liquids, likewise fall in the range $10^{-5}$ to $10^{-4}$ cm²/s; the higher values are due mainly to the higher temperatures involved. A rough initial estimate of $D = 10^{-5}$ cm²/s when liquid diffusivities are not known will therefore not be too far off the mark. The corresponding average value for gases is, as was seen, 10,000 times higher at $D = 10^{-1}$ cm²/s. However, because gases are much less dense than liquids and consequently have lower concentration gradients, the diffusion rates themselves are only about 10 times higher than the corresponding values in liquids.

**TABLE 3.3**

Diffusivities in Water ($T = 25°C$)

| Diffusing Species | Diffusivity, $cm^2/s \times 10^5$ | Molar Mass |
|---|---|---|
| *A. Gases* | | |
| Hydrogen | 4.8 | |
| Helium | 7.3 | |
| Methane | 1.8 | |
| Ammonia | 2.0 | |
| Carbon monoxide | 2.17 | |
| Oxygen | 2.42 | |
| Nitrogen | 2.0 | |
| Hydrogen chloride | 3.1 | |
| Carbon dioxide | 2.0 | |
| Sulfur dioxide | 1.7 | |
| *B. Liquids* | | |
| Methanol | 1.28 | |
| Ethanol | 1.24 | |
| Acetic acid | 1.26 | |
| Acetone | 1.28 | |
| Benzene | 1.02 | |
| *C. Ions* | | |
| $H^+$ | 9.3 | |
| $OH^-$ | 5.3 | |
| $NH_4^+$ | 2.0 | |
| $Na^+$ | 1.3 | |
| $Mg^{2+}$ | 0.71 | |
| $Cl^-$ | 2.0 | |
| $K^+$ | 2.0 | |
| $Fe^{3+}$ | 0.60 | |
| $Cu^{2+}$ | 0.71 | |
| $NO_3^-$ | 1.9 | |
| $SO_4^{2-}$ | 1.1 | |
| *D. Biological Substances* | | *Molar Mass* |
| Urea | 1.4 | 60 |
| Glucose | 1.3 | |
| Oxygen in blood | 1.4 | 32 |
| Oxygen in muscle tissue | 1.7 | 32 |
| Lysozyme (egg white) | 1.0 | 14,000 |
| Hemoglobin | 0.69 | 68,000 |
| Fibrinogen | 0.2 | 330,000 |
| Tobacco mosaic virus | 0.044 | 40,000,000 |
| *E. Environmentally Toxic Substances* | | |
| Chlorine | 1.45 | |
| Chlorobenzene | 0.91 | |
| 1,2,4-Trichlorobenzene | 0.76 | |
| Mercury | 2.9 | |
| DDT | 0.49 | |
| 2,4,2',4-Tetrachlorobiphenyl (a PCB) | 0.55 | |

**TABLE 3.4**

Diffusivities in Liquid Metals and Salts

| $T \, °C$ | Diffusing Species | Melt | $D \, (cm^2/s) \times 10^5$ |
|:---:|:---:|:---:|:---:|
| 1270 | Fe | Fe (4.6%C) | 10.0 |
| 40 | Hg | Hg | 2.0 |
| 600 | Zn | Zn | 5.0 |
| 450 | Sn | Pb | 2.0 |
| 906 | $Na^+$ | NaCl | 14.2 |
| 933 | $Cl^-$ | NaCl | 8.8 |
| 328 | $Na^+$ | $NaNO_3$ | 2.0 |
| 328 | $NO_3^-$ | $NaNO_3$ | 1.36 |

To demonstrate the use of liquid-phase diffusivities, we turn to the somewhat unusual case of ion migration in an electrolytic process, and the electrical current that results from it.

### Illustration 3.2: Electrorefining of Metals. Concentration Polarization and the Limiting Current Density

The final processing in the production of high-purity metals is often carried out electrolytically and is referred to as electrorefining. In this process the metal to be refined, such as copper or silver, has a typical initial purity of 95 to 99% and the aim is to reduce the impurity level to less than 0.1%. Conventional purification processes are often either inadequate or too expensive for this purpose. In electrorefining, the impure metal, e.g., copper, is placed in an electrolytic bath as an anodic plate that is paired with a cathode on which the purified metal is deposited electrolytically. The electrolyte typically consists of an aqueous solution of a salt of the metal to be purified, for example, copper sulfate, and the electrolytic cell is composed of an array of closely spaced alternating cathodes and anodes. A sample electrode pair and the configuration of the electrolytic cell are shown in Figure 3.2a.

When a potential is applied to the electrodes, two processes take place. At the anode, the metal, along with its impurities, dissolves as positive cations into the electrolytic bath. These ions migrate to the negative cathode where they are discharged as metal. The impurities, such as iron, usually require a higher potential to be deposited, and consequently remain in solution.

As a result of the applied voltage and the attendant migration of ions, a concentration gradient develops within the bath that ultimately leads to a linear steady-state profile shown in Figure 3.2b. A distinction must be made between the behavior of the cations, here exemplified by $Cu^{++}$, and the negative counterions represented by $SO_4^{--}$. For the former, the transport rate is made up of two components that act in the *same* direction (Figure 3.2c). Diffusional transport with a rate $N$, and transport due to the applied electrical potential, is represented by the rate $F_E$. We can then write

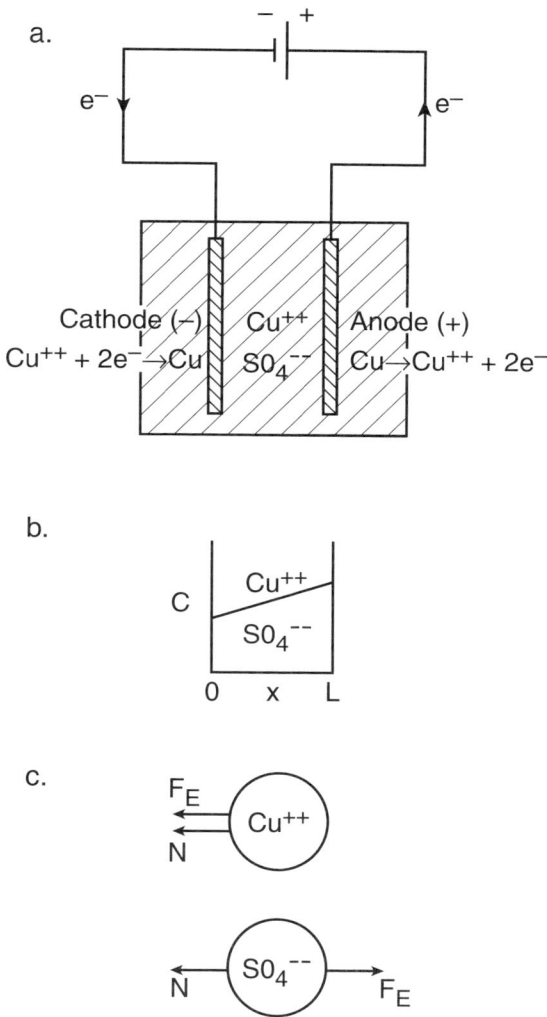

**FIGURE 3.2**
Electrorefining and concentration polarization: (a) sample electrode pair and electrolytic bath; (b) concentration gradients; (c) flux due to diffusion and electrostatic potential.

$$N_{Tot}/A = D_{Cu^{++}} \frac{dC_{Cu^{++}}}{dx} + F_E/A \qquad (3.4a)$$

Diffusional flux          Electrostatic flux

For the anions, the two transport mechanisms act in opposite directions (Figure 3.2c). $SO_4^{--}$ ions migrate toward the cathode by diffusion, as do the copper ions, but they are repelled there and driven back into solution by the negative charge of the electrode. The result is that there is no net flux of the anions. Diffusional transport is exactly balanced by electrostatic transport and we can write

$$0 = D_{SO_4^{--}} \frac{dC_{SO_4^{--}}}{dx} - F_E / A \tag{3.4b}$$

We note that $C_{SO_4^{--}} = C_{Cu^{++}}$, and for electrical neutrality to be maintained, we must have $D_{SO_4^{--}} = D_{Cu^{++}}$. Adding Equations 3.4a and 3.4b to eliminate the electrostatic flux, we obtain

$$N_{Tot} / A = 2 D_{Cu^{++}} \frac{dC_{Cu^{++}}}{dx} \tag{3.4c}$$

On integration by separation of variables there results

$$N_{Tot} / A = 2 \frac{D_{Cu^{++}} (C_{Cu^{++}}|_L - C_{Cu^{++}}|_0)}{L} \tag{3.4d}$$

The maximum flux occurs when the concentration at the cathode has dropped to zero. Noting that the concentration at the anode is then twice the initial electrolyte concentration, $C_{Cu^{2+}}^o$, we can write

$$(N_{Tot} / A)_{Max} = 4 \frac{D_{Cu^{++}} C_{Cu^{++}}^o}{L} \tag{3.4e}$$

Let us apply this equation to a particular condition. Suppose that the electrodes are spaced 1 cm apart and the initial concentration of copper sulfate is 0.1 molar. Using the diffusivity for $Cu^{++}$ listed in Table 3.3C we then obtain

$$(N_{Tot} / A)_{Max} = \frac{4 \times 0.71 \times 10^{-5} \times 10^{-4}}{1} \tag{3.4f}$$

$$(N_{Tot} / A)_{Max} = 2.84 \times 10^{-9} \; mol / cm^2 s \tag{3.4g}$$

*Comments:*
Equation 3.4e gives a good semiquantitative description of the behavior of electrolytic cells and their counterpart, the galvanic cells or batteries. It shows, first and foremost, that there is a limit to the rate of electrolytic deposition or production, which cannot be exceeded by increasing the

applied voltage. Similarly, there is a limiting maximum current that can be drawn from batteries. These limits are imposed by the diffusional processes, which accompany all electrochemical processes, be they galvanic or electrolytic in nature. Equation 3.4e also shows the need for close electrode spacing to achieve high production rates. In car batteries, for example, the plate spacing is of the order of 1 mm.

Deviations from Equation 3.4e occur mainly as a result of the concentration dependence of the diffusivities. This is the case particularly at high electrolyte concentration where electrostatic interactions reduce the ion mobility below the values listed in Table 3.3C. The latter apply to conditions at infinite dilution.

Equations of the form of Equation 3.4e can also be used to calculate the current associated with electrochemical processes. To do this, we use the conversion factor given by the so-called Faraday number $\Im$:

$$\Im = 96,520 \text{ C/mol electrons} \tag{3.5}$$

where C denotes coulombs or ampere-seconds.

Let us use this conversion and Equation 3.4e to calculate the current associated with the electrorefining of copper. We assume the unit to be composed of 20 cathodes and 20 anodes, each $1 \times 1$ m in dimension and spaced $L = 2$ cm apart. Diffusivity of the copper ions is $0.71 \times 10^{-5}$ cm²/s (see Table 3.3C) and the concentration $C$ is set at $10^{-3}$ mol/cm³. We obtain for the current $i$

$$i = 2\Im \times A \times D_{Cu^{++}} \times C / L \tag{3.6a}$$

$$i = 2 \times 96,520 \times (20 \times 100 \times 100) \times 0.71 \times 10^{-5}/2 \times 10^{-3} \tag{3.6b}$$

This yields

$$i = 137 \text{ A} \tag{3.6c}$$

Note that while currents in these operations are considerable, the applied voltage is quite small. Dissolution and deposition potentials almost exactly cancel each other and the only voltage drop that occurs is due to the Ohmian resistance of the electrolyte solution. This rarely amounts to more than a fraction of a volt.

The reader is directed to Practice Problem 3.4, which deals with the calculation of the size of an electrorefining plant.

### 3.1.3 Diffusion in Solids

In the diffusion through solids, several distinct cases arise that depend on the nature of the diffusing species and of the solid medium. The diffusing species can be gaseous or liquid in form, or, surprisingly, can also be a solid. For the solid medium, a distinction is made between consolidated media, such as polymers, and those that have a porous structure. We limit ourselves here to a discussion of three important cases of transport through solids, which we take up in turn: diffusion of gases through polymers and metals and through porous media, and the interdiffusion of solids.

### 3.1.3.1  Diffusion of Gases through Polymers and Metals

The diffusion of gases through polymers and similar consolidated media is viewed as a three-step process (Figure 3.3): (1) At the high-pressure interface, the gas dissolves or condenses in the solid matrix. (2) Following dissolution or condensation, the gas diffuses along a solid-phase concentration gradient $dC/dx$ in accordance with Fick's law to the low-pressure interface. (3) On arrival at the low-pressure gas–solid boundary, the dissolved gas is desorbed or released to the gaseous medium.

In Steps 1 and 3, the gas is assumed to be in instantaneous equilibrium with the neighboring solid matrix. The solid-phase concentration $C$ is related to the gas pressure by a linear relation, given as

$$C = Sp \qquad (3.7)$$

where $S$ is termed the solubility of the gas in the solid. The units used for $C$ are somewhat unconventional and are expressed as $cm^3$ gas at STP/$cm^3$ solid. Units for pressure $p$ are either in atmospheres or Pa, both of which are

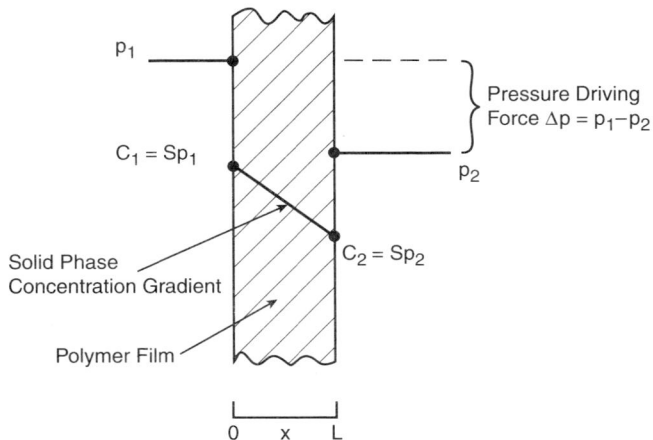

**FIGURE 3.3**
External gas pressures and internal solid-phase concentration in the diffusion of gases through polymers.

currently still in use. As a consequence of this dual usage, the solubility $S$ is reported either in units of cm³ (STP)/cm³ Pa or cm³ (STP)/cm³ atm.

A dual approach is also used in the formulation of the diffusional rate laws. In the first version, Fick's law is used with concentrations as defined above; i.e., we have

$$N(cc\,STP\,/\,s) = -DA\frac{dC}{dx} \tag{3.8a}$$

where $D$ is the diffusion coefficient in cm²/s.

The second and preferred version uses a pressure gradient and leads to the rate law

$$N(cc\,STP\,/\,s) = -PA\frac{dp}{dx} \tag{3.8b}$$

Here $P$ is the so-called permeability, which is most commonly expressed in units of cm³ (STP) × cm/cm² s or cm³ (STP) × cm/cm² s Pa. Several other additional units are still in current use and the translation from one set of units to another is a frequent necessity. To ease this task, we have provided in Table 3.5 a listing of the most commonly required conversion factors.

Permeability is related to the Fickian diffusivity $D$ through the solubility $S$. This is seen by substituting Equation 3.7 into Equation 3.8a and comparing the result with Equation 3.8b. We obtain the relation

$$P = DS \tag{3.9}$$

which can be used to calculate $D$ from independent measurements of permeability and the equilibrium solubility $S$.

In contrast to the diffusivities in gases and liquids, which cluster around values of $10^{-1}$ and $10^{-5}$ cm²/s, the diffusion coefficients and permeabilities we encounter here are not confined to a narrow numerical range. They are strongly material-dependent and range over several orders of magnitude. This is shown in Table 3.6, which lists values of $P$ and $S$ for six common gases and vapors in several commercial polymers.

Permeabilities typically vary over the range $10^{-17}$ to $10^{-10}$ cm³ STP × cm/cm² s Pa, values for $D = P/S$ over the somewhat narrower range of $10^{-9}$ to $10^{-5}$ cm³/s, with no apparent relation to species and material properties.

Even odder behavior is observed when we compare the mobility of different gases in the same material. Large molecules with a high molar mass, which we would intuitively expect to be less mobile than their lighter counterparts, can in fact have considerably higher permeation rates. Thus, in PVC, water vapor with a molar mass 4.5 times that of helium and twice its molecular size nevertheless has a permeability rate more than 100 times that of helium. This reinforces the notion that the movement of gas molecules

**TABLE 3.5**

Conversion Factors for Gas Permeabilities in Solids

| From | Multiplication Factors to Obtain $P$ in | | |
|---|---|---|---|
| | $\dfrac{cm^3\, cm}{cm^2\, s\, cm\, Hg}$ | $\dfrac{cm^3\, cm}{cm^2\, s\, Pa}$ | $\dfrac{cm^3\, cm}{m^2\, day\, atm}$ |
| $\dfrac{cm^3\, cm}{cm^2\, s\, cm\, Hg}$ | 1 | $7.5 \times 10^{-4}$ | $6.57 \times 10^{10}$ |
| $\dfrac{cm^3\, mm}{cm^2\, s\, cm\, Hg}$ | $10^{-1}$ | $7.5 \times 10^{-5}$ | $6.57 \times 10^{9}$ |
| $\dfrac{cm^3\, cm}{cm^2\, s\, atm}$ | $1.32 \times 10^{-2}$ | $9.87 \times 10^{-6}$ | $8.64 \times 10^{8}$ |
| $\dfrac{in.^3\, mil}{100\, in.^2\, day\, atm}$ | $9.82 \times 10^{-12}$ | $7.37 \times 10^{-15}$ | $6.45 \times 10^{-1}$ |
| $\dfrac{cm^3\, cm}{cm^2\, s\, Pa}$ | $1.33 \times 10^{3}$ | 1 | $8.75 \times 10^{13}$ |
| $\dfrac{cm^3\, mil}{cm^2\, day\, atm}$ | $3.87 \times 10^{-14}$ | $2.9 \times 10^{-17}$ | $2.54 \times 10^{-3}$ |

through polymers is a highly complex process, which does not exhibit the simple inverse relation to molecular mass and size seen in diffusion through gases and liquids. Much more elaborate theories, which are still in a state of development, are required to quantify this process.

The following example provides some practice in the use of solubilities and permeabilities.

## *Illustration 3.3: Uptake and Permeation of Atmospheric Oxygen in PVC*

Consider a polyvinyl chloride sheet 0.1-mm thick with a one-sided area of 1 m³. Its density is 1.1 g/cm³ and it is exposed to atmospheric air at 100 kPa. We wish to calculate (1) the uptake of oxygen from the air, and (2) the daily rate of permeation that prevails when one face of the sheet is in contact with the atmosphere, and the other is in contact with pure nitrogen at the same total pressure of 100 kPa.

### 1. Uptake of Oxygen

We start with Equation 3.7 and convert from $C$ (cc STP $O_2/m^3$) to mass $m_{O_2}$ of oxygen. This yields

$$m_{O_2} = S \times p_{O_2} \times V_{PVC} \times M_{O_2} / V_{O_2} \tag{3.10a}$$

**TABLE 3.6**

Permeabilities and Solubilities of Gases in Polymers
(*T* = 25°C unless otherwise indicated)

| Polymer | | H$_2$ | He | O$_2$ | CO$_2$ | N$_2$ | H$_2$O |
|---|---|---|---|---|---|---|---|
| Polyethylene | P | 7.4 | 3.7 | 2.2 | 9.5 | 0.73 | 68 |
| (low density) | S | 1.6 | 0.054 | 0.47 | 2.5 | 0.23 | |
| Polyethylene | P | | 0.86 | 0.30 | 0.27 | 0.11 | 9.0 |
| (high density) | S | | 0.028 | 0.18 | 0.22 | 0.15 | |
| Polystyrene | P | 17 | 14 | 1.9 | 7.9 | 0.59 | 1350 |
| (biaxially oriented) | S | | | | | | |
| Polyvinyl chloride | P | 1.3 | 1.5 | 0.034 | 0.12 | 0.0089 | 206 |
| (PVC-unplasticized) | S | 0.26 | 0.055 | 0.29 | 4.7 | 0.23 | 870 |
| Polyvinyldene chloride | P | | 0.233 | 0.0038 | 0.0022 | 0.00071 | 7.0 |
| (Saran) | S | | (30°C) | (30°C) | (30°C) | (30°C) | |
| Polytetrafluoroethylene | P | 7.4 | 9.0 | 3.2 | 7.5 | 1.0 | 6.8 (38°C) |
| (Teflon) | S | 4.9 | 1.1 | 2.1 | 9.2 | 1.2 | |
| Polychloroprene | P | 10 | | 3.0 | 19.0 | 0.88 | 683 |
| (Neoprene G) | S | 0.29 | | 0.74 | 8.2 | 0.36 | |
| Cellulose hydrate | P | 0.0046 | 0.00038 | 0.0016 | 0.19 | 0.0024 | 18,900 |
| (Cellophane) | S | | | | | | |
| Vulcanized rubber | P | 34 | | 15 | 98 | 5.3 | |
| | S | 0.40 | | 0.69 | 8.9 | 0.35 | |

*Note:* P in [cm$^3$ STP × cm/cm$^2$ s Pa]10$^{13}$; S in [cm$^3$ STP/cm$^3$ Pa]10$^6$.

Here $p_{O_2}$ equals the partial pressure of oxygen in the air (Pa), $V_{PVC}$ is the volume of the sheet (cm$^3$), and $M_{O_2}$ and $V_{O_2}$ are the molar mass and volume (STP) of the oxygen, respectively.

Using the values for solubility given in Table 3.6 we obtain

$$m_{O_2} = 0.29 \times 10^{-6} \, cm^3 \, STP / cm^3 \, Pa \times 21,000 \, Pa$$

$$\times (100^2 \times 0.01) cm^3 \times 32 \, g / cm^3 / 22,410 \, cc / mol \qquad (3.10b)$$

and consequently

$$m_{O_2} = 8.7 \times 10^{-4} \, g \qquad (3.10c)$$

i.e., approximately 1 mg.

### 2. Permeation Rate of Oxygen

The relevant expression for this case is Equation 3.8b, which after insertion of the appropriate numerical values yields

$$N(\text{cc STP / day}) = 0.034 \times 10^{-13} \, \text{cc STP cm / cm}^2 \, \text{s Pa}$$

$$\times \ 10^4 \, \text{cm} \times (21{,}000 \, \text{Pa / 0.01 cm})$$

$$\times \ 3{,}600 \, \text{s / h} \times 25 \, \text{h / day}$$ (3.10d)

$$N = 6.17 \, \text{cc STP/day}$$ (3.10e)

or equivalently

$$N'(\text{g / day}) = \frac{6.17}{22{,}410} \, 32 = 8.8 \times 10^{-3} \, \text{g / day}$$ (3.10f)

*Comments:*
The permeation rate of atmospheric oxygen in PVC is seen to be quite low, in spite of the considerable area of the sheet in question and its very small thickness. PVC would consequently be a good packaging material in cases where atmospheric oxygen is to be excluded.

### Illustration 3.4: Sievert's Law: Hydrogen Leakage through a Reactor Wall

Permeation of gases through metals initially follows the same mechanism that applies to diffusion through polymeric materials; i.e., the gas dissolves or condenses in the solid metal matrix. On dissolution, however, a number of diatomic gases such as hydrogen, oxygen, and nitrogen undergo dissociation into their component atoms according to the scheme:

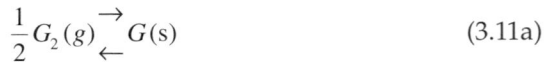

$$\frac{1}{2} G_2(g) \underset{\leftarrow}{\overset{\rightarrow}{}} G(s)$$ (3.11a)

As a consequence of this process, the equilibrium solid-phase concentration $C$ is no longer related linearly to gas pressure but depends instead on the square root of $p$. The linear relation (Equation 3.7), $C = Sp$, must consequently be replaced by the expression

$$C = Kp^{1/2}$$ (3.11b)

where $K$ is the equilibrium dissociation constant for the reaction (Equation 3.11a). A corresponding change must be effected in the definition of the permeability. Whereas previously we had defined $P$ as $P = DS$ (Equation 3.9), we now write

$$P = DK$$ (3.11c)

where $P$ has the new units of $\text{cm}^3$ STP cm/cm$^2$ s atm$^{1/2}$.

Expression 3.11b is known as Sievert's law and can be used to calculate the permeation rate of diatomic gases through metals.

The example considered here involves a tubular reactor made of steel that is to be used in the high-pressure hydrogenation of hydrocarbon vapor. The vessel is 3 m long, has an internal radius $r_i$ of 5 cm, and is to be operated at a pressure of 100 atm and a temperature of 450°C. The permeability $P$ at this temperature has a value of $8.4 \times 10^{-6}$ cm³ (STP) cm/cm² s atm$^{1/2}$.

We wish to calculate the wall thickness required to keep hydrogen losses within reasonable bounds. The reader may recall that a similar configuration and process had been considered in Illustration 1.2 to derive the diffusion rate through a hollow cylinder. That rate was given by the expression

$$N = D2\pi L \frac{C_i - C_o}{\ln r_o / r_i} \tag{1.7c}$$

where the subscripts $i$ and $o$ refer to inside and outside conditions.

For the case at hand, we replace $C$ by pressure using Equation 3.11b and eliminate $D$ by introducing the permeability $P$ given by Equation 3.11c. This leads to the result

$$N[\text{cm}^3 \text{(STP)} / \text{s}] = P2\pi L \frac{\left(\sqrt{p_i} - \sqrt{p_o}\right)}{\ln r_o / r_i} \tag{3.11d}$$

Assuming negligible external hydrogen pressure ($p_o = 0$), we obtain

$$N[\text{cm}^3 \text{(STP)} / \text{s}] = 8.4 \times 10^{-6} \, 2\pi 300 \frac{\sqrt{100}}{\ln r_o / r_i} \tag{3.11e}$$

The resulting diffusion rates as a function of external radius $r_o$ are tabulated below.

| $N$[cm³ (STP)/s] | $r_o$ (cm) | $r_o - r_i$ (cm) |
|---|---|---|
| 1.7 | 5.5 | 0.5 |
| 0.87 | 6 | 1 |
| 0.33 | 8 | 3 |
| 0.23 | 10 | 5 |
| 0.18 | 12 | 7 |
| 0.11 | 20 | 15 |

We note that, after an initial rapid drop, the diffusion rate tapers off asymptotically with an increase in wall thickness $r_o - r_i$ due to the logarithmic dependence on external radius. A wall thickness of 0.5 cm is clearly insufficient, as it leads to substantial hydrogen losses at a level of 1.7 cc (STP)/s. On the other hand, tripling the wall thickness from 5 cm to the inordinately high value of 15 cm merely brings about a reduction from 0.23 to 0.11 cc (STP)/s. A wall thickness of 5 cm therefore appears to be a reasonable compromise value. Ventilation will nevertheless have to be provided to prevent a dangerous buildup of hydrogen.

### *Illustration 3.5: The Design of Packaging Materials*

The use of packaging materials is today all-pervasive. Virtually every product that reaches the stores and the consumer is packaged at some stage during its passage from the manufacturing plant to its ultimate destination. Packaging in crates, cardboard boxes, or sacks and bags eases handling and provides protection from damage during transportation.

   Additional packaging may be provided to prolong shelf life. Food items are often packaged to maintain freshness and prevent the loss of moisture. In other cases, the opposite result is desired: Atmospheric moisture is the enemy and must be excluded by a suitable protective barrier. This arises in the packaging of moisture-sensitive items such as electronic components. Both instances require a barrier with low water permeability.

   Consider the case of a moisture-sensitive item that must be kept in an atmosphere of less than 10% relative humidity. This means that the partial pressure of moisture in the air cannot exceed 10% of the saturation vapor pressure. To protect the item against accidental excursions of the humidity of the surrounding air, it is proposed to package it in an appropriate material. Inspection of Table 3.6 shows that high-density polyethylene has a suitably low permeability to water vapor. We now stipulate that the thickness of the packaging material should be sufficient to protect the item against the accidental exposure to 95% humidity air of 1-h duration.

   It is not immediately clear at the outset how this information is to be obtained. Because the permeation process is an unsteady one, a good way to start is to set up an integral unsteady moisture balance, in which the rate of permeation into the package is balanced by the change in moisture content of the interior air. Some further thought will then reveal that the thickness $L$ being sought resides in the permeation gradient $\Delta p/L$. We can consequently write

$$\text{Rate of moisture in} - \text{Rate of moisture out} = \begin{array}{c} \text{Rate of change} \\ \text{of moisture} \\ \text{content} \end{array}$$

$$PA\frac{p_o - p_i}{L} - 0 = \frac{d}{dt}n_{H_2O} \qquad (3.12a)$$

where $p_o$ and $p_i$ are the constant exterior and varying interior partial water pressures and $n_{H_2O}$ equals the moles of water vapor contained in the package. As we have two dependent variables in $p_i$ and $n_{H_2O}$, a second expression relating the two is required. This relation is provided by the ideal gas law

$$n_{H_2O}(\text{mol}) = \frac{p_i V}{RT} \tag{3.12b}$$

Combining Equations 3.12a and 3.12b and introducing the molar conversion factor then yield

$$PA(p_o - p_i)/22.410 \, L = \frac{V}{RT}\frac{dp_i}{dt} \tag{3.12c}$$

where the factor 22.410 is used to convert cm³ (STP) contained in the permeability to units of mole.

Integrating by separation of variables, we obtain

$$\frac{(PA)(RT)}{22,410 \, VL}\int_0^t dt = \int_{p_i^o}^{p_i^f}\frac{dp_i}{p_o - p_i} \tag{3.12d}$$

and consequently

$$\frac{(PA)(RT)t}{22,410 \, VL} = \ln\frac{p_o - p_i^o}{p_o - p_i^f} \tag{3.12e}$$

where the superscripts $o$ and $f$ denote the initial and final moisture content of the interior air. The desired packaging thickness is obtained by solving for $L$:

$$L = \frac{(PA)(RT)t}{22,410 \, V \ln\frac{p_o - p_i^o}{p_o - p_i^f}} \tag{3.12f}$$

We set temperature at 25°C, area-to-volume ratio $A/V$ at 2, and the initial interior humidity at zero. Using a permeability of $9 \times 10^{-13}$ cm³ STP cm/cm² s Pa listed in Table 3.6 we obtain

$$L = \frac{(9\times10^{-13})\times(2)\times(8.314\times10^6)\times(298)\times(3600)}{22,410\times\ln\frac{0.9 \, p_{sat}}{0.9 \, p_{sat} - 0.1 \, p_{sat}}} \tag{3.12g}$$

$$L = 0.0061 \text{ cm} \tag{3.12h}$$

*Comments:*

This thickness of less than 0.1 mm is not unduly large. In fact, we could extend the exposure time to 10 h without exceeding acceptable thickness limits.

### 3.1.3.2  Diffusion of Gases through Porous Solids

The permeation process considered in the previous section was somewhat unusual. It required the prior dissolution of the gas in the solid matrix before it could make its way through the medium by a process of solid phase Fickian diffusion. A more conventional permeation process occurs when the solid involved is porous or is composed of loosely packed particles. Here the permeating gas enters the solid through the pore openings and continues its way through the porous passages, all the while remaining in the gas phase. Diffusion is strictly Fickian and no penetration of the solid matrix per se takes place.

To describe this process, two factors need to be taken into account. The first is the reduction in cross-sectional area available for diffusion, which reduces the diffusion coefficient by a factor equal to the void fraction $\varepsilon$ of the solid. Here $\varepsilon$ is expressed as the ratio of open cross-sectional area to total cross-sectional area and ranges from 0 to 1. The second factor is due to the tortuous nature of the porous pathway that is often associated with these media. The gas molecules are made to zigzag their way through the solid, rather than going straight through, occasionally coming to a complete halt in so-called dead-end pores (Figure 3.4a and Figure 3.4b). The net effect is to lengthen the diffusional pathway and consequently reduce the effective diffusion rate. This lengthening of the path and the effect of dead-end pores is accounted for through the so-called tortuosity factor $\tau$, which has a value greater than unity. Both of these effects are lumped into the diffusion coefficient, resulting in an effective and reduced diffusivity $D_e$; i.e., we have

$$D_e = \frac{D\varepsilon}{\tau} \tag{3.13a}$$

where $D$ remains the ordinary diffusivity applicable to free space. $D_e$ is used in conjunction with Fick's law, which retains its original form; i.e., we have

$$N/A = -D_e \frac{dC}{dx} \tag{3.13b}$$

Void fractions for many porous media typically vary over the range 0.1 to 0.5, with 0.3 a good average value. Tortuosity $\tau$ has a range of 1.5 to 10, with occasional excursions to higher values. A value of $\tau = 4$ gives a good initial estimate in the absence of precise data. The combined effect of the reduction in cross-sectional area and the lengthening of the diffusional pathway is to

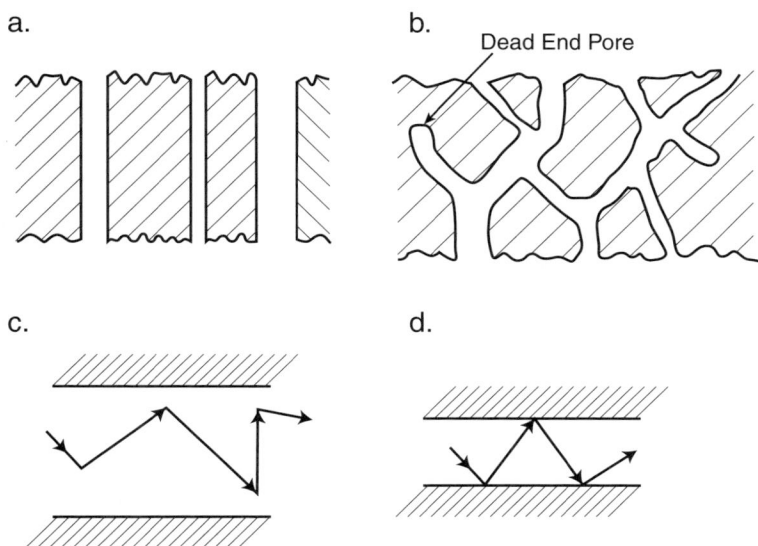

**FIGURE 3.4**
Diffusion in a porous solid: (a) straight-through pores; (b) tortuous pores with branching; (c) molecular diffusion; (d) Knudsen diffusion.

reduce the free-space diffusivity by approximately one order of magnitude. Thus, the diffusion of oxygen in air, previously set at 0.2 cm²/s (Table 3.1), now drops by a factor of 10 to approximately 0.02 cm²/s.

We have up to this point assumed that permeation in the pores follows the mechanism of gaseous diffusion in free space, i.e., that the gas molecules undergo repeated collisions with each other as they progress through the medium. However, when pore diameter drops below the value of the mean free path λ, the diffusional mechanism undergoes a change to what is termed *Knudsen diffusion*. Here the impeding collisions no longer occur between gas molecules, but rather between gas molecules and the wall of the pore itself. The pathway is still a random one, but it now zigzags between the walls of the pores, which deflect the gas molecules into a new direction (Figure 3.4d). Equation 3.2, which was given for diffusivity in free space, no longer applies. Although Knudsen diffusion still varies directly with temperature, and inversely with molar mass $M$, it now also depends on the radius of the pore itself. The relevant expression is given by

$$D_K(\text{cm}^2/\text{s}) = 9700 r_p (T/M)^{1/2} \tag{3.14}$$

where $r_p$ is the pore radius in cm and $T$ the absolute temperature in Kelvin. Knudsen diffusivities can be several orders of magnitude smaller than molecular diffusivities and are the controlling transport coefficient in many diffusional processes through porous media.

To illustrate diffusion through porous media, we consider two examples. The first deals with diffusion into and out of a leaf, which takes place through tiny, straight-through pores in the underside of leaves. This illustration, and the practice problem that accompanies it, introduces the reader to the fascinating world of plant physiology and the biophysical processes that sustain plant life, as well as our own. In the second example we consider diffusion in a porous catalyst pellet. Here the pore structure is no longer "straight-through" but becomes tortuous and often gives rise to Knudsen diffusion because of the small dimension of the pores. Transport into and out of catalyst particles plays an important role in determining the overall performance of the catalyst, and is taken up in more detail in Chapter 4.

### Illustration 3.6: Transpiration of Water from Leaves. Photosynthesis and Its Implications for Global Warming

The principal transport processes in a leaf, apart from the conveyance of nutrients, are the uptake of carbon dioxide from the air, and the release of oxygen and of water vapor to the atmosphere. The latter process is referred to as transpiration.

The cell structure of a typical leaf is shown in Figure 3.5. Transport takes place mainly through openings termed *stomatal pores* (*stoma*: Greek for *mouth*), which are concentrated at the underside of the leaf. Stomata have a typical length of 10 to 20 μm, an average radius of 5 to 10 μm and cover a fractional area ranging from 0.002 to 0.02 (0.2 to 2%). The remainder of the leaf surface is covered by a layer termed *cuticle,* which is essentially impermeable to gases (Figure 3.5).

The stomatal pores are flanked on either side by guard cells that control the size of the pore opening by expanding or contracting in response to external stimuli. In the dark, and at low external humidities, the guard cells are triggered to expand, partially closing the stomatal pores. Under these conditions, transport of gases through the pores is reduced or ceases entirely. The movement of the guard cells, which causes the losses, is brought about by a change in osmotic pressure of the cell fluids.

The interior of the leaves contains the *mesomorphic cells,* which are arranged in either a loosely packed, "spongy" configuration or in the denser "palisade" form. The intervening spaces are taken up by intercellular air. These mesomorphic cells contain smaller cells termed *chloroplasts,* which in turn carry chlorophyll, the principal substance responsible for photosynthesis, i.e., the conversion of carbon dioxide into organic compounds and oxygen. This process is addressed in greater detail in Practice Problem 3.7.

In the present illustration we consider the transport of water vapor from the interior of the leaf through the stomatal pores into the surrounding atmosphere. To obtain an assessment of the maximum possible moisture loss, we assume the interior of the leaf to be saturated with water vapor.

In principle, the external resistance will depend on wind conditions, which vary with time as well as with location. Extensive studies have shown that

a.

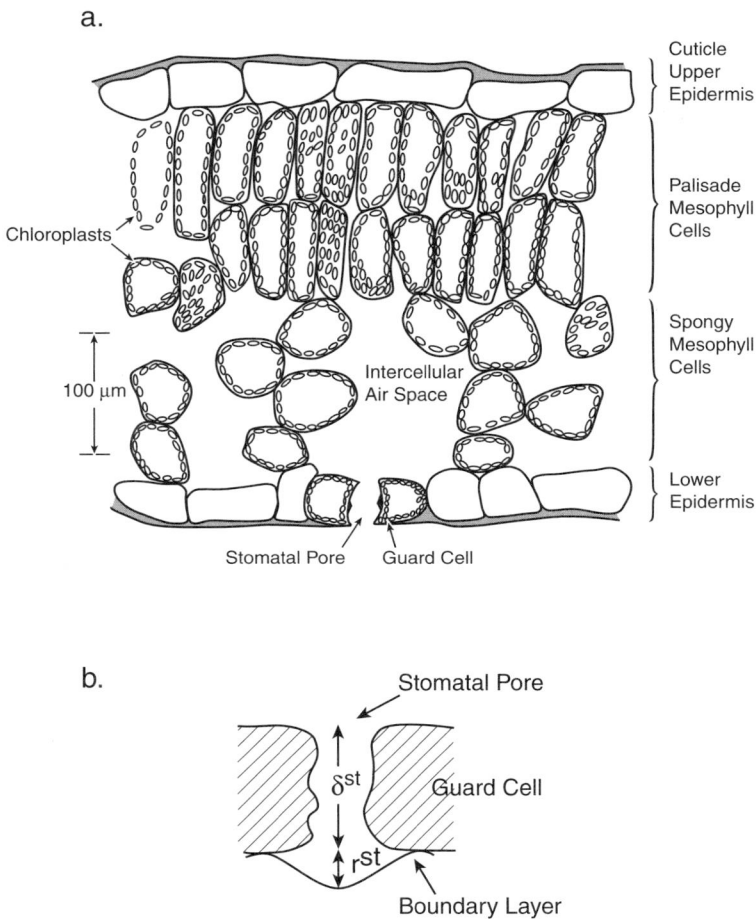

Cuticle
Upper
Epidermis

Palisade
Mesophyll
Cells

Chloroplasts

Spongy
Mesophyll
Cells

100 μm

Intercellular
Air Space

Lower
Epidermis

Stomatal Pore    Guard Cell

b.

Stomatal Pore

$\delta^{st}$    Guard Cell

$r^{st}$

Boundary Layer

**FIGURE 3.5**
Mass transport in a leaf: (a) leaf cell organization; (b) transport in a stomatal pore.

this resistance can be accounted for in an approximate fashion, by adding a distance equal to the average stomatal radius, $r^{st}$, to the length of the pore, $\sigma^{st}$. The total effective length then becomes $\ell^{st} = \sigma^{st} + r^{st}$ (Figure 3.5b).

Let us use these concepts to calculate the water loss from a typical garden or city tree of modest size. We assume a stomatal length and radius of $\sigma^{st} = 15 \times 10^{-6}$ m and $r^{st} = 7.5 \times 10^{-6}$ m, with a fractional open area of 1% ($\varepsilon = 0.01$), and set the leaf area at 25 cm² and the leaf population at 10,000 leaves per tree. External humidity is assumed to be 50% at a temperature of 25°C, with a corresponding saturation water vapor $p_{sat}$ pressure of 23.8 mmHg. Diffusivity $D$ of water vapor in air is $2.6 \times 10^{-5}$ m²/s (Table 3.1). The relevant transport equation is then given by

$$N = D\varepsilon A \frac{\Delta C}{\Delta x} = D\varepsilon A \frac{C_{sat} - 0.5 C_{sat}}{\sigma^{st} + r^{st}} \tag{3.15a}$$

We first calculate the saturation concentration $C_{sat}$, which is given by

$$C_{sat} = \frac{n_{sat}}{V} = \frac{p_{sat}}{RT} = \frac{(23.8/760)1.01 \times 10^5}{8.31 \times 298} = 1.28 \, mol/m^3 \tag{3.15b}$$

Substitution of this and other numerical values into Equation 3.15a then yields

$$N = 2.6 \times 10^{-5} \times 0.01 \times 10^4 \times 25 \times 10^{-4} \frac{1.28 - 0.5 \times 1.28}{15 \times 10^{-6} + 7.5 \times 10^{-6}} \tag{3.15c}$$

$$N = 0.185 \, mol/s \tag{3.15d}$$

This is the maximum possible moisture loss under the stipulated conditions. Actual losses may be considerably lower due to less than saturation values in the interior and due to evaporative cooling, which lowers the interior water vapor pressure and hence diminishes the driving force.

*Comments:*

The first impression of this model is its striking simplicity in the face of fairly complex circumstances. The fact that the interior of the leaf was taken to be at saturated water levels is acceptable in view of the larger surface area of cells and capillaries from which water emanates. The use of an equivalent resistance for the boundary layer equal to the radius of the stomata is not immediately transparent but becomes more reasonable on closer scrutiny of the configuration involved. Because no water vapor emanates from the leaf except at the stomata, the humidity will be constant and equal to the external level over the impermeable portion of the leaf. Concentration cannot change abruptly to the level prevailing at the pore mouth, and we must therefore expect that concentration contours in the form of humps will develop around the opening (Figure 3.5b). Based on these concepts, early workers postulated an equivalent resistance equal to the average pore radius. This was an inspired approximation, which was later confirmed in more elaborate studies. These more refined results, while much closer conceptually to the actual situation, nevertheless did not result in substantial changes in the original approximation. The use of the equivalent length $r^{st}$ is now firmly entrenched in the literature.

In Practice Problem 3.7 we examine the reverse process of $CO_2$ uptake by the leaf. The transport of carbon dioxide into the leaf interior is a more complex phenomenon involving several resistances as well as the reaction steps of photosynthesis. It is nevertheless possible to arrive at some simple results that illuminate this hugely important process with implications for global warming. Practice Problem 3.9 balances these results against the daily emissions of an average car.

## Illustration 3.7: Diffusivity in a Catalyst Pellet

Catalysts almost invariably consist of porous particles with a substantial internal surface area, which typically varies from a few square meters per gram to a few tens or even 100 $m^2/g$. For a reaction to occur, the reacting species must diffuse into the interior of the particle to reach the reactive surface. Conversely, the product gases, once formed, must in turn diffuse out in order to maintain a steady state. Both the diffusional process and the local reaction rate play a role in determining the overall rate of conversion of reactants. When diffusion is fast, which is the case for small particles with large pores, the reaction becomes the rate-determining step. Conversely, when the particle is large and the pores small, the rate of diffusion becomes the dominant factor. This interplay of diffusion and reaction is examined in greater detail in Chapter 4 (Illustration 4.9). It is shown there that the impeding effect of diffusional resistance can be expressed in terms of an effectiveness factor $E$, which varies in value from zero to one. When diffusion is fast, reaction is the dominant process and $E$ approaches one. Conversely, low diffusion rates and fast reactions lead to small values of the effectiveness factor.

The evaluation of $E$ requires a knowledge both of the diffusivity and the reaction rate constants. Our task in the present illustration is to estimate the diffusivity of a reacting species, given certain physical parameters of the catalyst particle.

Let us consider the diffusion of oxygen in a silica-alumina cracking catalyst with an average pore radius of 24 Å = $24 \times 10^{-8}$ cm and a void fraction $\varepsilon$ of 0.3 $cm^3/cm^3$. The tortuosity factor is not known, and we consequently use an average value of $\tau = 4$. Because pore radius is much smaller than the mean free path, which is of the order of $10^{-5}$ cm at atmospheric pressures, it is suspected that Knudsen diffusion may be operative. We turn to Equation 3.14 and obtain, for a reaction temperature of 420 K,

$$(D_K)_{pore} = 9700 \times 24 \times 10^{-8} (420/32)^{1/2} \tag{3.16a}$$

$$(D_K)_{pore} = 8.5 \times 10^{-3} \text{ cm}^2/\text{s} \tag{3.16b}$$

This value is more than 10 times lower than the molecular diffusivities of 0.20 $cm^2/s$. We conclude therefore that Knudsen diffusion is indeed the operative mode of diffusion. To calculate the effective diffusivity for the entire particle, we draw on Equation 3.13a and obtain

$$D_e = D_{pore}\varepsilon/\tau = 8.5 \times 10^{-3} \times 0.3/4 \tag{3.16c}$$

$$D_e = 6.4 \times 10^{-4} \text{ cm}^2/\text{s} \qquad (3.16\text{d})$$

This is the effective diffusivity to be used in assessing catalyst performance.

### 3.1.3.3  Diffusion of Solids in Solids

The diffusion of solid ions, atoms, or molecules through solid matrices, while seemingly not possible, can and does in fact take place. It does so by a mechanism akin to that which described liquid-phase diffusion: Local density fluctuations lead to a momentary opening or vacancy into which a neighboring particle can displace itself (see Figure 3.1b). Thus, a diffusional flow occurs, which follows Fick's law, as do more conventional diffusional processes.

A short compilation of important diffusion coefficients is presented in Table 3.7. Several features are of note here. The first is the strong dependence on temperature that we see in the diffusivities of boron in germanium, and the self-diffusion of Si. A 10 to 20% increase in absolute temperature can lead to a rise in diffusivity of several orders of magnitude. The temperature dependence is clearly an exponential one and is commonly expressed by an Arrhenius-type relation:

$$D = D_o \exp(-E_a/RT) \qquad (3.17)$$

where $E_a$ is an activation energy and $D_o$ is a preexponential factor. The implication here is that the solid-phase diffusion is to be viewed as an activated process in which the diffusing particle has to attain a threshold activation energy $E_a$ before displacement can take place.

A second feature seen in Table 3.7 is the strong influence of impurities on the magnitude of diffusivities. An impurity level of only 1% can result in a dramatic increase in the diffusion coefficient of several orders of magnitude (see the diffusion of carbon in iron). The reason for this lies in the local change in packing of the molecules with an attendant increase in the probability of a vacancy opening up. While the magnitude of solid–solid diffu-

TABLE 3.7

Approximate Diffusivities of Solids in Solids

| Diffusing Species | Solid Matrix | $D$ (cm²/s) | $T$ (K) |
|:---:|:---:|:---:|:---:|
| C | Fe | $6 \times 10^{-6}$ | 1667 |
| C | Fe + 2% Cr | $5 \times 10^{-5}$ | 1667 |
| C | Fe + 1% Mn | $4 \times 10^{-4}$ | 1667 |
| B | Ge | $1 \times 10^{-16}$ | 1000 |
| B | Ge | $3 \times 10^{-14}$ | 1110 |
| B | Ge | $4 \times 10^{-13}$ | 1176 |
| Si | Si | $7 \times 10^{-15}$ | 1429 |
| Si | Si | $3 \times 10^{-12}$ | 1667 |

sivities generally falls below values of $10^{-10}$ cm$^2$/s, much higher levels comparable to those that prevail in liquids are also encountered. An example of this is seen in the diffusion of carbon in solid iron (Table 3.7).

The reader may be puzzled about the relevance of solid–solid diffusional processes. The low values that pertain and the universal nature of the system would appear to make this a topic of mainly academic interest. In fact, the diffusion of solids within solids is highly important in a number of disciplines. In metallurgy, solid-phase diffusion is induced to desegregate local accumulations of alloy components and to relieve stresses caused during casting. This is implemented by maintaining the cast form just below its softening point for prolonged periods of time. The resulting diffusional process is an unsteady one that uses as its initial condition the concentration profile that was derived in Illustration 2.8.

Solid-state diffusion also plays an important role in the manufacture of semiconductors. To produce the junctions needed in these devices, a dopant such as boron is deposited on the surface of the semiconductor crystal, e.g., silicon or germanium, and is subsequently made to diffuse into the interior. This process, termed *drive-in diffusion*, is again carried out at elevated temperatures. An analysis of it and some relevant calculations appear in Chapter 4, Practice Problem 4.4. The diffusivity required in these calculations is derived in the short illustration given below.

### Illustration 3.8: Diffusivity of a Dopant in a Silicon Chip

We wish to calculate the diffusivity of the dopant boron in silicon at a temperature of 1150°C, using tabulated values for the Arrhenius constants of $D_o = 1.5 \times 10^{-4}$ m$^2$/s and $E_a = 357$ kJ/mol.

We substitute these values and other pertinent numbers in Equation 3.17 and obtain

$$D = 5.1 \times 10^{-4} \exp(-357 \times 10^3/8.314 \times 1383)$$

$$D = 1/6 \times 10^{-17} \text{ m}^2/\text{s}$$

## Practice Problems

3.1. *A Fermi Problem: Estimation of Gas Diffusivities*
   Make an order-of-magnitude estimate of the diffusion coefficient in simple gases other than He and $H_2$, given that the velocity of sound at atmospheric pressure is of the order 300 m/s and the liquid to gas densities for most gases are in the approximate ratio of 1:1000. (*Hints:* Sound propagates approximately at the same speed as that

of molecular motion. The diameter of a typical small molecule is of the order of 1 Å = $10^{-8}$ cm.)

3.2. *Pressure Dependence of Diffusivity in Gases*

a. Explain why molecular diffusivity varies inversely with pressure.

b. Why is Knudsen diffusivity independent of pressure?

c. Is Knudsen diffusion more likely to occur at high pressures or low pressures?

3.3. *Estimation of the Diffusivity of DDT in Water*

Use the Wilke–Chang equation (Equation 3.3) to estimate the diffusion coefficient of DDT in water at 25°C. The viscosity of water at this temperature is 0.894 cp (centipoises). Compare the result with the value tabulated in Table 3.3E. (*Hint:* The formula for DDT is given by 1-trichloro-2,2-bis(*p*-chlorophenyl)ethane. It contains two phenol rings and five chlorine atoms.)

3.4. *Electrorefining of Copper*

Copper is to be refined electrolytically at the rate of 10 kg/h using a cell with 2-cm plate spacing and a 1-molar $CuSO_4$ electrolyte.

a. What is the minimum electrode area required to carry out the process?

b. If the electrode dimensions are 2 m × 2 m × 1 cm, what is the minimum length of the electrolytic cell?

*Answer:* a. 309 $m^2$

3.5. *Diffusivity in Polymers*

Although permeability is the most commonly employed transport coefficient for polymers, occasional use is also made of diffusivities.

a. Calculate the range of diffusivities for the substances listed in Table 3.6. How do they compare with the diffusivity of gases in liquids?

b. Repeat the calculation for Illustration 3.3, Part 2, using a diffusivity and solid-phase concentrations instead of permeability and a partial-pressure driving force.

3.6. *Performance of Saran Wrapping*

A package is made of three ears of corn by placing them on a Styrofoam® tray and enclosing them with Saran Wrap® 0.01-mm thick. The exposed area is 400 $cm^2$. It is desired to keep moisture losses below 1% of the total water content estimated at 100 g per package for a shelf life of 10 days. Does the wrapping meet these requirements? Assume a constant internal saturation vapor pressure of 23.8 mmHg (25°C) and zero external moisture.

## 3.7. *Photosynthesis: $CO_2$ Uptake by an Average Tree*

Photosynthesis involves the uptake by plants of carbon dioxide from the surrounding air and its conversion to organic compounds and oxygen

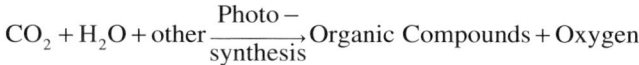

$$CO_2 + H_2O + \text{other} \xrightarrow[\text{synthesis}]{\text{Photo}-} \text{Organic Compounds} + \text{Oxygen}$$

It is the largest synthetic process on Earth and results in the fixing of approximately $7 \times 10^{13}$ kg carbon per year. Transport of $CO_2$ into the leaves is considered to be the rate-determining process in this chain of events and involves a series of steps, which are represented as a sum of resistances. This is akin to the concept we used in Section 1.4 of Chapter 1 to represent transport according to the two-film theory. The Fickian rate process is then expressed in the form

$$N/A = \frac{\Delta C}{\Sigma R} \tag{3.18a}$$

where $\Delta C$ is the concentration driving force and $\Sigma R$ represents the sum of resistances. In a leaf the major resistances are three in number and are due to (1) the stomata, (2) the mesophyll cells, and (3) the chloroplasts (see Figure 3.5). The range of values of resistances is tabulated in Table 3.8. We wish here to calculate the maximum daily $CO_2$ uptake by the modest-sized tree used in Illustration 3.6. To this end we choose the lower values of the resistances listed in Table 3.8 and set the chloroplast resistance at 200 sm$^{-1}$. We further neglect the carbon dioxide produced by the tree by respiration to maintain its own life. This is usually less than 10% of the total carbon dioxide uptake. Because photosynthesis is rapid compared with the transport process, the carbon dioxide concentration within the chloroplasts is set at zero. These are the approximations we use to calculate the carbon dioxide uptake.

*Answer:* 15 mol $CO_2$/day

**TABLE 3.8**

Representative Values of Resistances for $CO_2$ Diffusion into Leaves

| Component | Resistance sm$^{-1}$ |
|---|---|
| Crops, open stomata | 170–830 |
| Trees, open stomata | 500–2500 |
| Mesophyll cells | |
| Estimation | 600 |
| Measurements | 200–800 |
| Chloroplasts | |
| Estimation | <500 |
| Measurements | <400 |

3.8. *Choice of a Packaging Material*

Packaging materials often come in the form of laminates, which are composed of individual film barriers each of which is desired to meet a particular requirement. These include the reduction of light or heat transmission and the retention or exclusion of certain gases and vapors. Most packaged foods continue, after packaging, to undergo certain metabolic changes, including respiration. They are, in a sense, still "alive." These changes are usually undesirable and packaging materials must be designed to reduce these effects to a minimum. Consider the case of respiration, which is analogous to breathing in humans and has to be minimized as far as possible. Which of the materials listed in Table 3.6 would perform best in this respect?

3.9. *Carbon Dioxide Emissions from Cars: How Many Trees Will Compensate the Output of One Vehicle?*

The task in this problem is to compute the daily production of carbon dioxide by a vehicle that is driven an average of 100 km daily (36,500 km per year) and has a gasoline mileage of 15 km/l. Assume the fuel to be pure octane ($C_8H_{18}$) with a specific gravity of 0.8, which is quantitatively burned to carbon dioxide. Compare the calculated result with that given for Practice Problem 3.7.

3.10. *Temperature Dependence of Solid–Solid Diffusivities*

The self-diffusivities of sodium in solid sodium chloride were determined at two temperatures using radioactive sodium, with the following results:

| D, cm²/s | T, K |
|---|---|
| $4.5 \times 10^{-5}$ | 400 |
| $9.2 \times 10^{-5}$ | 667 |

Use these results to derive an analytical expression for the temperature dependence of sodium diffusivity.

# 4

## *More about Diffusion: Transient Diffusion and Diffusion with Reaction*

The process of diffusion is central to much of what we describe as mass transfer. It manifests in a variety of ways and needs to be looked at repeatedly, even at this introductory level, in order to grasp its full ramifications. This was done by first introducing the reader, in Chapter 1, to the notion of the rate of diffusion, enshrined in Fick's law, and to the linear driving force mass transfer rates derived from it. In Chapter 2 we demonstrated how these rate expressions are incorporated into mass balances leading to models of various mass transfer processes. Chapter 3 took up the topic of diffusivities, the all-important proportionality constants in Fick's law, and examined them both at the molecular and macroscopic level.

This chapter returns to the subject of diffusion per se and examines what happens when the rate of diffusion varies with both time and distance (Section 4.1) and when diffusion occurs simultaneously with a chemical reaction (Section 4.2). These are more advanced topics, which in the case of Section 4.1 lead to partial differential equations, notably Fick's equation given in Chapter 2 (Equation 2.18c). We do not attempt to solve it here, which would merely distract us from the main task, and confine ourselves instead to a presentation of the more important results in either analytical or graphical form. These are then used to solve a range of practical problems, a task that is far from trivial in spite of the appearance it gives of applying a set of convenient "recipes." Section 4.2 is confined to steady-state processes in which the state variable varies only with distance. Hence no partial differential equations arise here. We do, however, have to deal with ordinary differential equations, which sometimes require going beyond the elementary separation-of-variables technique seen in previous chapters by using the so-called $D$-operator method. This procedure is outlined in the Appendix at the end of the text.

## 4.1   Transient Diffusion

In transient diffusion, the concentration of a species varies, as we have seen, with both time and distance. The underlying process of diffusion may take place in isolation or it may be accompanied by a chemical reaction, by flow, or by both reaction and flow. These more complex cases are not taken up here, and we limit ourselves instead to the consideration of purely diffusive processes. Furthermore, with one or two exceptions, the treatment is confined to a *single* spatial coordinate represented by the Cartesian $x$- (or $z$-) axis or by the radial variable $r$. The last is used in formulating diffusion in a sphere, or in the radial direction of a cylinder. Fick's equation for these three cases can be deduced from the general conservation equation (Equation 2.24a) and Table 2.3. They are as follows:

Cartesian Coordinate in One Dimension

$$\frac{\partial C}{\partial t} = D\frac{\partial^2 C}{\partial x^2}$$
(2.18a)

Spherical Coordinate

$$\frac{\partial C}{\partial t} = D\left[\frac{\partial^2 C}{\partial r^2} + \frac{2}{r}\frac{\partial C}{\partial r}\right] = D\left[\frac{1}{r^2}\frac{\partial}{\partial r}\left(r^2\frac{\partial C}{\partial r}\right)\right]$$
(2.18b)

Cylindrical Coordinate (Radial Direction)

$$\frac{\partial C}{\partial t} = D\left[\frac{\partial^2 C}{\partial r^2} + \frac{1}{r}\frac{\partial C}{\partial r}\right] = D\left[\frac{1}{r}\frac{\partial}{\partial r}\left(r\frac{\partial C}{\partial r}\right)\right]$$
(2.18c)

The alternative formulations given on the right are sometimes found useful in the solution of steady-state problems.

Associated with these equations are a host of boundary and initial conditions. One can, for example, specify the concentration on the surface, or the flux prevailing there. Alternatively, there may be a film resistance at the surface, in which case the rate of mass transfer must be equated to the rate of diffusion at the surface.

Initial conditions may also be complex. Suppose, for example, that we wish to solve Fick's equation for transient diffusion into a sphere. The simplest case here is to assume that the sphere is initially "clean," i.e., contains no solute. But what if it is not? We would then have to specify an initial concentration distribution $C_{t=0} = f(r,\theta,\varphi)$ and this distribution would have to be entered into the solution process as an initial condition. Evidently there are an infinite number of such distributions; hence Fick's equation for this case will have an unlimited number of solutions.

Yet another complication arises if the diffusion process is triggered by a mass *source*. A puff of smoke emanating from a chimney, for example, constitutes such a source. If it lasts for only an instant, we speak of an *instantaneous source*. If the emanation persists, we speak of a *continuous source*. The existence of such sources must be incorporated into the Fickian model and leads to what we term *solutions of source problems*. Because of their importance, particularly in an environmental context, we address this topic separately in some detail.

This short discussion will have alerted the reader to the dismaying fact that Fick's equation has an unlimited number of solutions, which are intimately linked to the initial and boundary conditions and to the presence of sources. To keep our deliberation to manageable proportions, we propose to address the following limited but, nevertheless, highly important and instructive cases:

1. *Diffusion due to instantaneous and continuous sources emitting into infinite and semi-infinite domains.* Both point sources as well as area sources of infinite extent are addressed. These are referred to as *Source Problems*.

2. *Transient diffusion in a semi-infinite domain, into a solid bounded by parallel planes, in a sphere, and in the radial direction of a cylinder.* The domain is assumed to be initially uniformly loaded or uniformly clean, and to have a constant surface concentration. We term these *Nonsource Problems*. Occasional departures from the stated assumptions are noted as they occur.

### 4.1.1 Source Problems

We start our deliberations by considering a source located at the origin of a three-dimensional Cartesian coordinate system. At time $t = 0$, the source releases a substance of mass $M_p$ kg. The release is instantaneous, and leaves immediately; i.e., it is of infinitesimally short duration. Thereafter the released substance diffuses into the three-dimensional infinite space surrounding the source, giving rise to time-dependent three-dimensional concentration profiles $C(x, y, z, t)$. The three coordinates $x, y, z$ can be combined into a single radial distance $r$ anchored at the origin, where $r^2 = x^2 + y^2 + z^2$. The resulting concentration profiles are given by the following expression:

#### 4.1.1.1 Instantaneous Point Source Emitting into Infinite Space

$$C = \frac{M_p}{8(\pi Dt)^{3/2}} \exp(-r^2 / 4Dt) \qquad (4.1)$$

where $D$ is the diffusivity.

**TABLE 4.1**

Solutions to Source Problems

1. Instantaneous Point Source Emitting into Infinite Space

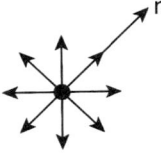

$$C = \frac{M_p}{8(\pi Dt)^{3/2}} \exp(-r^2/4Dt)$$

$M_p$ in kg

2. Instantaneous Point Source on an Infinite Plane Emitting into Half Space

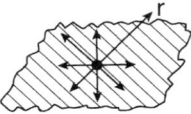

$$C = \frac{M_p}{4(\pi Dt)^{3/2}} \exp(-r^2/4Dt)$$

$M_p$ in kg

3. Instantaneous Infinite Plane Source Emitting into Infinite Space

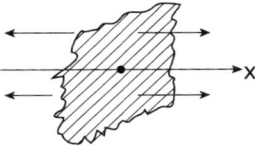

$$C = \frac{M_A}{2(\pi Dt)^{1/2}} \exp(-x^2/4Dt)$$

$M_A$ in kg/m$^2$

4. Instantaneous Infinite Plane Source Emitting into Half Space

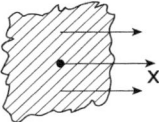

$$C = \frac{M_A}{(\pi Dt)^{1/2}} \exp(-x^2/4Dt)$$

$M_A$ in kg/m$^2$

5. Continuous Point Source Emitting into Infinite Space

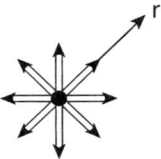

$$C = \frac{M_{CP}}{4\pi Dr} \, \text{erfc} \, \frac{r}{2\sqrt{Dt}}$$

$M_{CP}$ in kg/s

6. Continuous Point Source on a Semi-Infinite Plane Emitting into Half Space

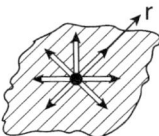

$$C = \frac{M_{CP}}{2\pi Dr} \, \text{erfc} \, \frac{r}{2\sqrt{Dt}}$$

$M_{CP}$ in kg/s

7. Continuous Infinite Plane Source Emitting into Half Space

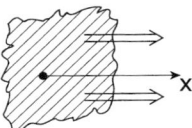

$$C = M_{CA} \left[ \left( \frac{t}{\pi D} \right)^{1/2} \exp(-x^2/4Dt) - \frac{x}{2D} \, \text{erfc} \, \frac{x}{2\sqrt{Dt}} \right]$$

$M_{CA}$ in kg/m$^2$s

Table 4.1 contains a number of solutions of both instantaneous and continuous source problems in infinite, semi-infinite, or half space. Let us examine two of these tabulated solutions.

The case of emissions into half space (see Item 2 of Table 4.1) can be obtained from the fully infinite case by arguing that the mass that would otherwise have gone into the lower half space is reflected upward, thus doubling the concentration. We obtain the following:

### 4.1.1.2    *Instantaneous Point Source on an Infinite Plane Emitting into Half Space*

$$C = \frac{M_p}{4(\pi Dt)^{3/2}} \exp(-r^2 / 4Dt) \tag{4.2}$$

The case of a continuous source is another example of interest. We would expect intuitively that the solution to this problem could in principle be obtained by integrating over time the result obtained for an instantaneous source. This is indeed the case and leads to the emergence of a special type of integral, termed the *error function* erf $x$, which is defined as

$$\text{erf } x = \frac{2}{\pi^{1/2}} \int_0^x \exp(-u^2)\,du \tag{4.3a}$$

The error function can be viewed as a partial area under the Gaussian distribution curve $(2/\pi^{1/2})\exp(-u^2)$. It has the value 0 at $x = 0$, and a value of 1 at $x = \infty$. The latter case corresponds to the full area under the distribution curve. A related expression is the *complementary error function* erfc $x$, which is defined as

$$\text{erfc } x = 1 - \text{erf } x \tag{4.3b}$$

and has the values erfc$(0) = 1$ and erfc$(\infty) = 0$.

The error function integral has to be evaluated numerically and can be found tabulated in texts on diffusion or conduction, or in mathematical tables. An abbreviated listing appears in Table 4.2, and some important properties of the function are summarized in Table 4.3.

With these definitions in place, we can now proceed to present the solution to the continuous point source problem. It is given by the following (see Item 5 of Table 4.1).

**TABLE 4.2**

Values of the Error Function

| $x$ | $\text{erf}(x)$ |
|---|---|
| 0 | 0 |
| 0.05 | 0.05637 |
| 0.1 | 0.11246 |
| 0.15 | 0.16800 |
| 0.20 | 0.22270 |
| 0.25 | 0.27632 |
| 0.30 | 0.32863 |
| 0.35 | 0.37938 |
| 0.40 | 0.42839 |
| 0.50 | 0.52050 |
| 0.60 | 0.60386 |
| 0.70 | 0.67780 |
| 0.80 | 0.74210 |
| 0.90 | 0.79691 |
| 1.0 | 0.84270 |
| 1.2 | 0.91031 |
| 1.5 | 0.96611 |
| 2.0 | 0.99532 |
| 2.5 | 0.999593 |
| 3.0 | 0.999978 |
| $\bullet$ | 1.00000 |

**TABLE 4.3**

Properties of the Error Function

1.  $\text{erf}(0) = 0$
2.  $\text{erf}(\infty) = 1$
3.  $\text{erf}(-x) = -\text{erf}(x)$
4.  $1 - \text{erf}(x) = \text{erf}(x)$ (complementary error function)
5.  $$\frac{d}{dx}\text{erf}\ x = \frac{2}{\sqrt{\pi}}e^{-x^2}$$
6.  $$\int_x^\infty \text{erfc}\ x\ dx = \frac{1}{\sqrt{\pi}}e^{-x^2} - x\ \text{erfc}\ x$$
7.  Approximation for small $x$: $\text{erf}\ x \cong 2\pi^{-1/2}x$
8.  Approximation for large $x$:
    $$\text{erfc}\ x \cong \pi^{-1/2}\ e^{-x^2}\left[\frac{1}{x} - \frac{2}{x^3} + ...\right]$$

### 4.1.1.3   Continuous Point Source Emitting into Infinite Space

$$C = \frac{M_{cp}}{4\pi Dr} \operatorname{erfc} \frac{r}{2\sqrt{Dt}} \tag{4.4}$$

For emissions into half space, the concentration given by Equation 4.4 is again doubled (see Item 6 of Table 4.1).

To obtain a sense of the spreading concentration waves that result from the emissions of sources, we have sketched the concentration profiles and histories associated with point and continuous sources (Figure 4.1). In the case of instantaneous point sources, the pulse emitted at $t = 0$ gradually spreads about the origin and diminishes in strength; i.e., concentration decreases with the passage of time (see Part 1 of Figure 4.1). Continuous sources result in a similar spreading of concentrations about the origin, but here there is a continuous increase in the level, which ultimately reaches a plateau (see Part 4). Of interest is the appearance of maxima at

### a. Point Source

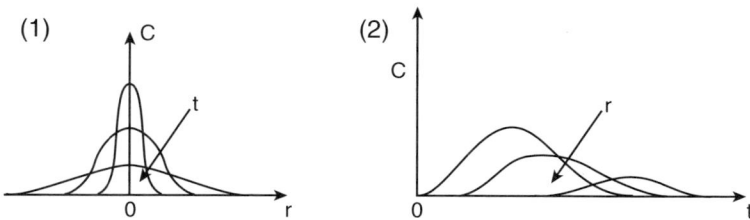

### b. Continuous Source

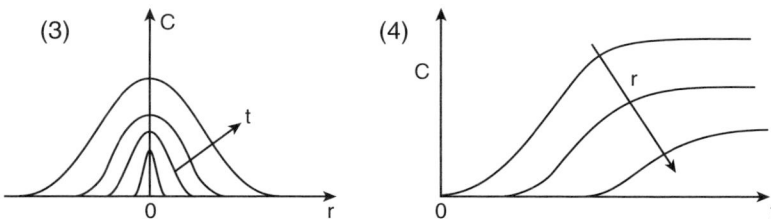

**FIGURE 4.1**
Concentration profiles and histories in source problems (infinite domain).

a particular location of the domain. In the case of the instantaneous source, this maximum occurs at finite times (see Part 2), while the concentrations from continuous emissions attain their maximum only in the limit of $t \rightarrow \infty$ (see Part 4 ). We address these maxima in Illustration 4.1 and in Practice Problem 4.2.

### Illustration 4.1: Concentration Response to an Instantaneous Point Source: Release in the Environment and in a Living Cell

The problem considered here involves calculating the concentrations that arise when material released by an instantaneous point source diffuses into a semi-infinite domain. In the environment, this could come about, for example, when a tank containing pressurized liquefied gas is ruptured and releases the contents, which instantly evaporate into the surrounding air.

Let us set $M_p$ at 1 kg (e.g., a small propane cylinder) and assume a typical diffusivity in air of $10^{-5}$ m²/s. We use the data to calculate the concentrations that result at a distance of $r = 100$ m from the source after 10, 100, and 1000 days, respectively. Substituting values into Equation 4.2, we obtain for $t = 10$ days

$$C = \frac{1}{4(\pi \times 10^{-5} \times 3{,}600 \times 24 \times 10)^{3/2}} \exp\left(-\frac{100^2}{4 \times 10^{-5} \times 3{,}600 \times 24 \times 10}\right)$$

and similarly for $t = 100$ and 1000 days. The results are tabulated below:

| T (days) | C (kg/m³) |
|----------|-----------|
| 10 | 0 |
| 100 | $1.4 \times 10^{-34}$ |
| 1000 | $6.3 \times 10^{-9}$ |

The data show a very slow, gradual rise in concentration. A subsidiary question, which we wish to address before commenting on the results, focuses on the time it takes for the concentration to reach its maximum value. That maximum occurs, as was shown in Figure 4.1a, Part 2, at some finite point in time, which is *independent of the amount released*. This makes it a convenient quantity for establishing the time scale of these events.

We write Equation 4.2 in the form

$$C(r,t) = \frac{u(t)}{v(t)} \tag{4.5a}$$

and require $\partial C / \partial t = 0$ or, alternatively,

$$(du/dt)v - u(dv/dt) = 0 \tag{4.5b}$$

We obtain

$$\exp(-r^2/4Dt)(r^2/4Dt^2)t^{3/2} - \exp(-r^2/4Dt)\frac{3}{2}t^{1/2} = 0 \qquad (4.5c)$$

and consequently

$$t_{Max} = \frac{1}{6}(r^2/D) \qquad (4.5d)$$

$$t_{Max} = \frac{1}{6}(100^2/10^{-5}) = 1.87 \times 10^8 \, s = 2164 \, days$$

Substituting this value into Equation 4.2 then yields

$$C_{Max} = 1.32 \times 10^{-7} \, kg/m^3 \qquad (4.5e)$$

The striking feature of these results is the extremely slow progress of the diffusion process. It takes almost 6 years for the concentration to attain its maximum at the modest distance of only 100 m. Evidently, these numbers do not give a realistic picture of what would happen in an actual accident. Wind and other air currents would intervene to disperse the material at an immensely higher rate. Quantifying the effect of wind, even when it is unidirectional, is a more complex task, which we address in Illustration 4.3.

Let us next consider the instantaneous release from a point source at the center of a living cell 100 μm in radius. Although the geometry here is a finite one, the initial stages of the diffusion process can be viewed as taking place into an infinite medium. If we assume $D = 10^{-10}$ m²/s, typical of a protein, then the time it takes for the maximum concentration to reach $r = 10$ μm equals one-half the value given by Equation 4.5d:

$$t_{Max} = \frac{1}{12}\frac{(10 \times 10^{-6})^2}{10^{-10}} = 0.083 \, s$$

i.e., attainment is near-instantaneous in spite of the much lower diffusivity. Note the vast difference in time scale for the two processes (6 years vs. less than a second), which is entirely due to the quadratic dependence on $r$.

### Illustration 4.2: Net Rate of Global Carbon Dioxide Emissions

The release of carbon dioxide into the atmosphere and its attendant effect on global warming have been widely publicized in recent years. A warning call came in the late 1980s and early 1990s when it was established that carbon dioxide concentrations in the air, then at a level of 370 ppm, were growing at a rate of 2 ppm per year. What we wish to do here is to use this information to calculate the net rate of global carbon dioxide emissions, i.e., the amount in excess of that consumed by plant life, which is a *minimum* measure of the amount by which emissions would have to be reduced. (Why minimum?)

The tools required for this task reside in Item 7 of Table 4.1; this information describes the continuous emission from an infinite plane, here taken to be the surface of Earth, into a semi-infinite domain, which is the atmosphere.

We start by noting that at ground level, $x = 0$, the expression given there reduces to the form

$$C = M_{CA}\left(\frac{t}{\pi D}\right)^{1/2} \tag{4.6a}$$

where $M_{CA}$ is the rate of emission in $kg/m^2\,s$, the unknown being sought here. The diffusivity $D$ of carbon dioxide in air at 25°C is $2.2 \times 10^{-5}\,m^2/s$, the air density $1.18\,kg/m^3$.

We proceed as follows: Solving for $t$ we obtain, for the interval of observation of 1 year,

$$t = \frac{C^2\pi D}{M_{CA}^2} = 3,600 \times 24 \times 365 = 3.15 \times 10^7\,s \tag{4.6b}$$

The result can be recast in the form

$$M_{CA}^2 = \frac{\pi D(C_2^2 - C_1^2)}{t} = \frac{\pi D(C_2 + C_1)(C_2 - C_1)}{3.15 \times 10^7} \tag{4.6c}$$

where $C_1$ and $C_2$ are the concentrations at the start and the end of the year.

Before introducing numerical values, the data, given in ppm ($m^3\,CO_2/m^3$ air) must be converted to $kg/m^3$. This is done by multiplying ppm values by the density of carbon dioxide, which is 44/29 times that of air, i.e., 1.18 $\times$ 44/29 = 1.79 $kg/m^3$. We obtain

$$M_{CA} = \left[\frac{\pi \times 2.2 \times 10^{-5}(370 + 372)(372 - 370)10^{-12}\,1.79^2}{3.15 \times 10^7}\right]^{1/2} \tag{4.6d}$$

$$M_{CA} = 1.0 \times 10^{-10}\,kg/m^2\,s \tag{4.6e}$$

The average radius of Earth is 6370 km. Hence the global net amount of carbon dioxide emitted is given by

$$M_{Tot} = 1.0 \times 10^{-10}\,4\pi(6.37 \times 10^6)^2 \tag{4.6f}$$

$$M_{\text{Tot}} = 5.1 \times 10^4 \text{ kg } CO_2/s \qquad (4.6g)$$

*Comments:*

This is still a staggering amount, considering that it only represents the *excess* carbon dioxide produced over and above what is removed by photosynthesis. The gross rate of carbon dioxide production is thus considerably higher. Some measure of what plant life can accomplish to alleviate this effect was given in Practice Problem 3.7.

### Illustration 4.3: Finding a Solution in a Related Discipline: The Effect of Wind on the Dispersion of Emissions

It has been indicated in Illustration 4.1 that the effect of air currents on the emissions from sources is considerably more difficult to quantify than purely diffusional processes. The standard literature dealing with solutions to Fick's equation (Crank, 1978) contains no mention of this problem, and although the solution may be lurking in the general literature, locating it is not an appealing task. It is often more rewarding in these cases to consult the related literature on conduction, which is considerably more voluminous and therefore a more promising source of solutions. It is the art of extracting such solutions, and translating them back into the original context, that we wish to practice here.

The analogy between diffusion and conduction rests on the fact that Fick's law of diffusion and Fourier's law of conduction are identical in form (see Table 1.1). When Fourier's law is applied to transient conductions by means of an energy balance, it results in the celebrated Fourier equation, which is analogous in form to Fick's equation (Equation 2.18c). Thus, for unsteady conduction in one-dimensional Cartesian coordinates we have

$$\frac{\partial T}{\partial t} = \alpha \frac{\partial^2 T}{\partial x^2} \qquad (4.7)$$

with $T$ taking the place of concentration in Fick's equation and $D$ replaced by thermal diffusivity $\alpha$. We can consequently expect its solutions to be likewise identical in form to those that apply to mass diffusion.

A search of the standard work on solutions to Fourier's equation reveals that it contains a section entitled "Moving Sources of Heat" (Carslaw and Jaeger, 1959, p. 266). Described in this section is a problem in which "heat is emitted at the origin for times $t > 0$ at the rate $q$ heat units per unit time, and that an infinite medium moves uniformly past the origin with a velocity $U$ parallel to the $x$-axis." This situation corresponds to the case of a continuous-point mass source emitting into half space in which the air moves in the $x$ direction with a velocity $U$.

The full solution to this problem is rather complex and is not reproduced here. It does, however, reduce to a simple expression for the limiting steady state case $t \rightarrow \infty$, which is of greater interest and which is given below in the form presented by Carslaw and Jaeger:

$$v = \frac{q}{4\pi KR} \exp[-U(R-x)/2\kappa] \tag{4.8a}$$

where $v$ = temperature (K), $q$ = heat emitted (J/s), $K$ = thermal conductivity (J/sm²K), $R^2 = x^2 + y^2 + z^2$, and $\kappa$ = thermal diffusivity (m²/s).

Translation into the corresponding mass diffusion case is straightforward: We replace $v$ by $C$ (kg/m³), $q$ by $M_{cp}$ (kg/s), and both $K$ and $\kappa$ by diffusivity $D$ (m²/s). The result is then given by

$$C = \frac{M_{cp}}{4\pi DR} \exp[-U(R-x)/2D] \tag{4.8b}$$

and represents the maximum concentration attained at steady state in response to the continuous point source.

A quick inspection of this equation shows that the wind velocity has an enormous effect on the concentration response. With a modest wind velocity of $U$ = 10 cm/s, a typical diffusivity in air of $10^{-5}$ m²/s, and $(R - x)$ of the order $10^2$ m, the exponential term immediately reduces to zero. Air movement would have to be reduced to the order of the diffusivity, $10^{-5}$ m/s, before any significant concentration levels arise.

The case of overriding interest here is the concentrations in the direction of the wind and along the positive $x$-axis. This would represent locations of maximum steady-state exposures. We have, in this case, $R = x$ and hence

$$C = \frac{M_{cp}}{4\pi Dx} \tag{4.8c}$$

This simple relation shows that the steady-state concentrations diminish in proportion to the downwind distance $x$ from the point source and that this decrease is the same, irrespective of the velocity of the wind.

Setting the emission rate $M_{cp}$ at $10^{-3}$ kg/s, for example, and distance $x$ at 100 m, we obtain for the concentration at that point

$$C = \frac{10^{-3}}{4\pi 10^{-5} 100} = 0.08 \text{ kg}/\text{m}^3 \tag{4.8d}$$

and 1/10 that value at a distance of 1 km. These concentrations drop off sharply in the immediate vicinity of the $x$-axis because of the exponential term in Equation 4.8b. The exposure zone is consequently confined to an exceedingly narrow strip along the line of sight of the source and in the direction of the wind.

*Comments:*

Locating solutions in related disciplines is not an easy matter. It requires a wide knowledge of what goes on elsewhere, but the results can be rewarding and often lead to new insights.

## 4.1.2 Nonsource Problems

Nonsource problems are by far the most prevalent type of problem involving diffusional processes, and almost all the material contained in standard monographs on diffusion is devoted to this topic. A parallel situation exists in the related field of heat conduction.

Mention was already made of the variety of geometries and boundary conditions that can arise in these problems, and we proposed to limit ourselves to the semi-infinite, parallel plane, spherical, and cylindrical geometries subject to constant initial and surface concentrations. We start with the simplest of these geometries, the semi-infinite medium, and follow this with a discussion of the other three principal geometries.

### 4.1.2.1 Diffusion into a Semi-Infinite Medium

Consider the case of the semi-infinite medium $x > 0$ in which the concentration $C_o$ is uniform throughout and which is exposed at time $t \geq 0$ and the position $x = 0$ to a constant surface concentration $C_s$. The solution to this problem can be given as a terse analytical expression and takes the form

$$\frac{C - C_s}{C_o - C_s} = \text{erf} \frac{x}{2\sqrt{Dt}} \qquad (4.9a)$$

For a medium initially devoid of solute, $C_o = 0$, the equation reduces to

$$\frac{C}{C_s} = 1 - \text{erf} \frac{x}{2\sqrt{Dt}} = \text{erfc} \frac{x}{2\sqrt{Dt}} \qquad (4.9b)$$

Both of these expressions make frequent appearances in the literature. They contain, as do some source problems, an error function, but lack the preexponential factor we have seen there. As a result, the concentration distributions that arise in this case are a function of only *one* dimensionless parameter, $x / 2\sqrt{Dt}$. It follows from this that:

1. The distance of penetration of any given concentration is proportional to the square root of time.

2. The time required for any point to reach a given concentration is proportional to the square of its distance from the surface and varies inversely with the diffusivity.

We demonstrate the use of these simple relations with the following example.

### Illustration 4.4: Penetration of a Solute into a Semi-Infinite Domain

1. Suppose a spill of a solvent has occurred (a) on land and (b) into a water basin. How much longer will it take a particular concentration to penetrate the same distance in water that it does in air? Since the concentration is the same in both cases, we must have, in accordance with Equation 4.9,

$$\left[ \frac{x}{2\sqrt{Dt}} \right]_{air} = \left[ \frac{x}{2\sqrt{Dt}} \right]_{water} \tag{4.10a}$$

$$t_{water} = \frac{D_{air}}{D_{water}} t_{air} \tag{4.10b}$$

Now, diffusion in air is of the order $10^{-5}$ m²/s, that in water of the order $10^{-9}$ m²/s (see Tables 3.1 and 3.3). Consequently, the time of penetration in water is 10,000 times longer than that in air.

2. If it takes a particular concentration 100 h to penetrate a distance of 1 m, how long will it take the same concentration to advance 10 m? Here again we are dealing with identical concentrations, which by virtue of Equation 4.9 leads to

$$\frac{x_1}{2\sqrt{Dt}} = \frac{x_2}{2\sqrt{Dt}} \tag{4.10c}$$

and consequently

$$t_2 = t_1 \left( \frac{x_2}{x_1} \right)^2 \tag{4.10d}$$

or

$$t_2 = 100 \left( \frac{10}{1} \right)^2 = 10^4 \text{ h} \tag{4.10e}$$

Hence it takes 100 times longer to penetrate from a distance of 1 m to a distance of 10 m.

These simple examples show that the case of diffusion into a semi-infinite medium can yield rapid answers in a relatively straightforward fashion. Furthermore, the geometry is not trivial. It can often be used to approximate finite geometries, particularly if the diffusion process is a slow one, as it is in liquid or solid media. Penetration will then be confined to short distances from the surface, at least initially, and the medium can consequently be regarded as a semi-infinite one for the short period under consideration (note the similarity to Illustration 4.1).

In diffusion problems, we often seek to calculate the *cumulative* amount of material that has entered or left a medium, rather than a particular concentration level or the time it takes to attain that level. In the case of accidental spills, for example, it is often of greater interest to know the time required for *complete* evaporation or dissolution of the material, rather than the detailed concentration transients. To obtain this information, some mathematical manipulations of the distribution Equation 4.9 are required, and these are discussed in the following illustration.

### Illustration 4.5: Cumulative Uptake by Diffusion for the Semi-Infinite Domain

Here we calculate the amount of material per unit area that has diffused into a semi-infinite medium to a certain point in time $t$. We proceed in two steps by first calculating the *rate of diffusion* at the base plane and then *integrating* that rate over time.

Calculation of the rate requires taking the derivative of the error function in Equation 4.9a. Let us consider the special (and usual) case of zero initial concentration $C_o = 0$. The need is then to evaluate

$$N/A = -D\frac{\partial C}{\partial x}\Big|_{x=0} = C_s\frac{\partial}{\partial x}\operatorname{erf}\frac{x}{2\sqrt{Dt}}\Big|_{x=0} \tag{4.11a}$$

Now from Item 5 of Table 4.1 we have

$$\frac{d}{dx}\operatorname{erf} x = \frac{2}{\sqrt{\pi}}\exp(-x^2) \tag{4.11b}$$

and consequently

$$\frac{\partial}{\partial x}\operatorname{erf}\frac{x}{2\sqrt{Dt}} = \frac{2}{\sqrt{\pi}}\frac{1}{2\sqrt{Dt}}\exp\left(-\frac{x^2}{4Dt}\right) \tag{4.11c}$$

Combining Equations 4.11a and 4.11c we obtain, for this intermediate step,

$$N \, / \, A = -D \frac{\partial C}{\partial x}\bigg|_{x=0} = \frac{DC_s}{\sqrt{\pi Dt}} \tag{4.11d}$$

This is the expression that now must be integrated with respect to time. We obtain immediately

$$M_t = 2C_s \left( \frac{Dt}{\pi} \right)^{1/2} \tag{4.11e}$$

where $M_t$ is the desired total mass per unit area that has diffused into the semi-infinite domain up to time $t$.

Note that $M_t$ varies directly with the surface concentration $C_s$ but only with the square root of time. In Practice Problem 4.3, use will be made of Equation 4.11e to estimate the time required for spilled solvent to evaporate into the atmosphere.

### 4.1.2.2 Diffusion in Finite Geometries: The Plane Sheet, the Cylinder, and the Sphere

It has previously been indicated that the diffusion into finite geometries leads to expressions of considerably greater complexity than was the case for the semi-infinite medium. The solutions typically take the form of infinite series, which not only are cumbersome to evaluate but also contain implicit parameters that cannot be conveniently extracted. It has become customary in these cases to represent the results graphically, which allows any of several parameters to be read off with ease. We present several of these for the convenience of the reader.

Figure 4.2 presents the concentration profiles that arise in a sheet of thickness $2L$, which is exposed to a surface concentration $C_s$ at time $t = 0$ and contains an initial concentration $C_1$. We can use these plots to calculate, for example, the time required for a certain concentration $C$ to reach a particular position $x$, or conversely, to calculate the prevailing concentration at a specified $x$ after the lapse of time $t$. The position $x = 0$ is of particular interest as it represents the midpoint of the sheet and is the farthest removed from the imposed surface concentration $C_g$. Note that the sheet is infinite in extent; i.e., there are no concentration variations in the $y$ and $z$ directions. The finite three-dimensional case evidently leads to more complex distributions, which cannot be plotted conveniently.

Figure 4.3 dispenses with the display of detailed profiles and presents, instead, the *average concentration* in the medium after the lapse of time. This quantity, like the total uptake that can be derived from it, is again of greater interest than the concentration distributions and can be used for a number of calculations of practical interest. The plots of this quantity are given for a slab or sheet of infinite extent, for an infinitely long cylinder, and for the sphere.

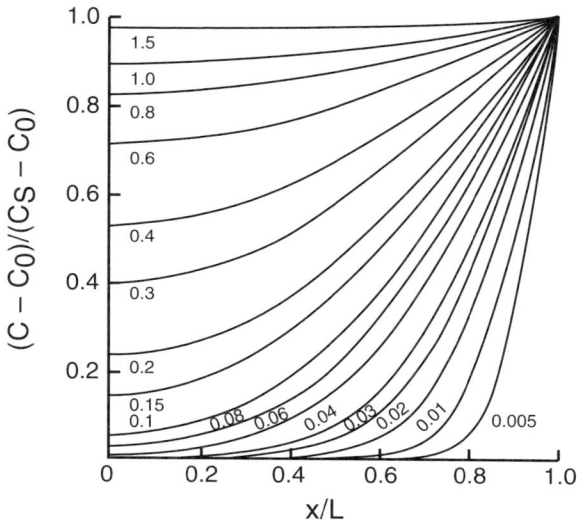

**FIGURE 4.2**
Concentration distributions at various times in the sheet $-L < x < L$ with initial uniform concentration $C_o$ and surface concentration $C_S$. Numbers on curves are values of $Dt/L^2$.

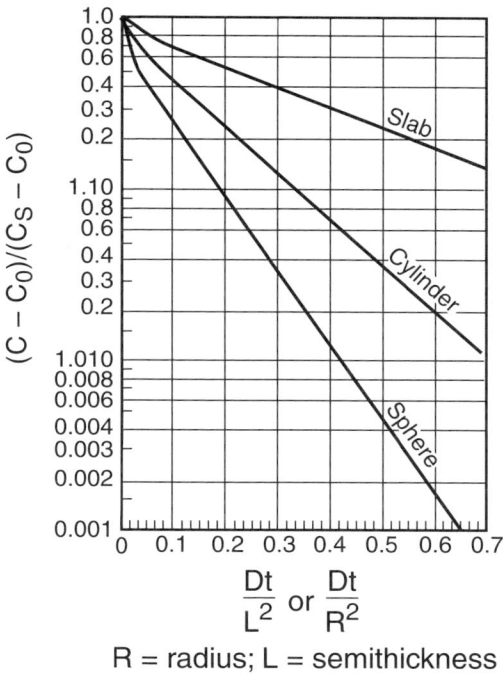

**FIGURE 4.3**
The relative change in average composition for the basic shapes.

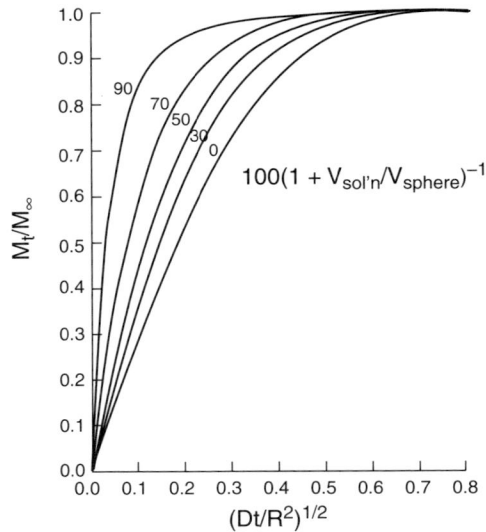

**FIGURE 4.4**
Fractional diffusional uptake and release in a sphere as a function of dimensionless time in a well-stirred solution of limited volume.

Finally, Figure 4.4 presents the cumulative uptake of material by a sphere contained in a solution of limited volume. This case arises when material must be deposited in a porous sphere or impurities are to be removed from a solution by adsorption. The parameter shown expresses the ratio of volumes of the sphere and the solution and also represents the amount of solute taken up by the sphere. For adsorption, the parameter must be adjusted to accommodate the partition coefficient, or Henry's constant $H$. We note that the plot can also be used for the reverse processes of leaching or desorption. This is illustrated with a number of examples.

### Illustration 4.6: Manufacture of Transformer Steel

To make a transformer steel with the proper hysteresis characteristics, it has to be loaded with silicon up to a certain prescribed content. This is accomplished by exposing a steel sheet that is low in silicon content and of 2 mm thickness to an atmosphere of $SiCl_4$ that dissociates to $Si(g)$ and $Cl_2(g)$. The silicon gas dissolves in the steel up to 3 wt% at equilibrium. The treatment is to be carried out at 1255 K. Silicon diffusivity in steel at this temperature is $8.2 \times 10^{-13}$ m$^2$/s.

The task here is to calculate the time necessary to achieve a target average concentration of 2.85 wt% silicon in the steel. We draw for this purpose on Figure 4.3, whose ordinate values contain the average concentration $\bar{C}$ being sought.

We start by noting that we can set the initial concentration $C_o = 0$ in view of the low silicon content of the steel to be treated. The surface concentration

$C_s$ required for the ordinate value is taken to be 3 wt%; i.e., it is assumed that at time $t = 0$, the steel surface is at equilibrium with the surrounding silicon atmosphere. We obtain, for the ordinate of Figure 4.3,

$$\frac{\overline{C} - C_s}{C_o - C_s} = \frac{2.85 - 3.0}{0 - 3.0} = 0.05 \tag{4.12a}$$

Linear extrapolation of the slab line of Figure 4.3 yields

$$\frac{Dt}{L^2} \cong 1.13 \tag{4.12b}$$

from which there results

$$t = (1.13)\frac{(1 \times 10^{-3})^2}{8.2 \times 10^{-13}} = 13.8 \times 10^6\,\text{s} = 383\,\text{h} \tag{4.12c}$$

Thus, some 16 *days* are required for the target concentration to be attained. This is entirely due to the low diffusivity of silicon in the steel, which is some three orders of magnitude lower than typical diffusivities in liquids. To reduce the inordinate length of treatment, the operation will likely have to be carried out at a higher temperature. Since the diffusivity in solids varies exponentially with temperature (see Equation 3.17) a modest increase in the latter will quickly lead to substantially higher diffusivities and thus shorten the duration of the treatment considerably.

Figure 4.3, which has been used here to calculate time $t$ required to achieve a certain concentration, can also be put to the task of extracting diffusivities from experimental data. This is shown in the following illustration.

### Illustration 4.7: Determination of Diffusivity in Animal Tissue

To determine the diffusivity of carbohydrates in animal tissue, a sample specimen 1 mm in thickness is soaked in sugar solution for a lengthy period of time, and then mounted on a holder and placed in a large, well-stirred bath of water. After exposure for 1 h, the sample is removed and the residual sugar content determined. It is found to have dropped to 20% of the initial concentration. We wish to use the data to calculate the diffusivity of sugar within the tissue.

The problem again calls for the use of Figure 4.3, with the ordinate value now given by $\overline{C} / C_0 = 0.2$. The corresponding abscissa value is 0.54 so that

$$\frac{Dt}{L^2} = 0.54 \tag{4.13a}$$

and consequently

$$D = 0.54L^2/t = 0.54(0.5 \times 10^{-3})^2/3600 \qquad (4.13b)$$

$$D = 3.8 \times 10^{-11} \text{ m}^2/\text{s}$$

### Illustration 4.8: Extraction of Oil from Vegetable Seeds

Vegetable oils can be extracted from their parent seeds by contacting them with solvent in a well-stirred tank. External film resistance can in these cases often be neglected and the principal events are confined to the interior of the seeds. Figure 4.4 can then be used to carry out pertinent calculations.

Suppose we desire to calculate the time required to extract 90% of the oil contained in oil-bearing vegetable seeds assumed to be spherical. The given data are as follows:

$$R = 0.25 \text{ cm}, D = 5 \times 10^{-6} \text{ cm}^2/\text{s}, V_{\text{Sol'n}}/V_{\text{Solids}} = 2$$

so that $100 (1 + V_{\text{Sol'n}}/V_{\text{Solids}}) = 33.3$.
From Figure 4.4 we obtain, for $M_t/M_\infty = 0.9$, $Dt/R^2 = 0.39$. Hence

$$t = 0.39R^2/D \qquad (4.14a)$$

$$t = 0.39 (0.25)^2/5 \times 10^{-6} = 4.9 \times 10^3 \text{ s} = 1.36 \text{ h} \qquad (4.14b)$$

## 4.2   Diffusion and Reaction

Processes in which diffusion is accompanied by a chemical reaction arise frequently and in a variety of different contexts. All catalytic reactions, in which the catalyst resides within a porous matrix, are necessarily accompanied by diffusional transport of the reactants and products into and out of this catalyst particle. In noncatalytic gas–solid and liquid–solid reactions, diffusion occurs not so much within the solid particle but rather through a gas or liquid film, or through ash layers surrounding the reacting core. Here again diffusion is coupled with reaction.

Reactions accompanied by diffusion also occur in fluid systems. The atmosphere is one vast reacting reservoir in which gaseous pollutants such as the nitrogen oxides (the famous Nox) or sulfur oxides (the equally famous Sox) diffuse into the air and undergo reactions with atmospheric oxygen. In gas–liquid systems, the liquid phase often contains a reacting component that interacts with the gas diffusing into it. Here, too, reaction is linked to diffusion.

Evidently, in each of the systems mentioned, a host of different reactions are possible. A reacting solid particle, for example, may involve the combustion of a fuel or the calcining of calcium carbonate (limestone) to calcium oxide and carbon dioxide. Reactants involved in catalytic reactions are almost infinite in their variety, as are those participating in atmospheric reactions.

To convey a flavor of these events without overwhelming the reader with a mass of details, we limit ourselves to the following cases.

### 4.2.1   Reaction and Diffusion in a Catalyst Particle

In this example, we consider the diffusion of a reactant into a porous catalyst particle where it undergoes a first-order reaction. The model here is a second-order ODE, which is solved by the $D$-operator method given in the Appendix and yields the concentration profile of the reactant within the particle. This is interesting but not immediately useful information for the design of catalytic reactors. To transform the result into a tool for engineering use, we derive the overall rate of reaction in the pellet with diffusional resistance and divide the result by the reaction rate that would prevail in the absence of a diffusional resistance. This ratio, known as the *catalyst effectiveness factor E*, can be viewed as the efficiency of the catalyst in converting the reactant to product. When diffusion is very fast, or the reaction slow, the interior reactant concentration will be nearly that prevailing at the surface. The effectiveness factor will then be nearly 1 and catalyst efficiency will be approximately 100%, the maximum it can attain. With increasing diffusional resistance, the reactant concentration within the particle begins to fall below that prevailing at the surface, causing the reaction rate to decrease. This is reflected in a lower effectiveness factor $E$, which is now below 1; the corresponding catalyst efficiency is less than 100%. We thus obtain a good sense, through the value of $E$, of how well the catalyst is performing, in coping with the diffusional resistance. It also enables us to take countermeasures to raise the efficiency, for example, by increasing porosity or reducing particle size, both of which have the effect of diminishing diffusional resistance.

There is a second important reason for introducing the concept of an effectiveness factor. In the ordinary course of events, concentrations within a catalytic reactor packed with catalyst particles will vary both axially in the direction of flow and radially within the catalyst pellets. The model mass balance for such a system would consequently lead to a PDE. By using an effectiveness factor we reduce the PDE to an equivalent set of two ODEs, one the pellet mass balance in the radial direction, and the other the reactor mass balance in the direction of flow. The reaction rate, which previously varied in two directions $r_A(r, z)$, is now a function of the axial distance only. We replace $r_A(z, r)$ by $Er_{Ai}(z)$; $r_{Ai}$ is the so-called intrinsic reaction rate, which is measured experimentally on a fine powder and excludes diffusional effects. The latter are lumped into the effectiveness factor, which now acts as a fractional efficiency on the intrinsic rate $r_{Ai}$. It is this product of $Er_{Ai}(z)$ that is used in the reactor mass balance.

### 4.2.2 Gas–Solid Reactions Accompanied by Diffusion: Moving-Boundary Problems

The topic addressed in this example is that of a solid particle that undergoes a continuous reaction with a gas, building up in the process a layer of porous solid product, which we denote by the general term *ash*. Reactant diffuses through a growing layer of ash to the surface of the shrinking particle where the reaction takes place. Such systems of two phases, in which the phase boundary undergoes a continuous movement due to some physical or chemical event, are referred to as *moving-boundary problems*. Examples of this type of behavior are numerous and important. In addition to reacting systems, moving-boundary problems arise in operations involving phase change such as evaporation, condensation, freezing and melting, crystal growth and dissolution, metal or polymer casting, and the freeze-drying of foods.

The state variables in these processes, such as temperature or concentration, are in principle functions of both distance and time, leading to PDEs that are usually coupled and nonlinear. To reduce the model to a manageable set of ordinary differential and algebraic equations, the following assumptions are made:

1. The "core" contained by the moving front, such as a burning fuel particle, has uniform properties and can be treated as an unsteady compartment.
2. The movement of the front itself is sufficiently slow that the transport gradients outside the core attain a quasi-steady state. This condition, which we have encountered before in Illustration 2.4, has the effect of eliminating time as a variable, with a consequent simplification of the model equation.
3. The processes involved — transport and reaction — are dominated by a rate-controlling slow step.

Thus, although both time and distance are retained as variables, distance (expressed through the changing size or mass of the core) becomes a dependent variable for the core-unsteady balance but is retained as an independent variable for the external, quasi-steady-state balance. Time is an independent variable as well, but appears only in the core balance.

A systematic way of modeling these systems is to start with unsteady balances about the core, followed by a consideration of the quasi-steady-state process outside the moving boundary. It is good practice to keep track of the number of dependent variables, which must ultimately be matched by the number of equations. These procedures are demonstrated in Illustration 4.10.

### 4.2.3   Gas–Liquid Systems: Reaction and Diffusion in the Liquid Film

It is not uncommon practice in gas-absorption operations to employ a solvent that reacts with the solute being absorbed from the gas phase. The purpose of this practice is to promote the solute removal rate and to enhance the efficiency of the gas absorber. Acid gases such as $H_2S$ and $CO_2$ are often contacted with solvents containing an alkaline component such as potassium or sodium hydroxide, or an ethanol amine. Conversely, the absorption of a basic solute such as ammonia can be promoted by reacting it with an acidic solvent.

The stoichiometry of these liquid-phase reactions can be represented in the general form

$$A(g) + bB(\ell) \rightarrow \text{products} \qquad (4.15)$$

$B(\ell)$ may refer to pure liquid B, or, more commonly, to B dissolved in a liquid solvent.

In Illustration 4.11 we take up the case of a first-order reaction taking place in the liquid, which is fast enough to cause the reactant concentration to drop to zero within a relatively short distance from the interface, i.e., within what is conventionally regarded as a liquid film.

The resulting concentration profiles are shown in Figure 4.5. We note here that because of the rapid reaction in the liquid film, the profiles are highly nonlinear and can no longer be approximated by a linear driving force. We must resort to a full diffusional mass balance. This is undertaken in Illustration 4.11.

Here again, as in the case of the catalyst particle, the resulting concentration profile does not yield information of immediate interest, and is translated instead into a quantity of greater engineering usefulness. The quantity in question is the so-called enhancement factor $E_h$, which is defined as the ratio of mass transfer with reaction to the mass transfer rate without reaction. In contrast to the catalyst effectiveness factor, the value of $E$ is above rather than below unity. This is because the reaction continuously removes reactant, thus sharpening its gradient and in consequence enhancing the mass transfer rate.

### *Illustration 4.9: Reaction and Diffusion in a Catalyst Particle. The Effectiveness Factor and the Design of Catalyst Pellets*

We consider, in this example, the diffusion of a reactant $A$ into a flat-plate catalyst particle, where it undergoes a first-order reaction. The geometry, the associated boundary conditions, and the resulting concentration profile are shown in Figure 4.6a.

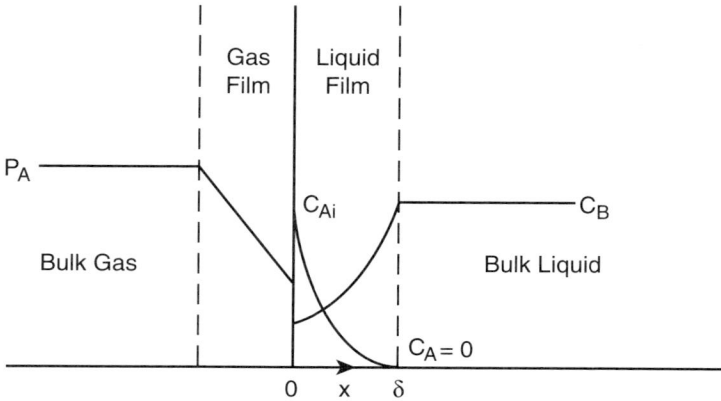

**FIGURE 4.5**
Concentration profiles in a gas–liquid reaction; fast first-order reaction in the liquid film, with reactant concentration dropping to zero at $x = \delta$.

Diffusion is assumed to take place unidirectionally through the largest exposed area, while the edges of the plane are taken to contribute an insignificant amount of flow. The surfaces of the particle are maintained at a constant concentration $C_{As}$, leading to the parabolic concentration profile shown in Figure 4.6a. The distribution is symmetric about the midplane and has a vanishing derivative at that point.

These two conditions, prevailing at the surface and the center plane, respectively, constitute the boundary conditions (BCs) for the model equation.

Figure 4.6b shows the difference element over which the mass balance is taken. We obtain

$$\text{Rate of } A \text{ in} - \text{Rate of } A \text{ out} = 0$$

$$-D_e A_C \frac{dC_A}{dx}\bigg|_x - \left[ \begin{array}{c} -D_e A_C \dfrac{dC_A}{dx}\bigg|_{x+\Delta x} \\ +k_r (C_A)_{avg} A_C \Delta x \end{array} \right] = 0 \qquad (4.16a)$$

where $D_e$ is the effectiveness diffusivity.

Dividing by $A_C \Delta x$ and letting $\Delta x \to 0$ we obtain

$$d^2 C_A / dx^2 - k_r C_A / D_e = 0 \qquad (4.16b)$$

This equation can be cast in a convenient nondimensional form by defining

$$\psi = C_A / C_{As} \qquad (4.16c)$$

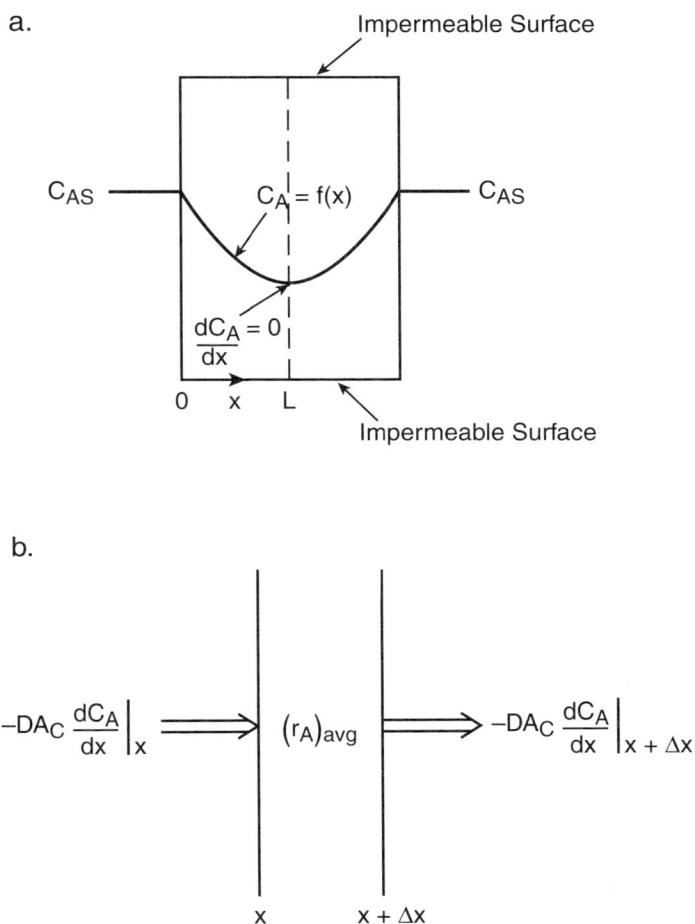

a.

b.

**FIGURE 4.6**
Diffusion and reaction in a flat-plate catalyst particle: (a) geometry and boundary conditions;
(b) difference element.

$$z = x/L \tag{4.16d}$$

$$\varphi = L(k_r/D_e)^{1/2} \tag{4.16e}$$

Equation 4.16b then becomes

$$d^2\psi/dz^2 - \varphi^2\psi = 0 \tag{4.16f}$$

This is the expression that now has to be solved for the dimensionless concentration ratio $\psi$. We draw for this purpose on the $D$-operator method given in the Appendix, which yields the solution:

$$\psi = C_1 \exp(\varphi z) + C_2 \exp(-\varphi z) \tag{4.16g}$$

$$\psi(0) = 1 \tag{4.16h}$$

$$\psi'(1) = 0$$

The integration constants obtained from these two conditions are

$$C_1 = \exp(-\varphi)[\exp(\varphi) + \exp(-\varphi)] \tag{4.16i}$$

$$C_2 = \exp(\varphi)/[\exp(\varphi) + \exp(-\varphi)\}$$

On introducing them into the solution (Equation 4.16g) we obtain

$$\psi = \frac{\exp[-\varphi(1-z) + \exp[\varphi(1-z)]}{\exp(\varphi) + \exp(-\varphi)} \tag{4.16j}$$

or alternatively

$$\frac{C_A}{C_{As}} = \psi = \frac{\cosh[\varphi(1-z)]}{\cosh \varphi} \tag{4.16k}$$

where we have used the hyperbolic function $\cosh \varphi = [\exp(\varphi) + \exp(-\varphi)]/2$ listed in the Appendix.

$\varphi$ is called the Thiele modulus, after one of the pioneers in the field, and expresses the combined effect of reaction rate constant $k_r$ and effective diffusivity $D_e$ on the concentration profile in the pellet. Large values of this parameter are associated with larger pellets and low diffusivities, and imply a strong pore diffusion resistance. This means that the reaction rate outpaces the transport rate, which is unable to adequately replenish the consumed reactant. A sharp drop in reactant concentration results and may in extreme cases lead to a vanishing concentration at some location in the pellet. Small values of $\varphi$, on the other hand, imply a low reaction rate and high diffusivities. Concentration profiles under these conditions will be nearly flat, and the pellet operates near its maximum effectiveness.

We now turn to the task of converting the concentration profile (Equation 4.16h), into the more useful effectiveness factor $E$, which had been defined as the ratio of the reaction rate with diffusional resistance to the reaction rate without diffusional resistance:

$$E = \frac{\text{rate with diffusional resistance}}{\text{rate without diffusional resistance at } C_{As}} \tag{4.17a}$$

Because the rate of reaction in the pellet must, at steady state, equal the rate of supply of reactant, we have alternatively

$$E = \frac{\text{rate of diffusion at } z = 0}{\text{rate of reaction at } C = C_{As}} \qquad (4.17b)$$

or

$$E = \frac{-D_e A_C (dC_A / dx)_{x=0}}{L A_C k_r C_{As}} \qquad (4.17c)$$

where the derivative is obtained from Equation 4.16k.

The result, obtained after some manipulation, is given by

$$E = -\frac{1}{\varphi^2} \left[ \frac{-\varphi \sinh \varphi (1-z)}{\cosh \varphi} \right]_{z=0} \qquad (41.7d)$$

or alternatively

$$E = (\tanh \varphi) / \varphi \qquad (4.17e)$$

A plot of this relation is shown in Figure 4.7. The important fact that emerges from it is that up to values of $\varphi = 0.5$, the pellet is very nearly at its maximum effectiveness, $E = 1$. This is precisely the region in which we want to operate and we can use this threshold value of $\varphi = 0.5$ to establish some of the properties a good catalyst pellet should possess.

Suppose it is desired to carry out a gas-phase reaction in a reactor packed with catalyst pellets. A reasonable size for the pellets that balances low-pressure drop and acceptable contact area is about 1 cm. Rate constants for many important gas-phase catalytic reactions are of the order $10^{-3}$ s$^{-1}$. The question then arises whether a typical pellet diffusivity $D_e$ satisfies the criterion $\varphi \leq 0.5$ or whether it has to be adjusted, along with pellet size, to yield an acceptable effectiveness. We have the requirement

$$\varphi = L(k_r / D_e)^{1/2} \leq 0.5 \qquad (4.17f)$$

A rough estimate of pellet diffusivity assuming molecular rather than Knudsen diffusion comes from the relation given in Chapter 3:

$$D_e = \frac{D\varepsilon}{4} \qquad (4.17g)$$

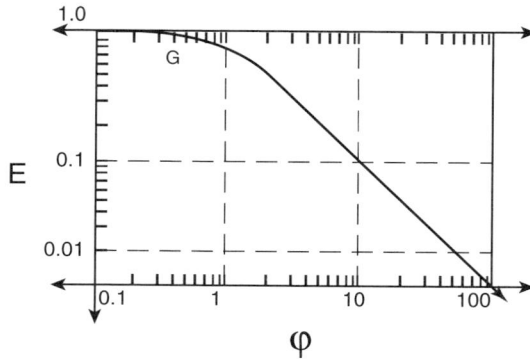

**FIGURE 4.7**
Catalyst particle effectiveness factor $E$ as a function of the Thiele modulus $\varphi$ for a flat plate and isothermal first-order reaction.

where $\varepsilon$ = porosity. If we set $\varepsilon = 0.4$ and use $D = 10^{-1}$ cm$^2$/s for the diffusivity of a typical gas, we obtain

$$\varphi = 0.5 \ (10^{-3}/10^{-2})^{1/2} = 0.16 \qquad (4.18)$$

Thus our criterion is fully met. In fact, we can drop $D_e$ by an order of magnitude or raise the rate constant by the same amount without violating the criterion. There is also some freedom to manipulate pellet size, if needed. All of this is revealed in simple fashion by an inspection of the Thiele modulus.

### Illustration 4.10: A Moving Boundary Problem: The Shrinking Core Model

The reader had previously been introduced to the concepts that must be applied in modeling the progress of a gas–solid reaction. They involve separating the system into a core, which reacts and continuously shrinks in size, and an external layer, which grows with time but is assumed to be at a quasi-steady state. This configuration is referred to as the *shrinking core model* and has associated with it the assumption of a quasi- or pseudo-steady state.

We illustrate these concepts with the following example: Suppose a solid particle undergoes a reaction according to the scheme

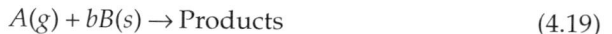

$$A(g) + bB(s) \rightarrow \text{Products} \qquad (4.19)$$

with rates $r_A$ and $r_B$, which are related by the stoichiometry of the reaction as follows

$$r_A = \frac{1}{b} r_B \qquad (4.20)$$

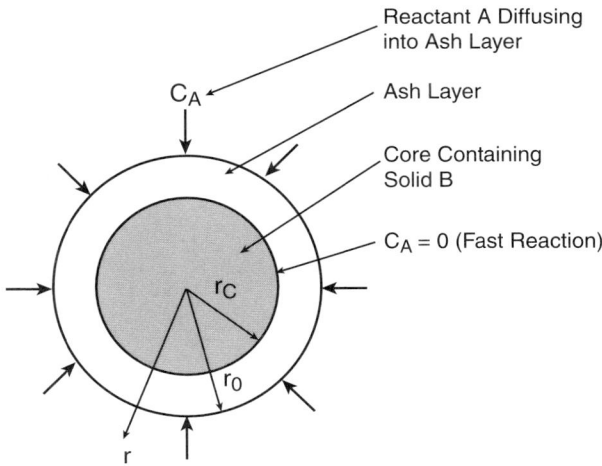

**FIGURE 4.8**
A reacting solid particle and its external ash layer.

The rate of reaction is assumed to be fast enough that all the reactant arriving at the surface of the particle is instantaneously consumed. Hence, $(C_A)_{r=r_C} = 0$ (Figure 4.8). On the other hand, the movement of the core front is, because of its high density, sufficiently slow that the external layer of solid products can be taken to be at steady state. The oxidative degradation of organic aerosols in the atmosphere is an example of such a reactor. The task is to determine the time dependence of the particle radius $r_C$ and hence that of its mass.

We start, as recommended, with an unsteady core balance and write

$$\text{Rate of } B \text{ in} - \text{Rate of } B \text{ out} = \begin{array}{c} \text{Rate of} \\ \text{change of } B \end{array} \qquad (4.21a)$$

$$0 - r_B = 4\pi r_C^2 \rho_B \frac{dr_C}{dt}$$

where we have placed the unknown reaction rate $r_B$ in the "out" column and denoted particle density by $\rho_B$.

This is followed by a quasi-steady-state mass balance in the product layer, which is composed as follows:

$$\text{Rate of } A \text{ in at } r - \text{Rate of } A \text{ out at } r_C = 0$$

$$4\pi r^2 D_e \frac{dC_A}{dr} - r_A = 0 \qquad (4.21b)$$

where $D_e$ is the effective diffusivity of reactant $A$ in the product layer. This last expression can be immediately integrated by separation of variables to yield

$$(4\pi D_e / r_A)\int_0^{C_A^{\,o}} dC_A = \int_{rC}^{rc} \frac{dr}{r^2} \qquad (4.21c)$$

or

$$(4\pi D_e C_A^{\,o} / r_A) = \frac{1}{r_C} - \frac{1}{r_o} \qquad (4.21d)$$

where $r_o$ = initial radius of particle.

We pause at this stage to take a brief inventory. We have, in Equation 4.20, Equation 4.21a, and Equation 4.21d, three equations in the three dependent variables $r_A$, $r_B$, and the core radius $r_C$. The model is consequently complete and we can proceed with the elimination of the unknown reaction rates $r_A$ and $r_B$ to arrive at the desired solution $r_C = f(t)$.

We obtain from the three aforementioned equations

$$-\frac{4\pi D_e b C_A^{\,o}}{1/r_C - 1/r} = r\pi r_C^{\,2}\rho_B \frac{dr_C}{dt} \qquad (4.21e)$$

which can again be integrated by separation of variables:

$$-\frac{D_e b C_A^{\,o}}{\rho_B}\int_0^t dt = \int_0^{rc}\left(r_C - \frac{r_C^{\,2}}{r_o}\right)dr_C \qquad (4.21f)$$

The final result is then

$$t = \frac{\rho_B r_o^{\,2}}{6bD_e C_A^{\,o}}\left[1 - 3\left(\frac{r_C}{r_o}\right)^2 + 2\left(\frac{r_C}{r_o}\right)^3\right] \qquad (4.21g)$$

*Comments:*

The illustration demonstrates how the use of clever simplifying assumptions and some inspired modeling can reduce the complexities of a process to manageable proportions. The assumptions would be violated only if the progress of core consumption were extremely rapid, in which case events in the outer layer become both time- and distance-dependent, leading to a PDE.

The time required for the reaction to be complete is obtained from Equation 4.21g by setting $r_C = 0$. This leads to

$$t_{\text{Tot}} = \frac{\rho_B r_o^2}{6bD_e C_A^o} \tag{4.21h}$$

where $t_{\text{Tot}}$ is seen to vary inversely with the effective diffusivity $D_e$ and the external reactant concentration $C_A^o$.

This was anticipated on physical grounds, as both these factors increase the rate of reactant consumption. Somewhat less expected is the direct dependence on the square of the initial particle radius, $r_o^2$. It might have been argued that the time necessary for total consumption should vary with the mass of the particle, i.e., with the radius cubed. That this is not so is neatly revealed by the model, and this is the case because the rate of consumption is dictated by, and hence proportional to, the surface area of the particle at any instant.

It will also be noted that a knowledge of the reaction rates is not required in this instance. This is a consequence of the assumption that the rate of reaction is very rapid and as a result ceases to affect the overall rate of the process. The rate-determining step here is the speed with which the core front recedes, which is very low given the high density $\rho_b$ of the particle. This, together with the low values of $D_e$ and $C_A^o$, leads to high values of the consumption time $t_{\text{Tot}}$.

### Illustration 4.11: First-Order Reaction with Diffusion in a Liquid Film: Selection of a Reaction Solvent

We consider the situation depicted in Figure 4.5, involving transport through a gas film to a liquid interface, followed by diffusion and reaction in the liquid film. The difference element over which the mass balance is taken is the same as that for the flat-plate catalyst pellet, and leads to the same differential equation and the same boundary conditions:

$$\frac{d^2 C_A}{dx^2} - k_r C_A / D_L = 0 \tag{4.16a}$$

$$CA(0) = CAi \tag{4.16b}$$

$$\frac{dC_A}{dx}(\delta) = 0$$

Where the treatment of the two cases differs is in the ultimate goal of the solution. The catalyst pellet required us to derive the ratio of *reaction rate with diffusion* to *reaction rate without diffusion*. For the liquid film the requirement is reversed: We compose the ratio of *diffusion rate with reaction* over *diffusion rate without reaction*. The results are exact inverses of each other. We obtain

For the catalyst pellet:

$$E = \frac{\tanh \varphi}{\varphi} \tag{4.16c}$$

For the liquid film:

$$E_h = \frac{Ha}{\tanh Ha} \tag{4.16d}$$

where $E_h$ is the *enhancement factor* and $Ha = \delta(k_r/D_L)^{1/2}$ is the Hatta number, named after one of the pioneers in the field. For the case of no reaction, Ha = 0 and $E_h$ = 1; that is to say, no enhancement occurs. Enhancement $E_h$ rises above 1 when Ha > 0 and continues to increase with further increases in the reaction-rate constants $k_r$. Values of $E_h$ can be read off Figure 4.7 by setting $\varphi$ = Ha and $E_h$ = 1/E.

Details of the derivation of Equation 4.16d are left to the exercises (Practice Problem 4.11). The intent here is to use the enhancement factor to make a rational selection of a reactive solvent. Let us assume a typical turbulent-flow film thickness of 0.1 mm (see Illustration 1.5) and a value of $D_L$ = 10$^{-5}$ cm$^2$/s, which is the commonly used order of magnitude of liquid-phase diffusivities (see Chapter 3). To find out at which value of Ha enhancement begins to exceed 1, we use the criterion Ha > 0.5, i.e., the inverse of that proposed for the catalyst pellet, $\varphi$ < 0.5. We obtain, for Ha = 0.5

$$k_r = D_L \left(\frac{Ha}{\delta}\right)^2 = 10^{-5}\left(\frac{0.5}{10^{-2}}\right)^2 = 0.025\,\text{s}^{-1}$$

Some other values are listed below:

| $k_r$ s$^{-1}$ | 2.5 | 10 | 25 | 100 |
|---|---|---|---|---|
| $E_h$ | 5 | 10 | 50 | 100 |

Thus, for the reaction to have a significant effect on mass transfer rate, the rate constant must be in excess of 0.025 s$^{-1}$. This is in fact the range of many liquid-phase reaction rates. The requirements become less stringent with an increase in film thickness. For $\delta$ = 1 mm, for example, rate constants can be lowered by a factor of 100 to achieve the same result.

Reactive solvents are routinely used in the scrubbing of acidic gases. An idea of what can be accomplished is reflected in Table 4.4.

**TABLE 4.4**

Enhancement Factor for the Absorption of Carbon Dioxide in Various Solvents

| Solvent | $E_h$ |
|---|---|
| 2N Potassium hydroxide | 6.25 |
| 2N Sodium hydroxide | 10.5 |
| 2N Ethanol amine | 62.5 |

## Practice Problems

4.1. *Emissions from a Chimney*

A 10-m-tall chimney emits a toxic substance at the rate of 10 kg/s. Calculate the prevailing concentration at the base of the chimney 100 h after the start of the emissions, assuming a diffusivity in air of $10^{-5}$ m²/s. (*Hint:* Use coordinates $z - z_o$, where $z_o$ = chimney height.)

*Answer:* $2.48 \times 10^{-24}$ g/m³

4.2. *A Simplified Model of a Nicotine Patch*

A nicotine patch attached to the skin of a smoker releases 1 µg of nicotine per second. The substance penetrates the skin and ultimately enters the blood vessels assumed to be 1 mm beneath the skin surface. Calculate the maximum concentration of nicotine attainable at the point of entry into the blood. (*Hint:* Assume the patch to be a continuous point source and set $D = 10^{-9}$ m²/s.)

*Answer:* 160 µg/cm³

*Note:* Nicotine patches usually last no more than 24 h, so the maximum given here is unlikely to be approached.

4.3. *Evaporation of a Solvent Spill*

A load of solvent is spilled over a large area, resulting in a layer of liquid with an area density of 1 kg/m². The solvent has a vapor pressure of 10 kPa and an average molar mass of 100. Diffusivity is estimated at $10^{-5}$ m²/s. Calculate the time of evaporation in the absence of any air currents. This is the maximum to be expected.

*Answer:* 136 h

4.4. *Doping of a Silicon Chip with Boron*
A junction in silicon is made by doping it with boron. This is done by first depositing a layer of boron on the chip (predeposition), followed by what is termed *drive-in diffusion*. If the deposition step requires 5 min, at what distance from the surface is the concentration of boron raised by $3 \times 10^{18}$ atoms/cm$^3$ during this interval? The density of pure boron is $5.1 \times 10^{20}$ atoms/cm$^3$, its diffusivity $5.8 \times 10^{-2}$ $\mu$m$^2$/h.

*Answer*: 0.271 μm

4.5. *Leaching of an Ore*
The leaching of ores to recover valuable mineral components is a commonly applied operation in the field of hydrometallurgy. The extraction of gold with cyanide solution is a familiar example. Suppose the finely ground ore can be thought of as plane flakes, and that it is desired to carry the process to the point where no more than 2% of the original material remains at the midplane of the flake, which has a thickness of 2 mm. Diffusivity is estimated at $10^{-9}$ m$^2$/s. Estimate the time of leaching.

*Answer*: 0.417 h

4.6. *Batch Adsorption of a Trace Substance*
When a diffusing solute partitions or adsorbs onto a solid matrix, we can often use standard solutions for nonsorbing solids to follow the course of adsorption by suitably modifying one of the solution parameters. For the case of adsorption by spherical particles from a well-stirred solution of limited volume, for example, the parameter $V_{Sol'n}/V_{Spheres}$ in Figure 4.4 is replaced by $V_{Sol'n}/KV_{Spheres}$, where $K$ is the partition coefficient or Henry's constant. Assume the following parameter values: $K = 10$, $V_{Sol'n}/V_{Spheres} = 10$, $D = 10^{-5}$ cm$^2$/s, $R = 0.46$ cm. What is the fractional saturation of the adsorbent after 1 h?

*Answer*: 0.74

4.7. *The Catalyst Pellet under Nonisothermal Conditions*
Derive the energy balance for a flat-plate catalyst pellet operating under nonisothermal conditions (first-order exothermic reaction). Give a plausible argument why the effectiveness factor can in this case exceed unity.

4.8. *Catalyst Pellets in the Form of Raschig Rings*
Catalyst pellets are on occasion cast in the form of hollow cylinders (Raschig Rings). Discuss the advantages and drawbacks of this geometry. How would you define the dimension needed to define the Thiele modulus?

4.9. *A Heat Transfer Counterpart to the Shrinking Core Model: Freeze-Drying of Food*

In the process considered here, it is desired to derive a model that would allow us to obtain relevant transport coefficients from freeze-drying rate data. The food to be dried, i.e., a slab of frozen poultry meat, has an initial (frozen) water content of $m_o$ kg. It is heated with an electric heater and, in the experiment in question, provided with thermocouples to measure surface temperature $T_g$ (Figure 4.9). Sublimation of the ice takes place in a vacuum chamber, and water loss is monitored by means of a spring balance. As sublimation progresses, the core ice front, assumed to be at the constant temperature $T_i$, recedes into the interior, exposing an ice-free matrix, which increases in thickness with time. Heat conduction through this matrix is assumed to be at a quasi-steady state so that a linear temperature gradient prevails at any given instant. Start in the usual fashion by first making mass and energy balances about the core, followed by an energy balance on the ice-free matrix. Use the fraction of ice removed, $f$, as the dependent variable.

*Answer:* $t/f = af + b$

4.10. *More about Gas–Liquid Reactions with Diffusion*

Show that for the system considered in Illustration 4.11, the rate of reaction is given by the expression

$$r_A = \frac{p_A}{\dfrac{1}{k_g} + \dfrac{H_A}{k_L E}}$$

where $H_A$ = Henry's constant.

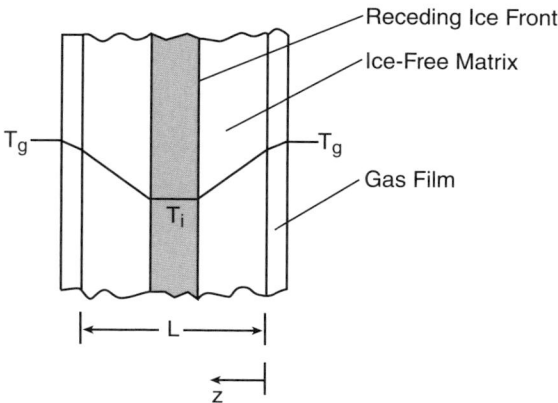

**FIGURE 4.9**
Freeze-drying of meat.

4.11. *The Enhancement Factor* $E_h$

Give a derivation of Equation 4.16d for the enhancement factor, using the procedure used to derive the catalyst effectiveness factor.

# 5

## A Survey of Mass Transfer Coefficients

Chapter 1 introduced the reader to the notion of a mass transfer coefficient and has shown the connection to what is termed film theory. In essence, this approach assumes the resistance to mass transfer to be confined to a thin film in the vicinity of an interface in which the actual concentration gradient is replaced by a linear approximation. The result is that the rate of mass transport can be represented as the product of a mass transfer coefficient and a linear concentration difference, or concentration driving force. Thus,

$$N_A/A = k_C\Delta C \tag{5.1}$$

It was further shown that individual transport coefficients could be combined into overall mass transfer coefficients to represent transport across adjacent interfacial layers. The underlying concept is referred to as two-film theory. Chapter 1 has been confined to simple applications of the mass transfer coefficient which is either assumed to be known, or is otherwise evaluated numerically in simple fashion.

This chapter seeks to enlarge our knowledge of mass transfer coefficients by compiling quantitative relations and data for use in actual calculations applied to practical systems. There are evidently a host of such systems, and our aim here is to convey the coefficients pertinent to these systems in an organized fashion.

When the system under consideration is in laminar flow, it is often possible to give precise analytical expressions of the transport coefficients. In most other cases, including the important case of turbulent flow, the analytical approach generally fails and we must resort to semiempirical correlations, arrived at by the device known as dimensional analysis, which involves the use of dimensionless groups.

To represent these facts in an organized fashion, we start our deliberations with a brief survey of the dimensionless groups pertinent to mass transfer operations. We next turn to transport coefficients that apply to systems in laminar flow and show how these coefficients are extracted from the solutions of the pertinent PDE models. This is followed by an analysis of systems in turbulent flow where the approach of dimensional analysis is used. We

describe the method and present the results obtained for some simple geometries, including flow in a pipe and around spheres and cylinders.

More complex geometries involving commercial tower packings are taken
up next. Such packings, used in separation and purification methods such
as gas absorption and distillation, have their own peculiar characteristics,
and our main source of information here is the data given in manufacturers'
catalogs. Even more complex conditions apply to mass transfer operations
carried out in stirred vessels. Configuration of the stirring mechanism and
the speed of stirring enter the picture here, leading to rather complex expressions.

Finally, we turn our attention briefly to transport in an environmental
context. The methodology used by environmentalists in determining transport coefficients has its own peculiarities, which are discussed and related
to the standard concepts used here.

## 5.1   Dimensionless Groups

The two principal dimensionless groups of relevance to mass transport are
the Sherwood and Schmidt numbers. They are defined as follows:

$$Sh = \frac{k_c \ell}{D} \tag{5.1a}$$

$$Sc = \frac{\mu}{\rho D} \tag{5.1b}$$

where $\ell$ is some pertinent dimension of the system, such as the diameter of
a pipe or a sphere, and $k_c$ is the mass transfer coefficient in units of m/s.

The Sherwood number can be viewed as describing the ratio of convective
to diffusive transport, and finds its counterpart in heat transfer in the form
of the Nusselt number. The Schmidt number is a ratio of physical parameters
pertinent to the system, and corresponds to the well-known Prandtl number
used in heat transfer. Added to these two groups is the Reynolds number,
which represents the ratio of convective-to-viscous momentum transport
and serves in essence to describe the flow conditions.

Two additional dimensionless groups, the Peclet number and the Stanton
number, are also used, although with lesser frequency. Both of these numbers
are composites of other dimensionless groups, which frequently occur in
unison. Thus, the Reynolds and Schmidt numbers often crop up combined
as a product, which leads to the Peclet number:

$$Pe = Re\,Sc = \frac{\ell v \rho}{\mu}\frac{\mu}{\rho D} = \frac{\ell v}{D} \qquad (5.1c)$$

The Stanton number is a combination of Sherwood, Reynolds, and Schmidt numbers, which likewise often appear in unison. It is defined as

$$St = Sh\,/\,Re\,Sc = \frac{k_c}{v} \qquad (5.1d)$$

The result here is a particularly simple one, and in essence represents the ratio of two velocities: the "velocity" of mass transfer $k_c$ in units of m/s and the velocity of flow, likewise in units of m/s. Table 5.1 summarizes these groups for both mass and heat transfer processes.

### *Illustration 5.1: The Wall Sherwood Number*

In dialysis and similar processes through permeable membranes, it has become convenient to replace the permeabilities, which were defined in Chapter 3, by an effective mass transfer coefficient, termed $k_w$, which equals the ratio of membrane diffusivity over membrane thickness. Thus,

$$k_w = \frac{D_m}{t_m} \qquad (5.2)$$

where $t_m$ = membrane thickness.

**TABLE 5.1**

Summary of Dimensionless Groups Used in Mass and Heat Transfer Processes

| Mass Transfer | Heat Transfer |
|---|---|
| Sherwood number $\ Sh = \dfrac{k_c \ell}{D}$ | Nusselt number $\ Nu = \dfrac{h\ell}{k}$ |
| Schmidt number $\ Sc = \dfrac{\mu}{\rho D}$ | Prandtl number $\ Pr = \dfrac{C_p \mu}{k}$ |
| Reynolds number $\ Re = \dfrac{\ell v \rho}{\mu}$ | Reynolds number $\ Re = \dfrac{\ell v \rho}{\mu}$ |
| Peclet number $\ Pe_m = \dfrac{\ell v}{D}$ | Peclet number $\ Pe_h = \dfrac{\ell v \rho C_p}{k}$ |
| Stanton number $\ St_m = \dfrac{k_c}{v}$ | Stanton number $\ St_h = \dfrac{h}{C_p v \rho}$ |

When flow in a tubular membrane device is laminar, with material diffusing from the flowing fluid into and across the membrane, the transport equations become distributed in both radial and axial directions and in consequence lead to PDEs. One of the boundary conditions for this PDE is given by the relation

$$k_w(C_w - C_{ext}) = -D\frac{dC}{dr}\bigg|_{r=R} \tag{5.3}$$

which in essence equates the rate of passage through the membrane to the rate of arrival of the dissolved solute at the membrane wall. It has become customary in models describing these events to combine the transport coefficients $k_w$ and $D$ with the tubular diameter $d$ into a dimensionless group termed the *wall Sherwood number*, $Sh_w$. Thus,

$$Sh_w = \frac{k_w d}{D}$$

This number can be viewed as the ratio of membrane transport to transport through the tubular fluid and has found extensive use in describing and correlating membrane transport processes.

With typical membrane and liquid diffusivities of $10^{-7}$ and $10^{-5}$ cm²/s, respectively, membrane thickness of $10^{-3}$ mm, and tubular diameter of 1 mm, a typical wall Sherwood number becomes

$$Sh_w = \frac{k_w d}{D} = \frac{(D_m/t_m)d}{D} \tag{5.4a}$$

$$Sh_w = \frac{(10^{-6}/10^{-4})}{10^{-5}}10^{-1} = 100 \tag{5.4b}$$

i.e., of the order 100. We have occasion to discuss membrane processes further in Chapter 8.

## 5.2 Mass Transfer Coefficients in Laminar Flow: Extraction from the PDE Model

Mass transport in laminar flow in a tubular geometry or around simple submerged shapes is generally modeled by PDEs because there is more than one direction of diffusional flow involved. These PDEs have been solved analytically for a number of cases and generally lead to fairly formidable expressions representing the concentration and velocity profiles in the geometries in question. As had been indicated on a number of previous occasions, concentration profiles, which represent the primary information obtained from the PDE model, are often not directly useful for engineering purposes. We have shown this, for example, in the case of diffusion and reaction in a catalyst pellet, Illustration 4.9, where the primary profiles were converted into the more useful quantity known as the effectiveness factor. In the present case the useful quantity we wish to extract from the primary information is an equivalent mass transfer coefficient.

Let us demonstrate its derivation using transport in a tube as an example. The situation here is one in which solute diffuses in the radial direction, either as a result of release from the wall, or in consequence of transport to and ultimately through a permeable wall. The model for this case has been presented in Section 2.8 and there referred to as the Graetz problem in mass transfer.

The concentration profiles that arise in this case are distributed in both the radial and axial directions, as diffusion in one direction is superposed on convective transport in the other direction. To obtain a mass transfer coefficient from this information, we perform a mass balance at the tubular wall, equating diffusional transport rate to an equivalent "convective" rate expressed by means of a mass transfer coefficient. Thus,

$$-D\frac{\partial C}{\partial r}\bigg|_{r=R} = k_c\left(C_m - C_{r=R}\right) \tag{5.5a}$$

To evaluate $k_c$, two quantities need to be obtained from the primary concentration profile. One is the derivative at the tubular wall given above, which is obtained by differentiating the solution $C(r,z)$; the second is the mean integral concentration $C_m$ in the flowing fluid. This latter quantity is obtained from the expression

$$C_m = \frac{\int_0^R C(r,z)2\pi r dr}{\pi R^2} \tag{5.5b}$$

Note that both of these items are part of the "information package" contained in a model that we had alluded to in Table 2.2.

With these two quantities in place, Equation 5.5a can be solved for $k_c$ and the latter tabulated. Similar calculations can be carried out for systems involving flow around simple geometries.

### 5.2.1    Mass Transfer Coefficients in Laminar Tubular Flow

In the case of tubular mass transfer coefficients, we distinguish between mass transfer in the so-called entry or Lévêque region, in which concentration changes are confined to a thin boundary layer $\delta(x)$ adjacent to the wall, and the so-called fully developed region, in which the concentration changes have penetrated into the fluid core. The situation is depicted in Figure 5.1, and represents a tubular wall coated with a soluble material of solubility $C_s$ dissolving into pure solvent.

Because of the thinness of the boundary layer, mass transfer in the entry region is very rapid, with Sherwood numbers in excess of 1000 attained near the tubular entrance (Figure 5.2). As we move away from the entrance in the downstream direction, the boundary layer gradually thickens and the Sherwood number diminishes with the one-third power of axial distance $x$. Eventually it levels off and attains a constant value as the fully developed region is reached (Figure 5.2). Table 5.2 lists some of the relevant Sherwood numbers obtained in ducts of various geometries and constant wall concentration.

For engineering calculations, it is often more convenient to deal with mass transfer coefficients that have been averaged over the entire length of the entry region. The result is expressed in terms of the dimensionless Reynolds

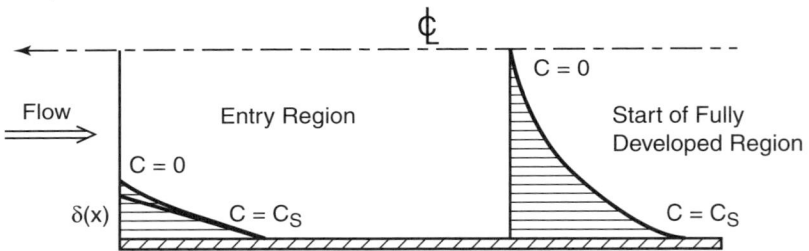

**FIGURE 5.1**
Mass transfer from a coated tubular wall into a flow of pure solvent.

**TABLE 5.2**

Mass Transfer Coefficients in Ducts of Various Geometries for Laminar Flow

| Duct Geometry | Entry Region | Fully Developed Region |
|---|---|---|
| Cylinder | $\mathrm{Sh} = 1.08\left(\dfrac{xD}{vd^2}\right)^{-1/3}$ | Sh = 3.66 |
| Parallel planes | $\mathrm{Sh} = 1.23\left(\dfrac{xD}{vd^2}\right)^{-1/3}$ | Sh = 7.54 |
| Square | — | Sh = 2.98 |
| Triangular | — | Sh = 2.47 |

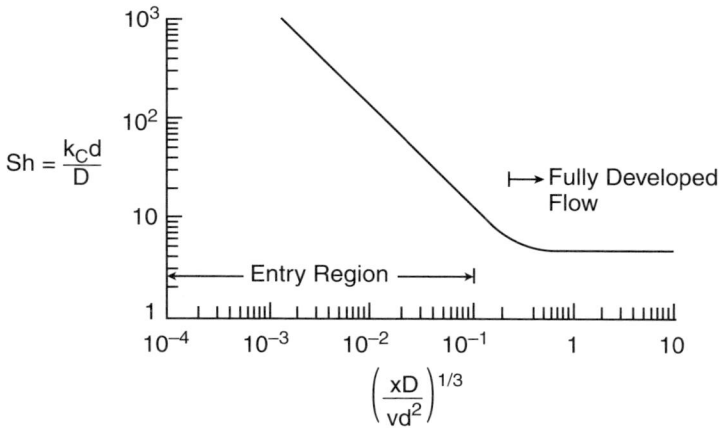

**FIGURE 5.2**
Sherwood numbers for laminar flow in a cylindrical duct.

and Schmidt numbers and takes the following form for laminar flow in cylindrical tubes

$$Sh = 1.86 \, (Re \, Sc \, d \, / \, L)^{1/3} = 1.86\left(\frac{vd^2}{DL}\right)^{1/3} \quad (5.6)$$

where $L$ is the length of pipe in question. The mass transfer coefficient here is a constant, rather than a function of distance, and can be directly incorporated in the usual tubular mass balances. This is shown in Illustration 5.2.

Mass transfer in the entry region is far from a rare event. Transport in the larger blood vessels lies entirely in the entry region, and so does mass transfer in commercial reverse osmosis desalination plants and other membrane processes (see Chapter 8). Entry lengths of many meters are not uncommon, particularly for low-solute diffusivities. The only requirement is that flow must be in the laminar regime (Re < 2000).

### 5.2.2 Mass Transfer Coefficients in Laminar Flow around Simple Geometries

In the case of mass transfer involving laminar flow about simple geometries, no distinction is made between regions near the leading front of the particle and portions farther downstream. Instead, we report transfer coefficients averaged over the entire particle. The results are expressed as correlations of the Sherwood or Stanton numbers as a function of Reynolds and Schmidt numbers.

Mass Transfer Coefficients in Laminar Flow around Simple Geometries

| Geometry | Correlation | Re |
|----------|-------------|-----|
| Flat plate | $St = 0.66(Re)^{-1/2} (Sc)^{-2/3}$ | $<10^5$ |
| Sphere | $Sh = 2.0 + 0.60 (Re)^{1/2} (Sc)^{1/3}$ | $<10^{-1}$ |
| Cylinder | $Sh = 0.43 + 0.53 (Re)^{1/2} (Sc)^{0.31}$ | $1 < Re < 10$ |

The correlation for three simple geometries, the flat plate, the sphere, and the cylinder, are tabulated in Table 5.3. The flat plate here is taken to be positioned parallel to the direction of flow, whereas the cylinder has its axis normal to the flowing medium.

### Illustration 5.2: Release of a Solute into Tubular Laminar Flow: Transport in the Entry Region

Consider the transport from a tubular wall coated with a soluble substance into a solvent in laminar flow. We set the diameter at $d = 1$ cm, velocity $v$ at 1 cm/s, and the length $x$ of the tube at 1 m. The solubility $C^*$ of the solute is 10 g/l and its diffusivity $D = 10^{-9}$ m²/s. The problems we wish to address are the following:

1. What is the boundary layer thickness at the exit of the tube?
2. What is the mean concentration at that position?

To answer question 1, we draw on the definition of a mass transfer coefficient, which was set out in the film theory given in Chapter 1, Equation 1.11b. The rearranged version of this expression has the form:

$$k_c = \frac{D}{z_{FM}} = \frac{D}{\delta} \qquad (5.7a)$$

from which the boundary layer thickness $\delta$ can be extracted.

To obtain $\delta$ we start by calculating the mass transfer coefficient $k_c$ for which we draw on Table 5.3. Note that, since $(xD/vd^2)^{1/3} = (1 \times 10^{-9}/10^{-2} \times 10^{-4})^{1/3} = 0.1$, the flow does in fact fall in the Lévêque region (Figure 5.2). We have

$$\frac{k_c d}{D} = 1.08\left(\frac{xD}{vd^2}\right)^{-1/3} \qquad (5.7b)$$

from which we obtain

$$k_c = \frac{1.08 \times D}{d}\left(\frac{vd^2}{xD}\right)^{1/3} = \frac{1.08 \times 10^{-9}}{10^{-2}}\left(\frac{10^{-2} \times 10^{-4}}{1 \times 10^{-9}}\right)^{1/3} \qquad (5.7c)$$

and consequently

$$k_c = 1.08 \times 10^{-6} \text{ m/s} \tag{5.7d}$$

Substitution of this value into Equation 5.7a and solving for $\delta$ yields the boundary layer thickness:

$$\delta = \frac{D}{k_c} = \frac{10^{-9}}{1.08 \times 10^{-6}} = 0.93 \times 10^{-3} \text{ m} = 0.93 \text{ mm} \tag{5.7e}$$

Thus, the boundary layer thickness is approximately 10% of the tubular diameter at the exit of a 1-m tube.

We next turn to the calculation of the concentration at the tubular exit. This requires setting up a differential solute balance over a tube segment and its subsequent integration over the entire length of the tube. For a finite tubular increment $\Delta x$ and volumetric flow rate $Q$, we can write

Rate of solute in − Rate of solute out = 0

$$\left[ \begin{array}{c} QC_{1x} \\ +k_c \pi d \Delta x (C * -C) \end{array} \right] - QC_{1x+\Delta x} \qquad = 0 \tag{5.8a}$$

where $C$ is the tubular concentration averaged over the entire cross section. Dividing by $\Delta x$ and letting $\Delta x \to 0$ yields

$$Q\frac{dC}{dx} - k_c \pi d(C * -C) = 0 \tag{5.8b}$$

or

$$v(d/4)\frac{dC}{dx} - k_C(C * -C) = 0$$

To keep the integration simple, we use the mean integral Sherwood number given by Equation 5.6, rather than the distance-dependent local coefficient listed in Table 5.2. We have

$$k_C = 1.86 \frac{D}{d}\left(\frac{vd^2}{DL}\right)^{1/3} \tag{5.8c}$$

$$k_C = 1.86 \left( \frac{vD^2}{dL} \right)^{1/3} \tag{5.8d}$$

$$k_C = 1.86 \left( \frac{10^{-2} \times 10^{-18}}{10^{-2} \times 1} \right)^{1/3} = 1.86 \times 10^{-6} \text{ m/s} \tag{5.8e}$$

This value is used in the solution of the ODE (Equation 5.8b) obtained by separation of variables.

$$\int_0^C \frac{dC}{C^* - C} = (4/vd)1.86 \times 10^{-6} \int_0^L dx \tag{5.8f}$$

$$\frac{C^* - C}{C^*} = \exp\left( -\frac{4 \times 1.86 \times 10^{-6}}{10^{-2} \times 10^{-2}} \times 1 \right) \tag{5.8g}$$

so that

$$1 - C/C^* = 0.93 \tag{5.8h}$$

and $C = 0.7$ g/l; i.e., the well-mixed solution is approximately 7% saturated at the exit. Note that the *core* concentration itself remains constant and equal to zero over the entire tube length, and average concentration changes very slowly. We use this fact to simplify the model for reverse osmosis taken up in Illustration 8.6.

---

## 5.3    Mass Transfer in Turbulent Flow: Dimensional Analysis and the Buckingham π Theorem

Both heat and mass transfer in turbulent flow are generally not amenable to analytical treatment. It is customary in these cases to resort to what is termed *dimensional analysis*. This device consists of grouping the pertinent physical parameters of the system into a number of dimensionless groups, thus reducing the number of variables that must be dealt with, and evaluating the undetermined coefficients experimentally. It is a powerful tool for arriving at a first qualitative description of complex systems and eases enormously the experimental work required to quantify the relationship.

### 5.3.1 Dimensional Analysis

The process of dimensional analysis can be carried out in three deceptively simple steps. Some comments are necessary to ensure the proper application of this scheme:

Step 1. *List all the variables that affect the system behavior.* This is by far the most important and difficult step and confronts the user with the task of deciding which independent variables to include in the analysis. There is no easy recipe for carrying out this step, but the following suggestions may be found useful:

a. Apply physical reasoning to identify the pertinent variables.
b. Use experimental observations to amplify this list and to verify the final result.
c. If no functional relationship of dimensional consistency results, reexamine the situation by adding or omitting parameters.
d. Make sure that only truly independent variables are included. Thus, if mass and acceleration are chosen as parameters, force cannot be added to the list since it is related to the former through Newton's law.

Step 2. *Write down the dimensions of these variables.* The basic dimensions that are commonly chosen in dimensional analysis are those of mass $(M)$, length $(L)$, time $(\theta)$, and temperature $(T)$. All other quantities are expressed in terms of these fundamental dimensions. Thus, force has the units of $ML\theta^{-2}$ by virtue of Newton's law. Joule (J) is not a fundamental dimension but is instead expressed as $ML^2\theta^{-2}$.

Step 3. *Combine the variables into a functional relationship involving dimensionless groups or some other dimensionally consistent form.* The ultimate aim of this step, and of the analysis as a whole, is to express system behavior in terms of the functional relationship:

$$F(\pi_1, \pi_2 \dots \pi_p) = 0 \qquad (5.9a)$$

where $\pi$ is the symbol commonly used for a dimensionless group. If $x_2 \dots x_p$ are independent variables, and $x_1$ the dependent variable, the dimensionless groups can be represented in the form

$$\pi_1 = \frac{x_1}{x_2^{a_1} x_3^{a_2} \dots} \qquad \pi_2 = \frac{x_i}{x_2^{\beta_1} x_3^{\beta_2} \dots} \quad etc. \qquad (5.9b)$$

Often, it is convenient to solve this relationship for the dependent variable. We then obtain

$$x_1 = x_2^{a_1} x_3^{a_3} \dots F(\pi_2, \pi_3 \dots \pi_n) \qquad (5.9c)$$

To aid the reader in implementing this analysis, a short list of the dimensions of important variables is presented in Table 5.4.

### 5.3.2 The Buckingham π Theorem

The application of dimensional analysis is relatively simple when the number of independent variables equals the number of fundamental dimensions. Straightforward application of Steps 1 through 3 then yields a result expressed in terms of a *single* dimensionless group. This case, however, is a relatively rare one. It is more common for the number of variables to exceed the number of dimensions, which means that more than one dimensionless group will be involved. The question then arises regarding how many such groups are to be sought out, and how they are to be composed. These problems were first addressed by Buckingham and resolved in his famous π theorem. Simply stated, that theorem reads as follows:

Given $p$ variables, $x_1, x_2, \ldots x_p$ are related to a physical phenomenon that can be expressed in terms of $r$ fundamental dimensions. Then these variables, which include the dependent variable, can be gathered into $p - r$ dimensionless groups π and cast in the functional form:

$$F(\pi_1, \pi_2, \ldots \pi_{p-r}) = 0 \qquad\qquad (5.10)$$

**TABLE 5.4**

Variables and Their Dimensions

|  | Variable | Symbol | Dimension |
|---|---|---|---|
| 1. Fundamental | Mass | $m$ | $M$ |
|  | Length | $\ell, d$ | $L$ |
|  | Time | $t$ | $q$ |
|  | Temperature | $T$ | $T$ |
| 2. Mechanical | Velocity | $v$ | $L\theta^{-1}$ |
|  | Acceleration | $a$ | $L\theta^{-2}$ |
|  | Density | $r$ | $ML^{-3}$ |
|  | Viscosity | $m$ | $ML^{-3}\theta^{-1}$ |
|  | Force | $F$ | $ML^{-1}\theta^{-2}$ |
|  | Pressure | $p$ | $ML^{-3}\theta^{-2}$ |
| 3. Thermal | Thermal conductivity | $k$ | $ML\theta^{-3}T^{-1}$ |
|  | Specific heat | $C_p$ | $L^2\theta^{-2}T^{-1}$ |
|  | Heat transfer coefficient | $h$ | $M\theta^{-3}T^{-1}$ |
| 4. Diffusional | Concentration | $C$ | $ML^{-3}$ |
|  | Rate of mass flow | $N$ | $M\theta^{-1}$ |
|  | Diffusivity | $D$ | $L^2\theta^{-1}$ |
|  | Mass transfer coefficient | $k_c$ | $L\theta^{-1}$ |

In other words, we have managed to replace a functional relation, which involves $p$ variables, with one that involves only $p - r$ variables. This constitutes a considerable saving.

To implement Buckingham's $\pi$ theorem, we retain the three steps we formulated earlier but amplify Step 3 in the following fashion:

Step 3a: Select a set of variables that equals the number of dimensions $r$ and does not include the dependent variable. Raise each variable to some unknown power $\alpha, \beta, \ldots$ and form a product of the result. This product is placed in the *denominator* of each dimensionless group $\pi$.

Step 3b: Place the remaining $(p - r)$ variables, which now include the dependent variable, in the *numerators* of the $(p - r)$ dimensionless groups. This results in the set of $\pi$ terms shown in Equation 5.10.

## Illustration 5.3: Derivation of a Correlation for Turbulent Flow Mass Transfer Coefficients Using Dimensional Analysis

The problem addressed here is that of expressing mass transfer coefficients that apply to turbulent flow conditions in a tube in terms of appropriate dimensionless groups. To implement Step 1, both fluid mechanical and transport properties must be taken into account. The former determine the degree of turbulence or ability to form eddies, and hence the rate at which mass is transported to or from the tubular wall, whereas the transport parameter determines the rate at which mass is conveyed through the film adjacent to the interface. It is proposed to use velocity $v$, density $\rho$, and viscosity $\mu$ as the fluid mechanical properties, as each of these parameters either promotes or resists the formation of eddies. Transport through the film is determined by only one parameter, the diffusivity of the conveyed species. In addition to these factors, we expect pipe diameter to play a role because it determines the distance over which the mass is to be transported and plays a role as well in the degree of turbulence generated in the system.

The reader will have noted that in choosing these parameters we had to proceed in a somewhat intuitive fashion. This is the difficult and challenging part of dimensional analysis we had alluded to in Step 1 of our outline. It requires physical insight as well as some inspired guesswork.

We can now proceed to Step 2 of the procedure and list the dimensions of these variables, making use of Table 5.4 for this purpose.

| Variable | Symbol | Dimension |
|---|---|---|
| Velocity | $v$ | $L\theta^{-1}$ |
| Density | $r$ | $ML^{-3}$ |
| Viscosity | $m$ | $ML^{-3}\theta^{-1}$ |
| Mass transfer coefficient | $k_c$ | $L\theta^{-1}$ |
| Diffusivity | $D$ | $L^2\theta^{-1}$ |
| Diameter | $d$ | $L$ |

It is seen that there is a total of six variables, including the dependent variable $k_c$. Hence, by Buckingham's theorem, there are $6 - 3 = 3$ dimensionless groups to be established.

As recommended in Step 3a and Step 3b we place $\rho, v, d$ in the denominator, and $k_c, \mu, D$ in each of the numerators. Thus,

$$\pi_1 = \frac{k_c}{\rho^{\alpha_1} v^{\alpha_2} d^{\alpha_3}} \Rightarrow \frac{L\theta^{-1}}{(ML^{-3})^{\alpha_1}(L\theta^{-1})^{\alpha_2}(L)^{\alpha_3}} \tag{5.11a}$$

$$\pi_2 = \frac{\mu}{\rho^{\beta_1} v^{\beta_2} d^{\beta_3}} \Rightarrow \frac{ML^{-3}\theta^{-1}}{(ML^{-3})^{\beta_1}(L\theta^{-1})^{\beta_2}(L)^{\beta_3}} \tag{5.11b}$$

$$\pi_3 = \frac{D}{\rho^{\gamma_1} v^{\gamma_2} d^{\gamma_3}} \Rightarrow \frac{L^2\theta^{-1}}{(ML^{-3})^{\gamma_1}(L\theta^{-1})^{\gamma_2}(L)^{\gamma_3}} \tag{5.11c}$$

We find by inspection, or by formally equating coefficients in the numerator and denominator,

$$\alpha_1 = \alpha_3 = 0 \qquad \alpha_2 = 1 \tag{5.11d}$$

$$\beta_1 = \beta_2 = \beta_3 = 1 \tag{5.11e}$$

$$\gamma_1 = 0 \quad \gamma_2 = \gamma_3 = 1 \tag{5.11f}$$

Hence, we obtain

$$\pi_1 = f(\pi_2, \pi_3) \tag{5.11g}$$

or

$$\frac{k_c}{v} = f\left(\frac{\mu}{\rho v d}, \frac{D}{v d}\right) \tag{5.11h}$$

Table 5.1 shows that we have here a first grouping in terms of the dimensionless Stanton, Reynolds, and Peclet numbers, i.e.,

$$St = F(Re, Pe) \tag{5.11i}$$

or, since $Pe = ReSc$,

$$St = F(Re, Sc)$$

A different set of dimensionless groups may be arrived at as follows: We divide $\pi_3$ by $\pi_2$, $\pi_1$ by $\pi_3$ and invert $\pi_2$, resulting in

$$\frac{k_c d}{D} = G\left(\frac{\rho v d}{\mu}, \frac{\mu}{\rho D}\right) \tag{5.11j}$$

or, equivalently,

$$Sh = G(Re, Sc) \tag{5.11k}$$

This is the more common form of dimensionless grouping seen in the literature and states that the Sherwood number Sh, which contains the mass transfer coefficient as the dependent variable, is a function of both Reynolds and Schmidt numbers.

Evidently, for relation 5.11k to be of practical use, it must be rendered quantitative. This is done by assuming that the functional relation is of a power form; i.e., we set

$$Sh = a\, Re^b\, Sc^c \tag{5.11l}$$

and evaluate the coefficients experimentally. This approach has proved itself successful in most undertakings of this kind. It will be noted that the amount of experimentation is considerably reduced by dimensional analysis, as the original six variables are replaced by three dimensionless groups.

We have compiled a list of the most frequently used correlations and tabulate them in Table 5.5. They include mass transfer correlations for turbulent flow about simple geometries as well as in tubes and in packed beds. The following illustration demonstrates the use of these correlations.

**TABLE 5.5**

Correlations for Mass Transfer Coefficients in Turbulent Flow

| Range | Correlation |
|---|---|
| 1. Flat plate | |
| Re > $10^6$ | $St = 0.036\,(Re)^{-0.2}\,(Sc)^{-0.67}$ |
| 2. Sphere | |
| Unlimited | $Sh = 2.0 + 0.60\,(Re)^{0.5}\,(Sc)^{0.33}$ |
| 3. Inside tubes | |
| Re > 20,000 | $Sh = 0.026\,(Re)^{0.8}\,(Sc)^{0.33}$ |
| 4. Packed bed of spheres | |
| Re > 50 | $St = 0.61\,(Re)^{-0.41}\,(Sc)^{-0.67}$ |

### Illustration 5.4: Estimation of the Mass Transfer Coefficient k$_Y$ for the Drying of Plastic Sheets

It is desired to establish the mass transfer coefficient that needs to be applied to estimate the time of drying of wet plastic sheets exposed to air in turbulent flow.

It is customary in drying operations to use humidity $Y$ as a driving force. This calls for the use of the mass transfer coefficient $k_Y$, which has been previously listed in Table 1.4. Because the correlations in Table 5.5 are given in terms of the coefficient $k_c$ (m/s), a two-step procedure is used to arrive at the desired result. In the first instance, we calculate $k_c$ from the correlation for a flat plate given in Table 5.5. This value is then converted into $k_Y$ using the appropriate conversion factor given in Table 1.4. The following are the data to be used in this connection:

Length of sheet $L$ = 15.2 m
Width of sheet $b$ = 1.52 m
Velocity of air $v$ = 15 m/s
Density of air $\rho$ = 1.12 kg/m³
Viscosity of air $\mu$ = 1.93 × 10⁻⁵ Pa s
Schmidt number Sc = 0.606
Vapor pressure at interface $p_A$ = 2487 Pa
Temperature of air $T_A$ = 44°C

We start by establishing that the flow conditions do in fact fall in the turbulent range. The Reynolds number, which is the criterion here, is given by

$$\text{Re} = \frac{Lv\rho}{\mu} = \frac{(1.52)(15)(1.12)}{1.93\times10^{-5}} = 1.32\times10^{7} \tag{5.12a}$$

i.e., it is in excess of 10⁶, which entitles us to use the correlation listed in Table 5.5. We have

$$\frac{k_c}{v} = 0.036\left(\frac{Lv\rho}{\mu}\right)^{-0.2}(\text{Sc})^{-0.67} \tag{5.12b}$$

Substitution of numerical values gives

$$\frac{k_c}{15} = 0.036(1.32\times10^{7})^{-0.2}(0.606)^{-0.67} \tag{5.12c}$$

from which there results

$$k_c = 0.0284 \text{ m/s} \tag{5.12d}$$

This agrees with the order of magnitude estimate of $10^{-2}$ m/s derived in Illustration 1.5.

Conversion to $k_Y$ requires the following relation given in Table 1.4:

$$k_Y = k_c \frac{M_B}{RT} p_{BM} = 0.0284 \frac{29 \times 10^{-3}}{8.31 \times 317} p_{BM} \tag{5.12e}$$

For the computation of $p_{BM}$ we note that, with water partial pressure $p_A$ set at 2487 Pa, we have for the inert air component $p_{B1} = 101300 - 2487 = 98813$ Pa and $p_{B2} = 101300$ Pa so that

$$k_Y = 0.0284 \frac{29 \times 10^{-3}}{8.31 \times 317} \frac{98813 - 101300}{\ln \dfrac{98813}{101300}} \tag{5.12f}$$

$$k_Y = 0.0313 \text{ kg H}_2\text{O/m}^2\text{s } \Delta Y \tag{5.12g}$$

This is the mass transfer coefficient to be used in modeling the air-drying of plastic sheets.

---

## 5.4   Mass Transfer Coefficients for Tower Packings

To this point in our narrative, we have confined ourselves to mass transfer in and around simple geometries such as channels of various types and exterior flow about flat plates, cylinders, and spheres.

In this section we turn our attention to more complex shapes represented by tower packings used in operations such as gas absorption, stripping, and distillation. The operation of such columns is addressed in more detail in Chapter 8.

Tower packings are used to fill the interior of large upright cylindrical shells in which two phases, usually a gas and a liquid or two liquids, are brought into intimate contact for the purpose of transferring or exchanging certain components contained in these phases. In gas absorption, for example, the aim is to remove an objectionable component from the gas stream or to recover a valuable component contained in it by contacting it with an appropriate liquid solvent. Both phases flow through the tower, usually in countercurrent fashion.

Packed towers and tower packings have been in use for more than 100 years. Some early examples of their application involve the production of sulfuric acid and the purification of coke oven gas. Prior to 1915, these towers were filled with coke, random-sized and -shaped quartz, broken glass, or broken crockery. Tower performance was unpredictable and no two towers would perform alike. The development in 1915 of the Raschig Ring made it possible for the first time to impart a degree of predictability and dependability to tower performance. These first uniformly shaped packings not only improved tower performance but also enabled engineers to translate the performance of one tower to others. Today, modern tower packings such as the Super Intalox Saddle and the Pall Ring greatly exceed the capacity and the efficiency of the early Raschig Ring. Figure 5.3 displays some of the shapes in current use.

The mass transfer characteristics obtained on packings are reported in several different ways. At the more fundamental level, we extract volumetric mass transfer coefficients from the experimental performance data. These coefficients, which we have encountered before in Illustrations 2.2 and 2.3, consist of the product of a film coefficient and the nominal specific surface area $a$ of the packing, expressed in units of m$^2$ per m$^3$ of packing. If we use the molar concentration-based coefficient $k_c$ or $k_L$, which has units of m/s, then the

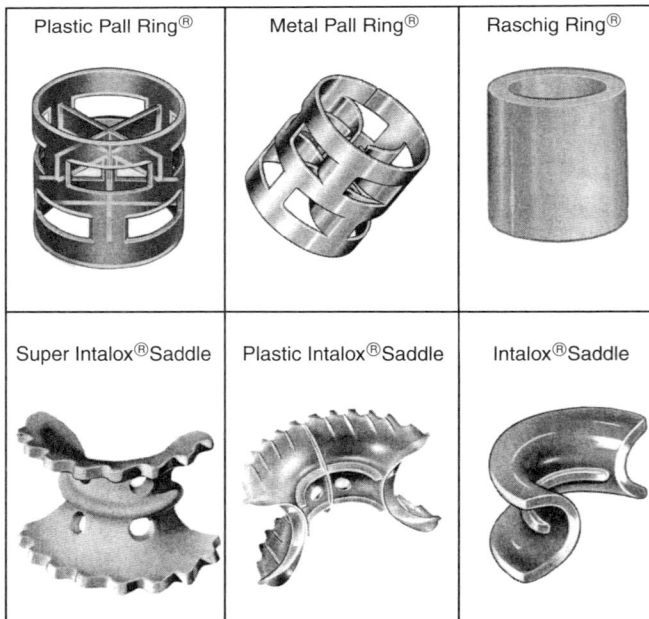

**FIGURE 5.3**
Sample tower packings.

volumetric coefficient given by the product $k_c a$ or $k_L a$ will have units of reciprocal seconds, which is the same as that of a first-order reaction rate constant.

Table 5.6 lists values of $k_c a$ and $k_L a$ extracted from the literature for four systems and a range of conventional packings and superficial flow rates per m² column cross section. These are all based on experimental performance data obtained over a range of gas and liquid flow rates.

The liquid flow rates listed in this table represent the range commonly used in gas–liquid operations. They fall between the extremes of too low a flow, which fails to thoroughly wet the packing, and an excessively high flow, which causes the tower to flood. Neither of these conditions is desirable and the range listed, typically $L = 1 - 10$ kg/m² s, provides a guideline for avoiding these extremes. The gas flow rates, which lie an order of magnitude lower at $G = 0.1 - 1$ kg/m² s, fall between two different extremes, that of insufficient flow to force the gas through the column and an excessively high flow, which would tend to entrain liquid out of the column. These two

**TABLE 5.6**

Mass Transfer Coefficients in Various Commercial Packings

| Packing | G, kg/m² s | L, kg/m² s | $k_c a$, s⁻¹ |
|---|---|---|---|
| *System: $CO_2$–aqueous NaOH ($k_c a$)* | | | |
| 1-in. Raschig ceramic | 0.61–0.68 | 1.4–14 | 0.14–0.33 |
| 2-in. Raschig ceramic | 0.61–0.68 | 1.4–14 | 0.09–0.31 |
| 1-in. Raschig metal | 0.61–0.68 | 1.4–14 | 0.16–0.32 |
| 1-in. Pall plastic | 0.61–0.68 | 1.4–27 | 0.16–0.33 |
| 2-in. Pall plastic | 0.68 | 1.4–54 | 0.13–0.33 |
| 1-in. Pall metal | 0.61–0.68 | 1.4–27 | 0.19–0.44 |
| 1-in. Intalox plastic | 1.2 | 2.7–27 | 0.28–0.43 |
| 2-in. Intalox plastic | 1.2 | 4.1–41 | 0.20–0.30 |
| ½-in. Intalox ceramic | 0.61–0.68 | 1.4–14 | 0.30–0.51 |
| 1-in. Intalox ceramic | 0.61–0.68 | 1.4–14 | 0.17–0.36 |
| 3-in. Intalox ceramic | 1.2 | 1.4–54 | 0.04–0.22 |

| Packing | G, kg/m² s | L, kg/m² s | $k_c a$ or $k_L a$, s⁻¹ |
|---|---|---|---|
| *System: $NH_3$–Water ($k_c a$)* | | | |
| 1-in. Raschig ceramic | 0.54 | 0.68–6.1 | 1.3–5.2 |
| 2-in. Raschig ceramic | 0.54 | 0.68–6.1 | 0.87–2.6 |
| 1-in. Berl ceramic | 0.54 | 0.68–6.1 | 1.7–4.3 |
| 2-in. Berl ceramic | 0.54 | 0.68–6.1 | 1.3–4.0 |
| 50-mm Pall plastic | 0.45–2.5 | 4.2 | 2.0–7.0 |
| *System: $CO_2$–water ($k_L a$)* | | | |
| 50-mm Pall plastic | 0.4–2.0 | 4.2 | 11 |
| *System: $O_2$–water ($k_L a$)* | | | |
| 1.5-in. Raschig ceramic | 0.054–0.54 | 2.7 | 0.14–1.4 |

undesirable conditions can be avoided by operating within the range indicated in Table 5.6. The mass transfer coefficients listed can be used as a guide for sizing packed columns.

Quantitative prediction methods for volumetric mass transfer coefficients that rely on empirical coefficients for each particular packing and packing size have been developed and can be found in the relevant literature. We do not often resort to complete predictions of this type, and it is more common to use relations that will extend known coefficients, such as those listed in Table 5.6, to a different set of conditions. This can be done in an approximate fashion using the following proportionalities:

For the gas film coefficient:

$$k_Ga(s^{-1}) \propto D_G^{0.67}(G\rho_G)^{0.8}(L/\rho_L)^{0.5} \qquad (5.13a)$$

For the liquid film coefficient:

$$k_La(s^{-1}) \propto D_L^{1/2}(L/\rho_L)^{0.75} \qquad (5.13b)$$

In Illustration 5.5, we use these relations to extend existing data to a different set of conditions.

A second and less fundamental way of expressing packing performance is through the concept of the height equivalent to a theoretical plate, or HETP. The theoretical plate, which is a concept we encounter in Chapter 7, is a hypothetical gas–liquid contacting device in which the two phases are brought into intimate contact and exit under equilibrium conditions. The number $N_p$ of such theoretical plates required for a particular separation performance is easily derived by means of graphical procedures to be described in Chapter 7. The total height $H$ of packed column needed is then established by the product of $N_p$ and the equivalent height of the theoretical plate; i.e., we have

$$H = N_p \times \text{HETP} \qquad (5.14)$$

For rough estimates of the HETP, the following recommendations, given in English units and nominal packing size $d_p$, hold:

1.  Pall Rings and similar high-efficiency random packings with low viscosity liquids

$$\text{HETP, ft} = 1.5d_p, \text{ in.} \qquad (5.15a)$$

2.  Absorption with viscous liquids in general

$$\text{HETP} = 5 \text{ to } 6 \text{ ft} \qquad (5.15b)$$

3. Vacuum service

$$\text{HETP, ft} = 1.5d_p, \text{ in.} + 0.5 \qquad (5.15c)$$

4. Small-diameter columns, $d_T < 2$ ft

$$\text{HETP, ft} = d_T, \text{ ft, but not less than 1 ft} \qquad (5.15d)$$

### Illustration 5.5: Prediction of the Volumetric Mass Transfer Coefficient of a Packing

The mass transfer coefficient for the system ammonia–water using 50-mm Pall Rings is known to have a value of 3.6 s$^{-1}$ at a liquid flow rate of 1.2 kg/m$^2$ s. We wish to calculate the coefficient that prevails at the same gas flow rate and a liquid throughput of 10 kg/m$^2$ s. Then from Equation 5.13a we have

$$\frac{(k_c a)_{new}}{(k_c a)_{old}} = \left(\frac{L_{new}}{L_{old}}\right)^{0.4} = \left(\frac{10}{4.2}\right)^{0.4} = 1.42$$

and consequently

$$(k_c a)_{new} = 3.6 \times 1.42 = 5.1 \text{ s}^{-1}$$

This compares with an experimental value of 5.3 s$^{-1}$.

## 5.5 Mass Transfer Coefficients in Agitated Vessels

Mass transfer involving tower packings, which we have considered in the previous section, is our first introduction to systems with complex and highly irregular geometries. The approach we had to take there was to make direct use of experimental data or else convert them by means of some simple empirical rates for use in similarly structured systems.

Agitated vessels represent yet another example of an unusual and not easily quantifiable geometry. The prominent irregularity here is the shape and size of the impeller and the geometry of its blades. Internal baffles, which are frequently used to enhance transport rates, are an additional unusual feature.

Agitated vessels find their use in a considerable number of mass transfer operations. At the simplest level, they are employed to dissolve granular or powdered solids into a liquid solvent in preparation for a reaction or other subsequent operations. The reverse process of precipitation or crystallization is likewise carried out in stirred vessels. Agitation is also used in leaching operations, or its reverse counterpart, adsorption, which is used to remove objectionable materials from a liquid solution or to recover valuable substances. Liquid extraction processes are often carried out in a batch mode using agitated tanks, as are a host of heterogeneous "stirred tank" chemical reactions. In all of these operations we are concerned with establishing mass transfer coefficients that determine the rate of transport to the continuous phase.

Fundamental work in this area dates to the 1940s and 1950s, and has been refined in subsequent decades. These studies have revealed that mass transfer coefficients in these systems can be correlated by the same combination of Sherwood, Reynolds, and Schmidt numbers we have encountered in simpler geometries, provided the former two are suitably modified to account for the altered system geometry and operation. These modifications are implemented as follows:

- For the Sherwood number, the dimensional length to be used is the vessel diameter $d_v$.

- For the Reynolds number, the dimensional length is represented by the impeller diameter $d_i$ and the dimensional velocity by the product $d_iN$, where $N$ represents the number of revolutions per unit time. All other parameters are used in the same fashion as before; i.e., $\mu$, $\rho$, and $D$ are the viscosity, density, and diffusivity of the continuous phase.

Using these modified dimensionless groups, it has been found possible to correlate a host of experimental data for a wide range of operations. The following correlations, tabulated in Table 5.7, have been found useful in predicting transport coefficients in the continuous phase of the stirred tanks.

**TABLE 5.7**

Mass Transfer Coefficients in Agitated Vessels

| System | Correlation |
| --- | --- |
| Solid–liquid baffled vessel Re $= 10^4 - 10^6$ | $\dfrac{k_c d_v}{D} = 1.46\left(\dfrac{d_i^2 N\rho}{\mu}\right)^{0.65}(Sc)^{0.33}$ |
| Solid–liquid unbaffled vessel Re $= 10^2 - 10^5$ | $\dfrac{k_c d_v}{D} = 0.402\left(\dfrac{d_i^2 N\rho}{\mu}\right)^{0.65}(Sc)^{0.33}$ |
| Liquid–liquid | $\dfrac{k_c d_v}{D} = 0.052\left(\dfrac{d_i^2 N\rho}{\mu}\right)^{0.833}(Sc)^{0.5}$ |

To illustrate the use of these equations, we consider the case of the dissolution of granular solids in a stirred tank. This is done in two steps: In Illustration 5.6, we derive the appropriate model to describe the process. The reader is then asked, in Practice Problem 5.7, to derive the mass transfer coefficient for the particular case of the dissolution of potassium hydroxide in a stirred tank and use it to compute the time of dissolution of the charge.

### Illustration 5.6: Dissolution of Granular Solids in an Agitated Vessel

The assumption made at the outset is that the concentration at the surface of the particles equals the saturation concentration $C_s$ of the solid material and that the mass transfer is driven by the linear potential $(C_s - C)$, where $C$ is the prevailing concentration in the liquid at any particular instant.

An initial unsteady mass balance over the solid phase leads to the following expression

$$\text{Rate of solid in} - \text{Rate of solid out} = \begin{array}{c} \text{Rate of change} \\ \text{of solid contents} \end{array}$$

$$0 \quad - \quad k_c A_s (C_s - C) \quad = \quad \frac{d}{dt} m \qquad (5.16a)$$

Note that both the surface area $A_s$ and concentration in the liquid $C$ vary with time or indirectly with the remaining mass $m$. For the area, which can be quite irregular, we stipulate that it varies with the two-thirds power of volume, so that

$$A_s = \alpha V^{2/3} = \frac{\alpha}{\rho^{2/3}} m^{2/3} = \beta m^{2/3} \qquad (5.16b)$$

where $\alpha$ is some shape factor and equals 4.83 for spherical particles.

To obtain an expression for the external concentration $C$, we apply a simple cumulative mass balance, which reads

$$\text{Initial solid} = \text{Solid leftover} + \text{Solid in solution}$$

$$m_o \quad = \quad m \quad + \quad CV \qquad (5.16c)$$

and consequently

$$C = \frac{m_o - m}{V} \qquad (5.16d)$$

Substituting Equations 5.16b and 5.16d into the original mass balance (Equation 5.16a), we obtain

$$-k_c \beta m^{2/3} \left( C_s - \frac{m_o - m}{V} \right) = \frac{dm}{dt} \qquad (5.16e)$$

which yields, after integration by separation of variables

$$At = \int_{m_o}^{m} \frac{dm}{m^{2/3}(m/V + C_s - m_o/V)} \qquad (5.16f)$$

where $A = k_c \beta = k_c \alpha / \rho_s^{2/3}$.

Evaluation of the integral is by numerical or graphical integration, which we do not address here. We consider instead the case where vessel volume $V$ and solubility $C_s$ are sufficiently high, that the term $(m_o - m)/V$ in Equation 5.16f can be neglected compared to $C_s$. This applies to the situation considered in Practice Problem 5.5. The result (Equation 5.16f) then becomes

$$t = \frac{3\rho_s^{2/3}}{\alpha k_c C_s} m_o^{1/3} \qquad (5.16g)$$

where $t$ is now the total dissolution time.

*Comments:*

There are several points of note in the final relation given. First, time of dissolution varies inversely with the mass transfer coefficient $k_c$ and the solubility $C_s$. This is in line with physical reasoning: High values of these coefficients imply a high mass transfer rate, which results in shorter dissolution times. A more startling result is the one-third power dependence on initial mass. This implies that an eightfold increase in the charge will increase dissolution time only by a factor of two. This was certainly not anticipated on physical grounds and is a direct consequence of the area–volume relation introduced in Equation 5.16b. We see here yet another example of the power of modeling to reveal the unexpected.

## 5.6   Mass Transfer Coefficients in the Environment: Uptake and Clearance of Toxic Substances in Animals: The Bioconcentration Factor

The last three decades have seen a dramatic increase in awareness of the effect of various toxic substances on animal life, in particular that of aquatic species. Among the identified culprits, the chlorinated hydrocarbons used

in pesticides, the polychlorinated biphenyls (PCBs) and various organome-
tallic compounds stand out.

Research in this area has focused, on the one hand, on the physiological
consequences of exposure to these substances, and on the other, on tracking
their fate both in the environment and in the affected wildlife and their
specific organs. It is the latter aspect that is considered here.

The fate of toxic substances within an animal is monitored in two ways:
first, by measuring the concentration changes during uptake from a controlled environment, and second by following the decline in concentration
after exposure has ceased. The latter process is termed clearance, elimination,
or depuration.

Interpretation of the experimental data is usually carried out by means of
compartmental models. The simplest of these, the one-compartment model,
yields the following results:

During uptake, the relevant mass balance over the animal body takes the
form

$$\text{Rate of toxin in} - \text{Rate of toxin out} = \begin{array}{c} \text{Rate of change} \\ \text{of toxic contents} \end{array}$$

$$k_{cu}A_uC_w \quad - \quad k_{cd}A_dC_a \quad = \quad V\frac{dC_a}{dt} \qquad (5.17a)$$

where $k_{cu}$ and $k_{cd}$ are the mass transfer coefficients in units of m/s for uptake
and depuration, $A_u$ and $A_e$ are the associated transfer areas, $C_w$ denotes
concentration in the water or other medium, taken to be constant, and $C_a$ is
the time-varying concentration within the animal or one of its organs.

Before integrating, it is customary to divide Equation 5.17a by volume,
with the results that $k_{cu}$ and $k_{ce}$ are converted into volumetric mass transfer
coefficients $k_u = k_{cu}A_u/V$ and $k_d = k_{cd}A_d/V$. Equation 5.17a consequently
becomes

$$k_uC_w - k_dC_a = \frac{dC_a}{dt} \qquad (5.17b)$$

The coefficients now have units of reciprocal time ($s^{-1}$) and are precisely
of the same type encountered in Section 5.4 in connection with mass transfer
in packed towers. We have termed them $k_da$ and $k_La$, where $a$ is the specific
surface area of the packing in units of $m^2$ per $m^3$ packing. The corresponding
term here is the ratio of transfer areas to body or organ volume $A/V$. The
advantage of this procedure is that both these factors, which are either
unknown or not known with precision, are lumped into a single empirical
transfer coefficient, which is determined experimentally.

The fact that these volumetric coefficients have units of reciprocal time has
led to the erroneous impression, and even statements, that the process is one
of chemical reaction, with $k_u$ and $k_e$ playing the role of first-order rate

constants. The mechanism of uptake is clearly one of mass transfer and that of elimination is probably a combination of reaction and transport. Mass transfer is therefore the key phenomenon that dominates these processes.

We now turn to Equation 5.17a and obtain, by separating variables and integrating,

$$C_a/C_w = (k_u/k_d)[1 - \exp) - k_d t)]  \qquad (5.17c)$$

This result expresses the toxin concentration in the animal, $C_a$, as a function of time. We note that as $t \to \infty$, a balance between uptake and elimination is obtained and a steady-state concentration ratio $(C_a/C_w)_{ss}$ results. This ratio is termed the bioconcentration factor (BCF) and expresses the magnification of toxic concentration in the animal over that prevailing in the surrounding water. It equals the ratio of the two mass transfer coefficients and is given by

$$\text{BCF} = (C_a/C_w)_{ss} = k_u/k_d  \qquad (5.17d)$$

Turning next to the elimination process, a mass balance similar to that performed for the uptake step leads to the following result

$$\text{Rate of toxin in} - \text{Rate of toxin out} = \frac{\text{Rate of change}}{\text{in toxin contents}}$$

$$0 \qquad - \qquad k_{Cd} A_d C_a \qquad = \qquad V\frac{dC_a}{dt}  \qquad (5.18a)$$

or equivalently

$$-k_d C_a = \frac{dC_a}{dt}  \qquad (5.18b)$$

which upon integration by separation of variables yields

$$C_a/(C_a)_{ss} = \exp(-k_d t)  \qquad (5.18c)$$

Here $(C_a)_{ss}$ is taken to be the steady-state concentration attained during uptake.

The uptake and elimination coefficients $k_u$ and $k_e$ can in principle be calculated by first extracting $k_d$ from Equation 5.18c using measured-clearance histories, and then substituting it into Equation 5.17c and performing a similar analysis of uptake-concentration histories. In actual practice, it is more common to perform an independent evaluation of $k_u$ by using the initial portion of the uptake process, which is unaffected by the relatively slow elimination process. Equation 5.17b then assumes the reduced form

$$k_u C_w = \frac{dC_a}{dt} \tag{5.19a}$$

which is integrated to yield

$$C_a = k_u C_w t \tag{5.19b}$$

This expression is used to calculate $k_u$ from a linear plot of the initial uptake data.

Table 5.8 presents uptake and elimination constants obtained on bluegill fish, exposed to an aqueous environment containing anthracene and benzopyrene, and ring doves, which were fed pellets containing various PCBs. This is supplemented, in Table 5.9, by the BCF found in various organs of the bluegill fish. Of note in the latter is the extraordinary magnification that takes place particularly in the gallbladder of this species, which attests to the dangerous effect of these toxins.

### Illustration 5.7: Uptake and Depuration of Toxins: Approach to Steady State and Clearance Half-Lives

Two topics are addressed here. The first involves the calculation of the time required to attain 95% of the ultimate steady state during toxin uptake. This conveys a sense of the speed with which this process occurs. In the second calculation, we seek to quantify the depuration process by calculating its half-life, i.e., the time required for the toxin concentration to drop to one half its original steady-state concentration. This again serves as an indicator of the speed with which depuration proceeds.

**TABLE 5.8**

Uptake and Depuration of Some Toxic Substances

|                | Toxin        | $k_u$ (h$^{-1}$) | $k_d$ (h$^{-1}$) |
|----------------|--------------|------------------|------------------|
| Bluegill fish  | Anthracene   | 36               | 0.04             |
|                | Benzopyrene  | 49               | 0.01             |
| Ring doves     | PCBs         | $(0.3–4.4)10^3$  | 0.094–0.24       |

**TABLE 5.9**

BCF in Various Organs of Bluegill Fish

|             | BCF         |              |
|-------------|-------------|--------------|
| Tissue      | Anthracene  | Benzopyrene  |
| Gallbladder | 1,800       | 14,000       |
| Liver       | 561         | 1,600        |
| Viscera     | 640         | 770          |
| Brain       | 555         | 90           |
| Carcass     | 42          | 30           |

*1. Approach to 95% of Steady State*

Here we make use of Equation 5.17c and Equation 5.17d, which upon division by each other yield

$$C_a / (C_a)_{ss} = 1 - \exp(-k_d t) \tag{5.20a}$$

and consequently

$$t = -\frac{1}{k_d} \ln[1 - C_a / (C_a)_{ss}] \tag{5.20b}$$

Applying this to the uptake of benzopyrene by bluegill fish listed in Table 5.8, we obtain

$$t = (-1 / 0.01) \ln[1 - 0.95] \tag{5.20c}$$

$$t + 300 \text{ h} \tag{5.20d}$$

*2. Half-Life t₁/₂ of Depuration*

The pertinent expression here is obtained from Equation 5.19c by setting $C_a / (C_a)_{ss} = 0.5$. Hence

$$0.5 = \exp(-k_d t) \tag{5.21a}$$

and consequently

$$t_{1/2} = \frac{\ln 2}{k_d} \tag{5.21b}$$

Applying this to the same data as before, we have

$$t_{1/2} = \frac{\ln 2}{0.01} = 69 \text{ h} \tag{5.21c}$$

To obtain a more direct comparison with the uptake case, we allow $C_a / (C_a)_{ss}$ to drop to 5% of its original value. This time there results

$$t = \frac{\ln 20}{0.01} = 300 \text{ h} \tag{5.21d}$$

*Comments:*

The surprising fact that emerges from these calculations is that the uptake and depuration processes proceed at the same speed. Both are long, drawn-out events taking place over a period of days. This is encouraging news for the uptake step because it implies no harmful effects for brief exposure times of, say, 10 min. The long depuration period, on the other hand, is disturbing because the animal will require many days in a clean environment to recover from its exposure. We have here a mix of good and bad news, which is often the norm in environmental events.

The BCF mentioned above has its benign counterpart in the effective therapeutic concentration (ETC), which has been encountered in Practice Problem 2.1. Both result when the rate of inflow of the material is exactly balanced by the rate of elimination. Thus, although the two substances are at opposite poles, one, toxic, the other, therapeutic, the mechanism by which they reach their plateau values are identical.

## Practice Problems

5.1. *The Heat Transfer Analogy*
Derive the heat transfer counterpart for laminar entry flow and for turbulent flow in a cylindrical pipe.

5.2. *Mass Transfer Coefficients in Terms of Shear Rate $\dot{\gamma}$*
In physiological flows, such as that of blood, it is customary to replace velocity by the shear rate $\dot{\gamma}$, which equals the velocity gradient at the vessel wall. Show that the $k_C$ in the entry region is then given by

$$k_C = 0.54\left(\frac{\dot{\gamma}D^2}{\lambda}\right)^{1/3}$$

(*Hint:* Use the parabolic velocity distribution for laminar flow as a starting point and derive the relation $\dot{\gamma} = 8v/d$.)

5.3. *Mass Transfer Regimes in Blood Flow: The Critical Blood Vessel Diameter*
Shear rates $\dot{\gamma}$ in physiological blood flow typically lie in the range $100 - 1000$ s$^{-1}$. Show that for proteins ($D = 10^{-10}$ m$^2$/s), mass transfer in the "larger" blood vessels, $d > 1$ mm, falls entirely in the developing (Lévêque) region, while for $d < 10^{-2}$ mm, the concentration profile is fully developed. (*Hint:* Use the relation $\dot{\gamma} = 8v/d$.)

5.4. *Mass Transfer in the Kidney: The Loop of Henle*
The Loop of Henle is part of an intricate system of permeable channels that carry raw urine through the kidney and to the bladder. During its passage, the urine exchanges water and solutes with the surrounding tissue to ensure that not too much or too little of either is withdrawn from the body. The *total amount* of salt reabsorbed into

the tissue depends in a complex way on flow velocity, and early workers attributed this to a flow-sensitive boundary layer within the tube: Show that this is not the case and show also that we can still have flow-sensitive reabsorption. Data: $d = 2 \times 10^{-3}$ cm, $D = 2 \times 10^{-5}$ cm²/s, $v = 10^{-1}$ cm/s, wall Sherwood number $\leq 2$. (*Hint:* Show that transport is in the fully developed region.)

5.5. *Controlled Release of Anticoagulants from Artificial Blood Vessels*
The use of artificial grafts to replace diseased blood vessels has by now become a commonplace operation. The danger of blood clot formation, which attends these replacements, can be largely eliminated by incorporating an anticoagulant in the graft, which is slowly released to the bloodstream. One such anticoagulant is the protein heparin, which effectively prevents the onset of coagulation. It has been estimated that to achieve this, a microenvironment with a heparin concentration of $C_s = 0.5$ µg/cm³ must be provided at the blood vessel surface. The problem to be addressed here is to calculate the release rate $N$ (µg/cm² s) required to maintain this concentration within 0.5 cm from the tubular entrance. Flow is laminar and mass transfer resides entirely within the entry region. Shear rate is set at 1000 s⁻¹ and the heparin diffusivity is $7.5 \times 10^{-7}$ cm²/s.

*Answer:* $5.2 \times 10^{-4}$ µg/cm² s

5.6. *Solution Mining of Potash*
One proposed method for harvesting underground deposits of potash (KCl or sylvite) is the technique called *solution mining*. The procedure consists of cracking the deposit open by pumping high-pressure water through an "injection well" drilled at one end of the deposit ("hydrofracturing"). A second well, the so-called production well, is drilled at the far end of the fracture. Water is then continuously pumped into the deposit through the injection well, dissolving potash as it passes through the open fracture and returning to the surface through the production well as a KCl solution. It is desirable in these operations to have an *a priori* estimate of the KCl concentration in the effluent, as this will determine the production rate of the mine. Consider in this connection the following data:

Fracture height $d = 1$ m

Fracture length $L = 500$ m

Fracture perimeter $P = 500$ m

Water velocity $v = 0.1$ m/s

As a result of the geothermal gradient, temperature near the fracture is estimated to be 50°C. The following physical parameters have been calculated with this factor in mind:

Density $\rho$: 985 kg/m$^3$

Viscosity $\mu$: $0.6 \times 10^{-3}$ Pas

Diffusivity $D$: $1.8 \times 10^{-9}$ m$^2$/s

The problem we set ourselves is to calculate the degree of saturation $C/C^*$ attained at the far end of the fracture. *Simplifying assumptions:* We assume isothermal operation even though the endothermic nature of the dissolution process will cause a drop in temperature and consequently of the solubility $C^*$. The variation in fracture height that occurs with time is neglected, as are the variations in physical properties due to the changing KCl concentration. The correlation for tubular flow listed in Table 5.5 is assumed to hold, with fracture height taking the place of tubular diameter.

*Answer:* $C/C^* = 0.047$

5.7. *Dissolution of Potassium Hydroxide in an Agitated Vessel*

In preparation for carrying out a hydrolysis reaction, a 1-molar aqueous solution of KOH at 100°C is to be produced using a baffled stirred tank. The vessel has a diameter $d_v$ of 4 m with 0.3-m-long impeller blades $d_i$ and a volume of approximately 50 m$^3$. The required amount of potassium hydroxide $m_o$ is 2500 kg ($\rho_s = 2000$ kg/m$^3$) and its saturated concentration 900 kg/m$^3$. In addition, we set the number of revolutions $N$ at 1 s$^{-1}$. Other data are as follows:

$$\mu = 0.28 \times 10^{-3} \text{ Pas}$$

$$\rho = 1000 \text{ kg/m}^3$$

$$D = 2 \times 10^{-9} \text{ m}^2/\text{s}$$

*Note:* These are only approximate averages of the actual time-varying values. It is desired to use this information to estimate the approximate time of dissolution of the charge.

Calculate the dissolution time.

*Answer:* 5380 s

5.8. *Diameter of a Packed-Gas Scrubber*

List some criteria you would use for choosing a packed tower diameter. Where in this chapter can you find some guidelines?

5.9. *Bioconcentration in a Two-Compartment Model*

Consider the two-compartment model shown in Figure 5.4. Toxins enter the first compartment from the water phase and are eliminated from it with first-order rate constants of $k_1$ and $k_2$. Simultaneously,

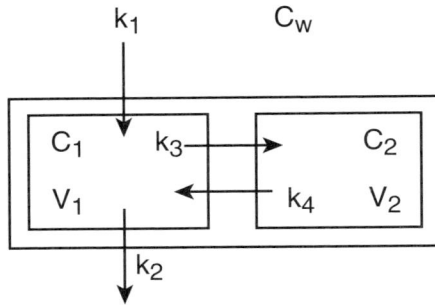

**FIGURE 5.4**
Two-compartment model for bioconcentration.

a reversible exchange of toxins takes place with an adjacent second compartment with first-order rate constants of $k_3$ and $k_4$, respectively. The first compartment may be viewed as the circulatory system; the second compartment represents the tissue, which ultimately, at steady state, equilibriates with the first unit. We define a relative mass for the first compartment, given by

$$f_1 = \frac{m_1}{m_1 + m_2}$$

Show that with this definition in place the total BCF for the two compartments is given by

$$\text{BCF}_{\text{Tot}} = C_{\text{Tot}}/C_w = f_1(k_1/k_2) + (1 - f_1)(k_1/k_2)(k_3/k_4)$$

(*Hint:* Set up the ultimate steady-state balances for the two compartments.)

# 6

## Phase Equilibria

Throughout the preceding chapters, it is evident that, along with transport coefficients, phase equilibria play a crucial role in determining overall mass transfer rates. In processes involving single-film resistances, equilibrium compositions or pressures constitute the anchor of the driving force responsible for the transport of mass. This is shown most vividly in Figure 1.5. The driving forces here are all of the form $C^* - C$ or $p^* - p$, where the asterisked quantities represent equilibrium compositions and partial pressure, and the plain symbols denote the same quantities in the bulk fluid. Evidently, transport will continue only as long as the two quantities are unequal, and will come to a halt when $C^* = C$ and $p^* = p$. The two phases are then said to be in equilibrium.

Phase equilibria also appear in processes involving two-film resistances in series, but their role here is somewhat more complex. They still appear as one of the two partners constituting the driving forces in the individual films. This comes about because, in two-film theory, the interface separating the two phases is postulated to be at equilibrium (see Figure 1.7). The use of individual coefficients and driving forces, however, is awkward because conditions at the interface are generally unknown. It then becomes convenient to express transport in terms of overall driving forces that bridge the interface and extend from one bulk concentration to the other. Driving forces still retain their previous form, $C^* - C$ and $p^* - p$, but the asterisked quantities are now *hypothetical* compositions and pressures denoting equilibrium with the second phase at a particular point of the system. This is described in detail in Section 1.5.

Yet another important aspect of phase equilibria rests on the fact that they determine certain maximum or minimum quantities associated with the process. Suppose, for example, that a liquid evaporates into an enclosure. By allowing the process to proceed to equilibrium, i.e., to full saturation, we are able to determine the *maximum amount of liquid* that will have evaporated, or, conversely, the *minimum mass of air* that can accommodate that amount of vapor. Suppose next that the same liquid, for example, water, adheres to a solid that is to be dried by passage of air over it. Then by using very low flow rates we can ensure that the air leaving the drying chamber is fully saturated and that consequently the air consumption is at a minimum.

The picture that emerges from this brief discussion is that the role of equilibrium is a very considerable one. Equilibrium, or the departure from it, determines the driving potential of the process and sets upper and lower limits to the enrichment attainable and the material inventory involved. Equilibrium can consequently be viewed as one of the two key players in mass transport; the other is the transport coefficients themselves. Because each of these factors has a distinct role to play, it has become customary to examine the component factors separately and subsequently combine the results for a complete description of the event. In this procedure we start by first ignoring the effect of transport resistance and allowing the two phases to come to equilibrium. The device in which this step is carried out is termed an *equilibrium* or *ideal stage*. Concentration or pressure changes that result from the procedure are noted and set aside. The effect of transport resistance is examined next. In the case of staged operations, this involves an assessment of *stage efficiency*, which is a measure of the effect of transport resistance on the amount transferred. When resistance is negligible, the efficiency is 100% and the process proceeds to complete equilibrium. When efficiency is 50%, only half of the attainable enrichment is obtained. In the final step, the results of this dual scrutiny are combined to arrive at an overall description of the process. These steps, as well as the underlying concept of an equilibrium stage, are taken up in the next chapter.

While some of the topics in the present chapter will be new to the reader, others may be known from previous courses in thermodynamics. They are repeated to provide a refresher and a link to subsequent chapters.

## 6.1   Single-Component Systems: Vapor Pressure

Most pure substances that the reader will be familiar with can exist in the solid, liquid, and vapor phases. The principal exceptions are high-molecular-weight solid compounds such as proteins, carbohydrates, and polymers that decompose before they can pass into the liquid or vapor phase. These will not be of concern here. The wider class of substances, which are addressed in this chapter, is capable of existing in all three phases. Their behavior is best illustrated by means of a pressure–temperature phase diagram; a representative example is shown in Figure 6.1a.

The diagram is divided into three regions representing the solid phase ($S$), the liquid phase ($L$), and the vapor phase ($V$). The dividing boundaries between these regions are the melting point or freezing point curve $A$, the sublimation curve $B$, which separates the solid and vapor phases, and the vapor pressure or boiling point curve $C$, which is the dividing line between liquid and vapor phases. It is principally the latter curve and the relation between vapor pressure and temperature that we are concerned with here. Also marked on the diagram are four specific pressure–temperature pairs,

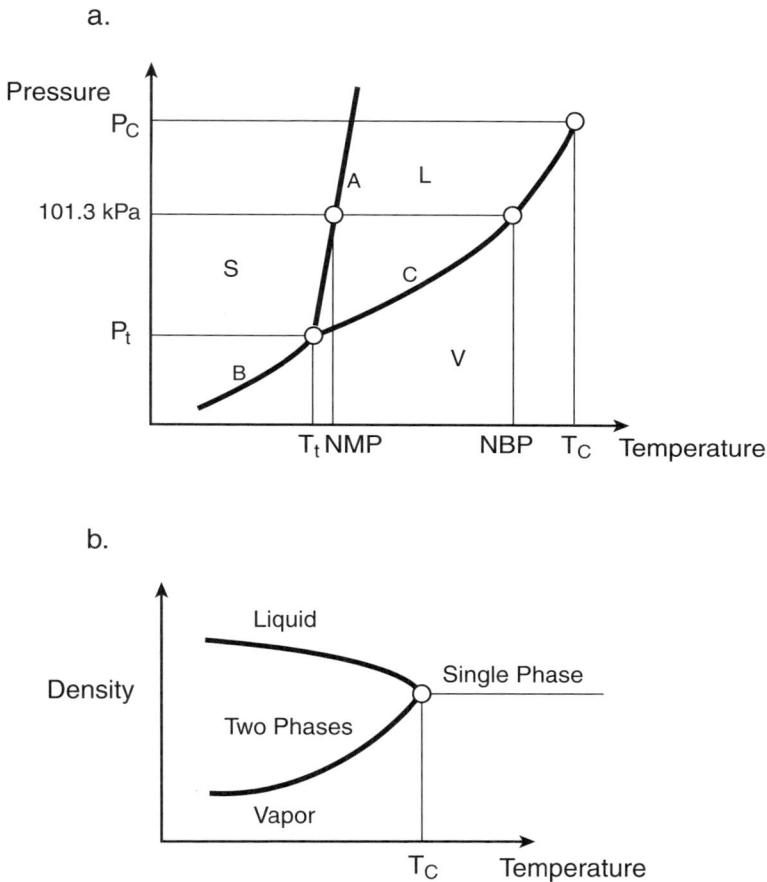

a.

b.

FIGURE 6.1
(a) Phase diagram of a pure substance; (b) approach to the critical point.

which characterize the system. NMP and NBP are the "normal" melting and boiling points, i.e., those that prevail at a pressure of 1 atm or 101.34 kPa. The prefix *normal* is often dropped and the shorter version used to denote conditions of 1 atm. The pair $P_t$-$T_t$ marks the *triple point* and represents the conditions under which all three phases can coexist. For water, the pressure and temperature values are 0.006 atm and 0.01°C, respectively.

Suppose, now, that a liquid sealed in a container and in equilibrium with its vapor is heated starting from room temperature. As the temperature is raised, liquid passes into the vapor phase, thus increasing the density of the latter, while the liquid itself undergoes expansion with an attendant decrease in density (Figure 6.1b). A point is eventually reached where the densities of the two phases become identical and the two phases merge into a single entity. The specific condition ($P_c$-$T_c$) at which this occurs is termed the critical point of the system. Beyond it, i.e., above the critical temperature $T_c$, a substance can exist only as a vapor and no liquefaction can occur, no matter how high the pressure is raised. For water, this point is reached at a temperature of 374.4°C and a corresponding pressure of 219.5 atm.

Note that while a substance can only exist as a gas above the critical temperature, its density is nevertheless high enough that it can also qualify as a quasi-liquid. This dual behavior has certain advantages, which are exploited in a process termed *supercritical fluid extraction*, which is taken up in Section 6.2.5.

We now turn to the consideration of the equilibrium between liquid and vapor. The aim is to establish a quantitative relation for the curve $C$ in Figure 6.1a, i.e., for the vapor pressure of the liquid in the interval between the triple point $T_t$ and the critical temperature $T_c$. The starting point is given by the well-known Clapeyron equation, which is the thermodynamic expression of phase equilibrium for a pure substance:

$$\frac{dP}{dT} = \frac{\Delta H}{T\Delta V} \tag{6.1a}$$

Here $\Delta H$ and $\Delta V$ denote the molar enthalpy and volume changes that occur during the passage from one phase to another. The equation is confined to pure substances, but is otherwise quite general and capable of expressing the transition between any two of the phases shown in Figure 6.1a. Thus, for a passage from the solid to the liquid phase, $\Delta H$ will represent the latent heat of fusion, while $\Delta V$ denotes the difference between liquid and solid molar volumes. The derivative $dP/dT$, which appears on the left side of Equation 6.1a, gives the slope at any point of the phase boundary curve $A$ shown in Figure 6.1a.

Equation 6.1a cannot be integrated in straightforward fashion since both $\Delta H$ and $\Delta V$ depend on temperature in a complex fashion. However, when one of the states is represented by the vapor phase, the exact Equation 6.1a can be simplified by introducing the following two approximations:

1.  The molar volume of the vapor phase is much greater than that of the liquid or solid phase. Thus,

$$V_v \gg V_{s,\ell} \tag{6.1b}$$

and consequently

$$\Delta V \doteq V_v \tag{6.1c}$$

2.  At low pressures, it may be assumed that the vapor phase behaves ideally, so that

$$V_v = \frac{RT}{P^o} \tag{6.1d}$$

On substituting these two simplifying relations into Equation 6.1a and rearranging, we obtain

$$\frac{1}{P^o}\frac{dP^o}{dT} = \frac{\Delta H}{RT^2} \tag{6.1e}$$

or alternatively

$$\frac{d\ln P^o}{d(1/T)} = -\frac{\Delta H}{R} \tag{6.1f}$$

Equations 6.1e and Equation 6.1f are known as the differential forms of the Clausius–Clapeyron equation. If we further assume that the latent heat of vaporization $\Delta H$ is constant over the interval in question, we obtain by integration

$$\log P^o = A - B/T \tag{6.1g}$$

where $A$ and $B$ $(= \Delta H/2.302R)$ are constants.

Expression 6.1g, known as the Clausius–Clapeyron equation, provides an excellent representation of the solid vapor pressure but is valid only over a limited temperature range of the liquid vapor pressure. It breaks down, in particular, near the critical point where vapor and liquid densities approach each other in magnitude. Various semiempirical modifications have been proposed as a result, among which the Antoine equation (1888) has proved to be particularly successful. It takes the form

$$\log P^o = A - B/(T + C) \tag{6.1h}$$

where $P^o$ is the liquid vapor pressure in mmHg, $T$ the temperature in degrees Celsius, and $A$, $B$, and $C$ are empirical constants. Values of $A$, $B$, and $C$ for some common organic liquids appear in Table 6.1. They have been computed from data measured up to several atmospheres and provide a precise representation of the vapor pressure over this range and beyond. To demonstrate the use of the Antoine equation, we consider the following, somewhat unusual case.

### Illustration 6.1: Maximum Breathing Losses from a Storage Tank

During a rise in ambient temperature, solvent-laden air in the headspace of storage tanks expands and is partially expelled into the atmosphere through

**TABLE 6.1**

Antoine Constants for Various Liquids

| Substance | A | B | C |
|---|---|---|---|
| Methyl chloride | 7.09349 | 948.582 | 249.336 |
| Methylene chloride | 7.40916 | 1325.938 | 252.616 |
| Chloroform | 6.95465 | 1170.966 | 226.252 |
| Carbon tetrachloride | 6.87926 | 1242.021 | 226.409 |
| Acetone | 6.11714 | 1210.595 | 229.664 |
| Diethyl ether | 6.92032 | 1064.066 | 228.799 |
| Methanol | 8.08097 | 1582.271 | 239.726 |
| Ethanol | 8.11220 | 1592.864 | 226.184 |
| *n*-Hexane | 6.88555 | 1175.817 | 224.867 |
| *n*-Octane | 6.91874 | 1351.756 | 209.100 |
| Benzene | 6.89272 | 1203.531 | 219.888 |
| Toluene | 6.95805 | 1346.773 | 219.693 |
| *o*-Xylene | 7.00154 | 1476.393 | 213.872 |
| Water | 7.96681 | 1668.210 | 228.000 |

a vent pipe. When the temperature drops, the process reverses itself and fresh solvent-free air enters the headspace. Subsequent cycles of rising and falling temperature cause a cumulative loss in solvent. An accurate calculation for this loss would require a knowledge of the time- and space-dependent concentrations and temperatures in the tank, and would thus call for the solution of PDEs (mass and energy balances) along with the appropriate equilibrium relation. The somewhat irregular geometry (tank and vent pipe) and the possibility of both conductive and free convective transport, plus uncertainties in the external heat transfer coefficient, make this a formidable problem to solve.

In this first elementary treatment of this problem, these complications are avoided by confining ourselves to the calculation of the *maximum* loss that can occur in the course of a single temperature cycle. This is achieved by assuming that the tank contents are well mixed and in thermal and phase equilibrium at the maximum temperature attained in a cycle. The result is an enormous simplification of the problem because we are now dealing merely with algebraic expressions representing the vapor pressure of the system and the appropriate gas laws.

Suppose that the stored liquid in question is benzene, and that the headspace of the storage tank is 100 m³. We assume that the temperature rises from 15 to 30°C in the course of a day and that, as a result, some 5% of the headspace air is ejected. The task is to calculate the maximum amount of solvent lost.

We start by computing the vapor pressure at the maximum temperature of 30°C using the Antoine constants for benzene listed in Table 6.1. Thus,

$$\log P^o = 6.89272 - \frac{1203.531}{30 + 219.888}$$

and consequently

$$P^o = 119.3 \text{ mmHg}$$

Substituting this value into the ideal gas law, we obtain

$$m = \frac{M P^o V}{RT}$$

or

$$m = \frac{78(119.3 / 760)1.013 \times 10^5 \times 5}{8.314 \times 303}$$

and therefore

$$m = 2.46 \text{ kg}$$

Although this amount represents a theoretical maximum, more-refined calculations have shown that these losses are indeed quite substantial and would in most cases be considered unacceptable. Provision is therefore often made to recover the escaping vapors by compression condensation or by adsorption, or to cover the tank by a floating top. An attendant benefit of this procedure is the avoidance of the adverse effect of such emissions on the environment.

## 6.2 Multicomponent Systems: Distribution of a Single Component

We take up the topic of multicomponent equilibria by drawing a distinction between systems in which several or all components are present in the two equilibrated phases, and those in which only one component plays a key role by distributing itself in significant amounts between the phases in question. Vapor–liquid equilibria of mixtures and other similar multicomponent systems involving the appearance of several solutes in each phase are the prime example of the former, while distributions of a single component occur in a number of different contexts, which we take up in turn below. They include the equilibrium of a single gas with a liquid solvent, or a solid (gas absorption and adsorption), and the distribution of solutes between a liquid solution and an immiscible solvent (liquid extraction) or solid (liquid-phase adsorption). We note that although more than one component may be present in both phases, their appearance does not affect the distribution of the principal component under consideration. Thus, in gas absorption, solvent vapor is inevitably present in the gas phase but does not interfere with the distribution of the main solute in any way. Similarly, in liquid extraction, the two solvents may not be perfectly immiscible, but this does not significantly affect the distribution of the solute.

### 6.2.1  Gas–Liquid Equilibria

Examples of gas–liquid equilibria abound both in the physical world surrounding us and within an industrial context. Gases of both a benign and toxic nature are taken up or released by bodies of water. The example of dissolved oxygen, which is essential to the sustenance of aquatic life, immediately comes to mind. On the industrial scene both valuable and objectionable gases are often selectively removed or recovered by gas absorption or gas scrubbing. We have already alluded to this process in Illustration 2.3 and Section 5.4, and more on this process appears in Chapter 8.

The phase equilibrium between a gas and a liquid solvent is usually expressed in terms of the amount absorbed or liquid-phase concentration as a function of gas pressure. A diagram of this relation appears in Figure 6.2. The concentration of the dissolved gas is seen to increase with pressure and it does so indefinitely; i.e., no limiting saturation value is attained. This is in contrast to gas–solid and liquid–solid adsorption equilibria in which the solid surface ultimately becomes saturated with solute. An increase in temperature, on the other hand, diminishes the solubility of the gas, and hence its concentration. One notes in addition that at the lower end of these diagrams the plot becomes linear. The slope of this linear portion is termed the Henry's constant $H$ and the phase equilibrium in this range is said to follow Henry's law, given by

$$p = HC \tag{6.2}$$

Here $p$ is the gas pressure usually expressed in kPa and $C$ the concentration of the dissolved gas (mol/l or mol/m$^3$). Henry's constants at 25°C for some common gases are displayed in Table 6.2. The validity of Equation 6.2 for these gases extends to several atmospheres, and in the case of permanent gases, such as $H_2$, $O_2$, and $CO_2$, to several tens of atmospheres. A good deal of useful information can therefore be gathered through the use of Henry's law and the associated Henry's constants. We demonstrate this with the following illustration.

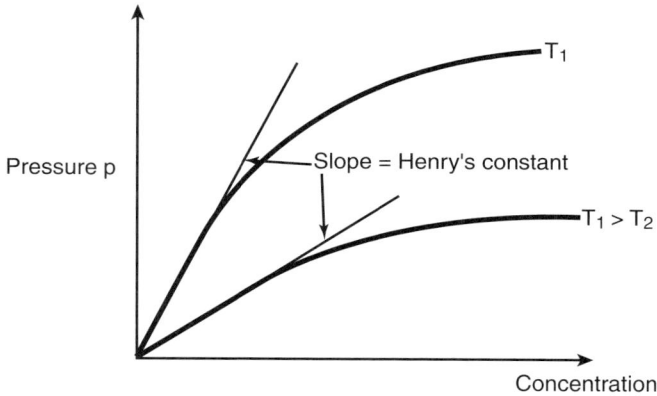

**FIGURE 6.2**
Gas–liquid equilibrium isotherm.

**TABLE 6.2**

Henry's Constants for Gases in Water at 25°C

| Gas | $H$ (kPa m$^3$ mol$^{-1}$) |
| --- | --- |
| Hydrogen | 130 |
| Helium | 260 |
| Carbon monoxide | 100 |
| Nitrogen | 150 |
| Oxygen | 79 |
| Carbon dioxide | 2.9 |
| Methane | 71 |
| Ethane | 53 |
| Propane | 66 |
| $n$-Butane | 80 |

## Illustration 6.2: Carbonation of a Soft Drink

It is common practice in the soft drink industry to carbonate drinks by dissolving a *fixed volume* of carbon dioxide in the liquid, rather than by applying a *prescribed pressure* to the contents. That volume is set at 3 to 5 times the volume of the liquid contents.

Consider a standard 1.5-l soft drink bottle with a headspace of 5%. The task is to calculate the pressure in the bottle after carbonation and the consumption of carbon dioxide in a plant bottling 10,000 containers per day. We assume a $CO_2$ charge equal to 5 l.

Taking account of the 5% headspace, and assuming a bottling temperature of 298 K together with a molar volume of STP of 22.4 l, this leads to a carbon dioxide volume of

$$V_{CO_2} = 5000 - 0.05 \times 5000 = 4750 \, l/m^3 \tag{6.3a}$$

or equivalently

$$C = (4750/22.4)(273/298) \text{ mol } CO_2/m^3 \tag{6.3b}$$

We apply this value to Henry's law (Equation 6.2) and use a Henry's constant of 2.90 taken from Table 6.1 to obtain

$$p = H \times C = 2.9(4750/22.4)(273/298) \tag{6.3c}$$

or

$$p = 563 \text{ kPa}$$

To this value has to be added the initial air pressure of 100 kPa, which brings the total pressure in the container to slightly above 6.5 atm.

To obtain this result in 10,000 bottles of 1.5 l each, one requires a $CO_2$ volume of

$$V_{CO_2} = 10,000 \times 1.435 \times 5 \, l \tag{6.3d}$$

where the headspace of 0.075 l has been subtracted from the total bottle volume of 1.5 l. The corresponding mass of $CO_2$ in the carbonated drink $(m_{CO_2})_d$ is given by

$$(m_{CO_2})_d = \frac{10,000 \times 1.425 \times 5}{22.4} (273/298)M \tag{6.3e}$$

where $M$ = molar mass of carbon dioxide = 44. Consequently

$$(m_{CO_2})_d = 128 \, kg \tag{6.3f}$$

The amount of carbon dioxide in the headspace is a small fraction of this value and is given by

$$(m_{CO_2})_h = 10,000 \frac{PV}{RT} M = 10,000 \frac{5.63 \times 10^5 \times 0.075 \times 10^{-3}}{8.314 \times 298} \times 44 \tag{6.3g}$$

$$(m_{CO_2})_h = 7499 \, g = 7.5 \, kg \tag{6.3h}$$

Thus, a total of 135.5 kg of $CO_2$ will be required in the daily operation of the plant.

## 6.2.2 Liquid and Solid Solubilities

A second example of a binary system in which one component is considered to be confined to one phase only involves the solubility of liquids and solids in a solvent. The confined species here is the solvent, which is in contact with a second phase containing a pure liquid or solid. At equilibrium the solvent phase has become saturated with the dissolved species or solute and no further dissolution takes place. The concentration corresponding to this state is termed the *solubility* of the liquid or solid in question.

The number of possible solute–solvent combinations is evidently quite large, and the corresponding number of required measurements infinite if temperature is considered an additional variable. It is customary therefore to deal with only a small number of solvents, principally water, and to perform measurements at a standard temperature of 25°C. Most reported data have been obtained within this framework.

A listing of the solubility in water of a number of solutes, mostly organic in nature, appears in Table 6.3. Sodium nitrate and glucose show high values of close to 50%, as expected, while solubilities of DDT and mercury are measured in parts per billion. The latter are nevertheless sufficiently significant to be of environmental concern. The reader will also note the relatively high solubility of diethyl ether (~7.5%), which is generally thought of as insoluble. Unusually high values are also shown by chloroform and benzene. Some caution should therefore be exercised before declaring a solvent "insoluble" in water. The amounts dissolved may in fact be considerable, and their presence in discharged process water must be duly accounted for. This is illustrated in the following example.

**TABLE 6.3**

Solubilities of Liquids and Solids in Water at 25°C

| Substance | Solubility (g/l water) |
|---|---|
| Chloroform | 11 |
| Carbon tetrachloride | 0.8 |
| Diethyl ether | 75 |
| $n$-Hexane | 0.15 |
| $n$-Octane | $6.6 \times 10^{-4}$ |
| Benzene | 1.8 |
| Toluene | 0.52 |
| $o$-Xylene | 0.18 |
| Naphthalene | $3.3 \times 10^{-2}$ |
| DDT | $1.2 \times 10^{-6}$ |
| Glucose | 820 |
| Potassium chloride | 360 |
| Sodium chloride | 360 |
| Sodium nitrate | 940 |
| Mercury | $3.0 \times 10^{-5}$ |

### *Illustration 6.3: Discharge of Plant Effluent into a River*

We consider here the case of process water saturated with benzene being discharged into a river. The question to be addressed is whether the diluting effect of the river flow is sufficient to reduce the effluent concentration to within permissible limits and, if not, how much of the offending substance has to be removed to meet environmental standards. A sketch depicting effluent and river flow is shown in Figure 6.3.

Let us consider an effluent discharge of 150 l/min and a river flow that varies seasonably from 23,000 to 50,000 l/s. Note that since the regulatory limit has to be met at all times during the year, the lower summer flow rate of 23,000 l/s must be used. The standard used here is that of the U.S. Environmental Protection Agency, which has set the maximum permissible level of benzene in drinking water at 0.05 mg/l or 5 ppb. The effluent is assumed to be saturated with benzene at the solubility level of 1.8 g/l (1.8 $\times$ 10$^6$ ppb) given in Table 6.3.

We commence with a mass balance around the juncture of effluent and river flow shown in Figure 6.3 and assume the contents of the envelope to be well mixed and to have attained a steady state. Thus,

Rate of benzene in − Rate of benzene out = 0

$$Q_e C_e - (Q_e + Q_r)C_r = 0 \tag{6.4a}$$

where the subscripts $e$ and $r$ refer to effluent and river, respectively, and $Q$ denotes volumetric flow rates. Substituting the given data into this equation, we obtain

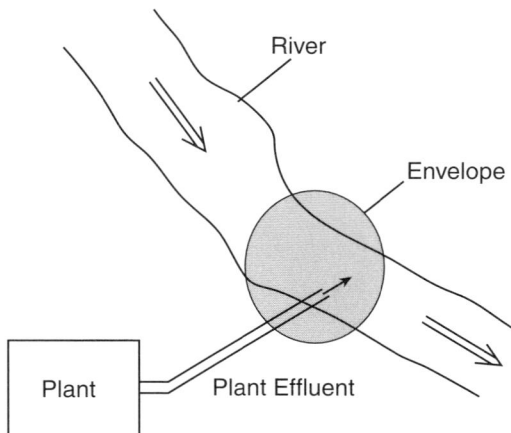

**FIGURE 6.3**
Discharge of a plant effluent into a river.

$$(150/60)1.8 \times 10^6 - [(150/60) + 23,000]C_r = 0 \qquad (6.4b)$$

Solving for $C_r$ yields a downstream river concentration of

$$C_r = \frac{4.5 \times 10^6}{23002.5} = 195 \, \text{ppb} \qquad (6.4c)$$

With the allowable concentration set at 5 ppb, the required fractional removal $R$ is given by

$$R = 1 - \frac{5}{195} = 0.974 \qquad (6.4d)$$

In other words, slightly more than 97% of the benzene in the plant effluent will have to be removed to meet the aforementioned standard. It is likely that an adsorption purification process using activated carbon can be used to achieve this goal. Such a process is taken up in Illustration 6.4.

### 6.2.3 Fluid–Solid Equilibria: The Langmuir Isotherm

We previously, in Section 6.2.1, considered the case in which a gas is absorbed into and comes into equilibrium with a liquid solvent. This process of absorption, in which the solute gas permeates the entire body of the liquid, differs from adsorption, which is essentially a surface phenomenon. Here the solute also penetrates the porous structure of the solid, and the process is therefore initially at least akin to absorption. Ultimately, however, the solute molecules come to rest on the walls of the porous structure and remain confined there in dynamic equilibrium with the surrounding pore space. Thus, while permeation of the solid structure does take place, the solute molecules are not uniformly dispersed but rather are localized on the internal surface of the solid matrix. We show this, as well as the differences between adsorption and absorption, in Figure 6.4.

There is a further distinction to be made between the two processes. A liquid has in principle an unlimited capacity for dissolving a gas, although that capacity diminishes asymptotically as gas pressure is increased (see Figure 6.2). In adsorption, the surface area available for accommodating solute molecules is limited and finite. Here an increase in pressure or solute concentration will ultimately lead to complete coverage by a "monolayer" or saturation of the surface. On reaching this state, no further adsorption can take place. This is indicated by the asymptotic saturation capacity shown by the isotherm in the phase diagram of Figure 6.5. An exception occurs when the solute gas is within reach of a state of condensation. The solute may then form multiple adsorbed layers or multilayers and ultimately fill the entire pore space by condensation ("capillary condensation"). This less

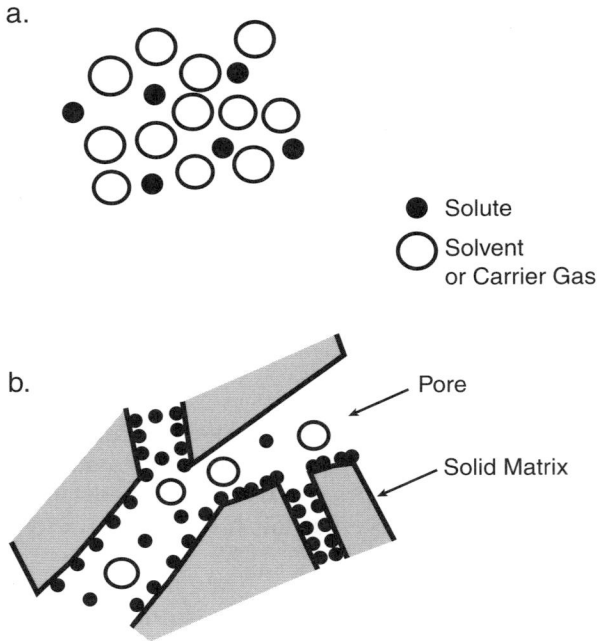

a.

● Solute

○ Solvent
  or Carrier Gas

b.

Pore

Solid Matrix

**FIGURE 6.4**
(a) Absorption: dispersion of solute molecules throughout a liquid solvent. (b) Adsorption: localization of solute molecules on pore walls.

frequent case leads to inflecting isotherm curves and is shown in Figure 6.6. Most fluid–solid equilibria, however, lead to monolayer coverage and are well-represented by the isotherms shown in Figure 6.5.

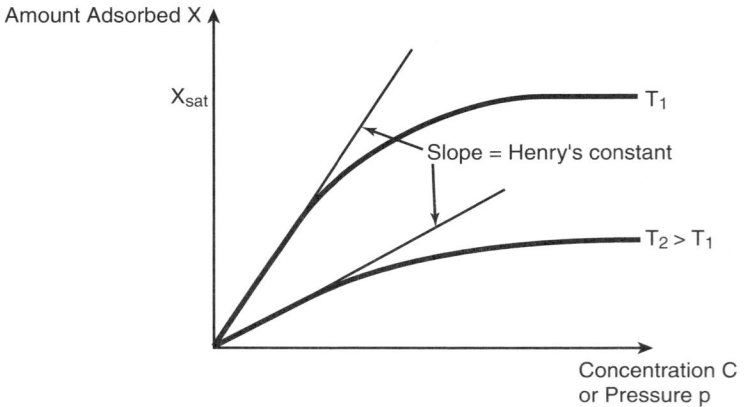

Amount Adsorbed X

$X_{sat}$

Slope = Henry's constant

$T_1$

$T_2 > T_1$

Concentration C
or Pressure p

**FIGURE 6.5**
The Langmuir adsorption isotherm.

One of the earliest attempts to derive a theoretical expression for these isotherms, and to date also the most successful one, is that due to Langmuir. In this derivation, Langmuir postulated that the adsorption equilibrium was the result of two rate processes, equal in magnitude but opposite in direction: The rate of adsorption, which was taken to be proportional to solute pressure or concentration and the available free surface area, and a desorption rate, which varied directly with the fractional surface coverage. The result of equating these two rate expressions can be expressed, after some manipulation, in the following form:

$$X = \frac{a'p}{1+b'p} \tag{6.5a}$$

or

$$X = \frac{a'C}{1+b'C} \tag{6.5b}$$

Here $a'$ and $b'$ are semiempirical constants, $p$ and $C$ are gas partial pressure and solute concentration, respectively, and $X$ represents the amount adsorbed. For general engineering purposes, it is often more convenient to replace $p$ and $C$ by a single fluid-phase concentration $Y$, expressed in units of kg solute/kg inert gas, or kg solute/kg solvent. Equation 6.5a and Equation 6.5b can then be coalesced into a single expression of the form

$$X = \frac{HY}{1+bY} \tag{6.5c}$$

where $a'$ and $b'$ are replaced by the new empirical constants $H$ and $b$. The amount adsorbed $X$ is generally expressed in units of kg solute/kg solid, or less frequently, as mol solute/g solid. Evidently, when dealing with pure solute gases, the fluid ratio $Y$ can no longer be used and we must revert to Expression 6.5a. The need to do this rarely arises in practice because in most practical applications the gas phase contains an inert, nonsorbable component such as air.

Let us now examine the asymptotic behavior of the Langmuir isotherm: At low values of the fluid-phase concentration $Y$, the term $bY$ becomes small compared to 1, and the Langmuir isotherm approaches the limiting linear form

$$X = HY \tag{6.5d}$$

$$Y \to 0$$

The slope of this line is often referred to as Henry's constant, and Expression 6.5d as Henry's law for adsorption, in analogy to the corresponding case of gas–liquid equilibrium (see Equation 6.2).

At the other extreme of high values of $Y$, $bY$ becomes the dominant factor and the Langmuir isotherm converges to the form

$$X_{Y \to 0} = H/b = X_{Sat} \qquad (6.5e)$$

Both these limiting cases nicely accord with the features shown in Figure 6.5.

We note that high values of $H$ correspond to high adsorbent capacities, in contrast to gas absorption where large values of Henry's constant are associated with *low* solvent capacities.

Adsorption Henry's constants are central to adsorptive purification processes of dilute streams and also reach, as will be seen in the next section, into areas of environmental concern. To acquaint the reader with their magnitude, we have compiled values of $H$ on carbon for some important trace solutes in aqueous solution, which are displayed in Table 6.4. Of note here is the extremely high value for PCBs, which dominates the table. The reader should be reminded, however, that this is partly offset by the extremely low solubility of PCBs.

Although the great majority of adsorbed solutes show Langmuir-type behavior, a considerable number of substances exhibit inflecting isotherms. This is particularly the case with vapors in the vicinity of saturation. Figure 6.6, which shows moisture isotherms on a variety of adsorbents, illustrates this type of behavior. Zeolitic sorbents are the only ones among them that display Langmuir-type behavior. They are most effective at low humidities where they show a substantial uptake of water. Silica gel and activated alumina have high uptakes at higher humidities, where they exceed zeolite capacities by factors of two or more. Carbon, because of its hydrophobic nature, has only a minimum affinity for water.

**TABLE 6.4**

Henry's Constants for Aqueous Solutions on Carbon at 25°C

| Solute | $H$ (kg $H_2O$/kg C) |
|---|---|
| Methyl chloride | 6.6 |
| Methylene chloride | 14 |
| Chloroform | 74 |
| Carbon tetrachloride | 360 |
| *n*-Pentane | 2,200 |
| *n*-Hexane | 10,400 |
| Benzene | 400 |
| Styrene | 600 |
| Chlorobenzene | 500 |
| PCBs (upper limit) | $1.5 \times 10^8$ |

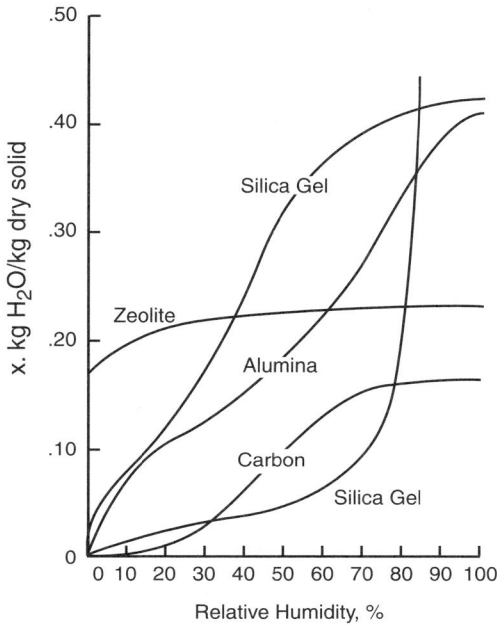

**FIGURE 6.6**
Moisture adsorption isotherms.

In what follows, we use the tabulations of Table 5.4 to explore a particular and highly useful limiting case of the adsorptive purification of an aqueous solution.

## Illustration 6.4: Adsorption of Benzene from Water in a Granular Carbon Bed

The purification of both potable and wastewater by adsorption is a wide-spread practice, which in the U.S. alone consumes over 100 million kg of activated carbon a year. Typically in such an operation, the water is passed through a fixed bed of the granular adsorbent, which becomes progressively saturated with the impurities. In the course of this process, a concentration profile develops within the bed, which takes the form of an S-shaped curve ranging in level from full saturation $q_s$ to essentially zero impurity content. This profile, which is shown in Figure 6.7a, progresses steadily through the adsorber until it reaches the end of the bed. An observer stationed at this position would at this point see the first traces of impurity break through. It would then rise in level, again in the form of an S-shaped curve, until the feed concentration $Y_F$ is attained. This is, however, not allowed to happen and the operation is instead terminated when the effluent concentration reaches a prescribed maximum permissible level, $Y_P$. This level and the associated concentration breakthrough curve are shown in Figure 6.7b. The

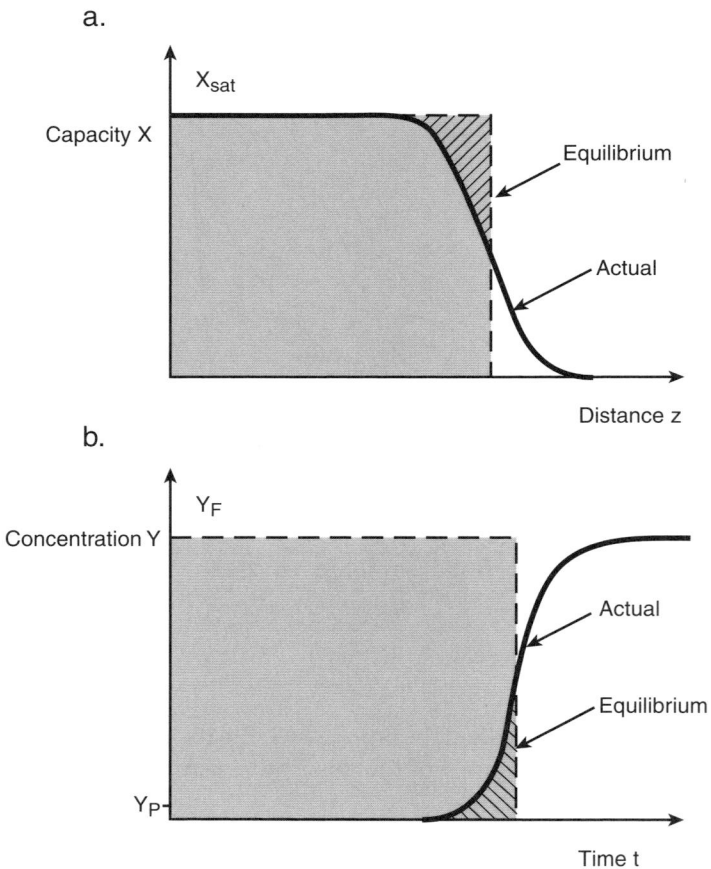

a.

b.

**FIGURE 6.7**
Adsorption in a fixed bed: (a) adsorbent concentration profiles; (b) fluid phase concentration breakthrough curve.

operation is continued by switching the feed stream to a second clean bed, held on hand while the spent adsorbent is either regenerated or replaced by a fresh charge of the material.

Adsorption in a fixed bed is a complex system to model. Concentrations evidently vary with distance, and although they ultimately attain a steady form of distribution, they also vary with time. The model would consequently consist of two mass balances, one for the fluid phase and a second for the solid phase, and both of these would be partial differential equations in time and distance, which generally have to be solved numerically. To avoid this complication, a procedure has come into use in which mass transfer resistance is neglected and the two phases are everywhere assumed to be in equilibrium. The concentration then propagates in the shape of a rectangular front, shown in Figure 6.7 and denoted "Equilibrium." The movement of this front and its dependence on flow rate and feed concentration can be analyzed by means of a simple cumulative mass balance, which takes the following form:

$$Y_F v \rho_f A_C t \quad = \quad X_F \rho_b A_C z \quad + \quad Y_F \rho_f A_C z \qquad (6.6)$$

| Amount introduced to time $t$ | Amount retained by adsorbent | Amount left in fluid |
|---|---|---|

Here $v$ and $A_C$ are the fluid velocity and cross-sectional area of the bed, respectively, and $\rho_f$ and $\rho_b$ are the fluid and bed densities.

The last term in this equation is generally negligible because the bulk of the impurity will reside in the adsorbent. It would otherwise not be a very efficient adsorbent. Equation 6.7a can then be recast in the form

$$z / t = \frac{\rho_f v}{\rho_b H} \qquad (6.7a)$$

where we have substituted Henry's constant for the concentration ratio $X_F / Y_F$.

Equation 6.7a can be used to calculate the time $t$ it takes the front to reach a particular position or, conversely, the position attained after a prescribed time interval. These quantities are, by necessity, limiting lower values because full saturation will in fact be retarded by the mass transfer resistance. However, in many instances the fluid flow is sufficiently slow that local equilibrium is attained, or will be nearly attained, during the interval of contact. We make use of this fact in Illustration 6.6 to analyze the contamination of soil that results from polluted groundwater.

This illustration addresses the problem of sizing a carbon bed to be used in the purification of a plant effluent. Our purpose here is best served by recasting Equation 6.7a in a form suitable for the calculation of bed requirements. This is done by first cross-multiplying the expression and then dividing and multiplying by the cross-sectional area $A_C$. We obtain

$$\frac{A_C z \rho_b}{A_C v \rho_f t} = \frac{(W_b)_{\text{Min}}}{G \times t} = \frac{(W_b)_{\text{Min}}}{v \rho_f t} = \frac{1}{H} \qquad (6.7b)$$

or equivalently

$$\frac{\text{Minimum bed weight}}{\text{Weight of fluid treated}} = \frac{1}{H} \qquad (6.7c)$$

Thus, the larger the Henry constant $H$, the smaller the required bed size. Note, however, that this is a *minimum* requirement, since we have assumed the phases to be in equilibrium.

Let us see how this works out in practice. We refer to the preceding example of an aqueous plant effluent saturated with benzene, 97% of which has to be removed before being discharged. The rate of discharge was 150 l/min.

Suppose that we wish to size a bed of granular carbon, which will stay on-stream for 6 months before breakthrough occurs. The amount of effluent treated in this period comes to

$$G \times t = 150 \times 60 \times 24 \times 180 = 3.9 \times 10^7 \text{ kg} \qquad (6.7d)$$

Using $H = 400$ for benzene, listed in Table 6.4, we obtain from Equation 6.7d

$$(W_b)_{\text{Min}} = G \times t / H = 3.9 \times 10^7 / 400 = 9.8 \times 10^4 \text{ kg carbon} \qquad (6.7e)$$

Now, the bed density $\rho_b$ for granular activated carbon is, typically, of the order 500 kg/m$^3$. If, therefore, a bed 3 m in diameter $d$ is chosen, the packed height $h$ of that bed is given by

$$h = \frac{(W_b)_{\text{Min}}}{(\pi d^2 / 4) \times \rho_b} = \frac{9.8 \times 10^4}{(\pi 3^2 / 4) \times 500} = 28 \text{ m} \qquad (6.7f)$$

Since adsorbent beds typically range up to 10 m in height, this figure is excessive. We therefore reduce the time on-stream to 1 month and arrive at the more reasonable bed height of approximately 4.5 m. This allows some slack for *actual* bed requirements, which will be somewhat higher because of the neglected mass transfer resistance.

### Illustration 6.5: Adsorption of a Pollutant from Groundwater onto Soil

Soils show a considerable sorptive affinity for pollutants that is principally brought about by the soil's carbon content (a result of decaying organic matter). That content typically ranges from 1 to 2% of the total mass. Adsorptive capacities per unit weight are consequently some 50 times lower than the carbon values shown in Table 6.4, and this figure was used to compose the Table 6.5 listing of Henry's constants for soils.

We consider here again an effluent saturated with benzene and assume that seepage into the groundwater has occurred. The task will be to calculate the stretch of soil that will have been contaminated after 10 days of exposure. Groundwater velocity is set at 1 mm/s and soil density at 2500 kg/m$^3$. Using a Henry's constant of 0.54 given in Table 6.5, we obtain from Equation 6.7a

$$z / t = \frac{\rho_f}{\rho_b} \frac{v}{H} = \frac{1000 \; 10^{-3}}{2500 \; 0.54} = 7.54 \times 10^{-4} \text{ m / s}$$

**TABLE 6.5**

Henry's Constants for Aqueous Solutions on Soil at 25°C

| Solute | $H$ (kg water/kg soil) |
| --- | --- |
| Methyl chloride | 0.033 |
| Methylene chloride | 0.071 |
| Chloroform | 0.37 |
| Carbon tetrachloride | 1.8 |
| *n*-Pentane | 11 |
| *n*-Hexane | 52 |
| Benzene | 0.54 |
| Styrene | 3.0 |
| Chlorobenzene | 2.5 |
| PCBs, upper limit | $7.3 \times 10^5$ |

and consequently

$$z = 7.4 \times 10^{-4}\, t = 7.4 \times 10^{-4} \times 3600 \times 24 \times 10$$

$$z = 640 \text{ m}$$

*Comments:*

The reader is reminded that the results obtained in both Illustration 6.4 and Illustration 6.5 represent *minimum* values, i.e., minimum carbon bed requirements and minimum penetration into the soil. Because of the mass transfer resistance, the concentrations on the percolating fluid run ahead of the rectangular front shown in Figure 6.6a, and penetrate deeper into the solid matrix than predicted. The corrections that must be applied often amount to no more than 20 to 30% of the ideal length, which is therefore a highly useful engineering estimate for the situation at hand. Higher corrections are needed in cases involving large particles or high fluid velocity.

Note that the rigorous PDE model would require, in addition to equilibrium data, the relevant transport parameter, i.e., the mass transfer coefficients within both the solid particle and the liquid. The solid-phase coefficient, in particular, requires fairly elaborate measurements and is usually unavailable to the general practitioner.

## 6.2.4 Liquid–Liquid Equilibria: The Triangular Phase Diagram

Liquid–liquid equilibria deserve our attention principally because of their widespread occurrence in industrial- and laboratory-scale extraction processes. They also play a role in assessing the effect of accidental spills of oil and organic solvents in lake and ocean waters.

On the industrial scene, the most prominent applications both in scale and number are seen in the petroleum industry. Liquid extraction is used here to separate petroleum fractions selectively and to purify or otherwise refine them. In the Edeleanu process, which is close to a century old, liquid sulfur dioxide is used to extract aromatics from various feedstocks. The removal of the ever-present sulfur compounds is accomplished by extraction with sodium hydroxide solutions. In addition, a wide range of organic solvents is used in the purification and refining of various lubricants.

Considerable use of liquid extraction is also made in the metallurgical industry for the separation and refining of metals, and in the food industry for the extraction of oils and fats and other edible products. The pharmaceutical industry uses liquid extraction to recover their end products, such as penicillin, from the reaction mixture. In general, whenever dealing with heat-sensitive materials, extraction at low temperatures is often the process of choice.

A number of the traditional solvents used in these processes, such as the chlorinated and fluorinated hydrocarbons, have in the last two decades come under increasing scrutiny and criticism because of health and environmental concerns. The effect of certain of these solvents on the ozone layer has received a good deal of publicity, as has their role as potential carcinogens. This has opened the door to the development of alternative processes, chief among them *supercritical fluid extraction*. We address this topic in some detail in Section 6.2.5.

Liquid–liquid equilibria differ from the previous cases we have considered in that the two phases will, with an appropriate shift in the relative concentrations of the three components, merge into a single homogeneous phase. Consider, for example, the two "immiscible" solvents, benzene and water, to which the solute ethanol is added. Initially, at low solute concentrations, two distinct phases are maintained, with ethanol distributed between them in certain concentration ratios. As the ethanol content is increased, the two solvents begin to show greater mutual solubility. This trend continues until ultimately, at sufficiently high ethanol concentrations, the two phases coalesce into a single homogeneous phase.

To obtain a valid representation of the entire domain of liquid–liquid phase behavior, it is necessary to take account of the concentrations of all three components. This is done by using the trilinear coordinate system, which leads to a construction known as a *triangular diagram*, shown in Figure 6.8.

Each side of this diagram is scaled from 0 to 100%. The apexes of the triangle represent the pure components $A$, $B$, and $C$, respectively. $C$ is usually taken to be the solute, and $A$ and $B$ represent the two mutually immiscible solvents. Any point on the side of the triangle denotes a two-component (binary) mixture, while the interior of the diagram represents the full complement of three-component (ternary) mixtures. Thus, the point $G$ on the side $AB$ represents 40% solute $C$ and 60% solvent $B$, while the interior point $M$ signifies a mixture containing 30% $A$, 40% $B$, and 40% $C$. Compositions are usually expressed as weight percent, or, less frequently, as mole percent.

Liquid–liquid equilibria are determined experimentally by equilibrating the two phases having a total concentration $M$ (Figure 6.9) and recording their compositions. These values are entered in the diagram and form the

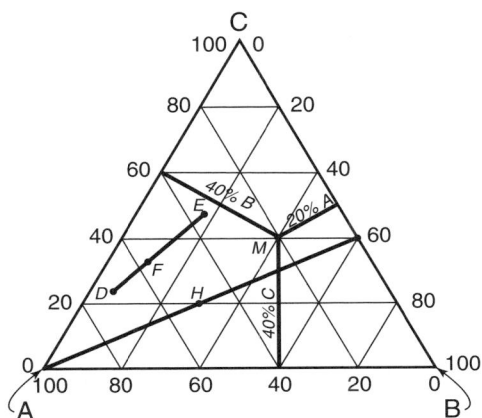

**FIGURE 6.8**
The triangular diagram.

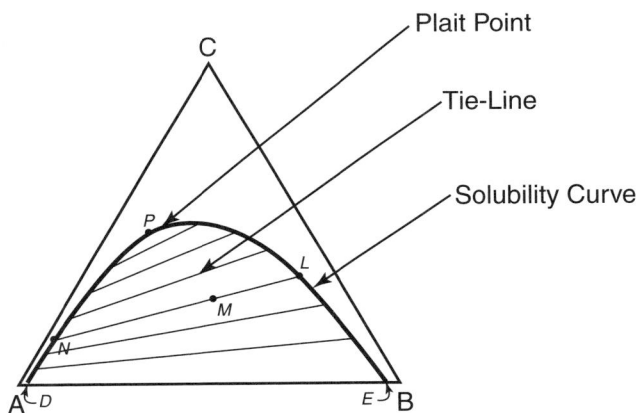

**FIGURE 6.9**
Representation of liquid–liquid equilibrium in trilinear coordinates.

end points of what is termed a *tie-line*, shown in Figure 6.9. There is in principle an infinite set of such tie-lines, only a few of which appear in the diagram. They are ordinarily not parallel and change their slope slowly with changing composition. As more solute $C$ is added to such a mixture, the mutual solubility of the two solvents $A$ and $B$ increases until the tie-line end points merge at the point $P$, known as the *plait point*. The line connecting all these points and passing through $P$ is known as the *binodal* or *solubility curve* and encloses all mixtures showing two-phase behavior. Compositions lying outside this solubility curve represent homogeneous, single liquid-phase solutions, while the end points $D$ and $E$ denote the mutual solubility of the two solvents. When these are completely immiscible, $D$ and $E$ coincide with the two apexes $A$ and $B$.

Representation of liquid–liquid equilibria in ternary diagrams is widely used in the graphical solution of extraction problems. This topic is taken up and discussed in considerable detail in the next chapter. Occasions arise, however, when it becomes convenient to use an alternative representation in rectangular coordinates, known as a *distribution curve*. This diagram, shown in Figure 6.10, consists of a plot of the solute weight fraction in the two phases against each other. Its construction is accomplished by transferring the compositions represented by the tie-line end points to rectangular coordinate axes, as shown in Figure 6.10.

Clearly, this type of representation provides only a limited picture of the entire domain of compositions. Neither of the two solvent weight fractions, for example, can be deduced. Various other constructional features of the triangular diagram, which will be taken up below, likewise cannot be translated into rectangular coordinates. The distribution diagram does, however, convey several key features of two-phase behavior. It shows the maximum solute concentration $(x_{CB})_{Max}$ beyond which no phase separation can take place, and it locates the plait point on the 45° diagonal, which denotes equality of phase concentrations. Compositions that lie above the diagonal are richer in solute content than those of the companion phase, while points below it denote a depletion in solute. The greater the distance from the diagonal, the greater the degree of enrichment or depletion. By noting these features we gain an immediate sense of the potential of a particular system for extractive enrichment.

To quantify these properties, it is useful to define a quantity known as the distribution coefficient $m$, which equals the solute mole-fraction ratio in the two phases and is given by

$$m = \frac{x_{CB}}{x_{CA}} \tag{6.8}$$

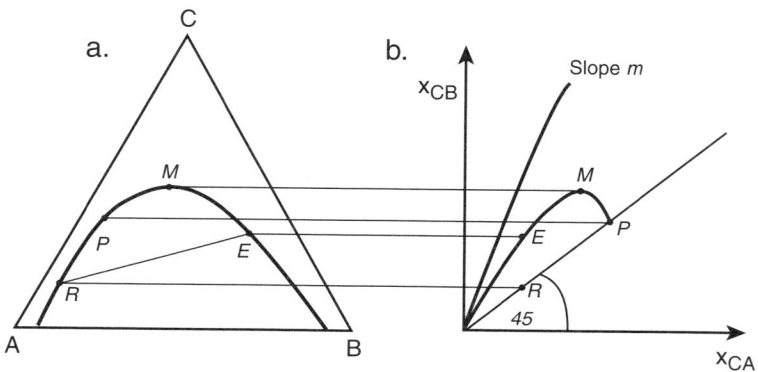

**FIGURE 6.10**
Translation from trilinear to rectangular coordinates: (a) triangular diagram; (b) distribution curve.

Values of $m$ greater than unity denote solute enrichment in the $B$ layer, and those less than unity denote enrichment in the solvent $A$. In the limit of low concentration, or infinite dilution of the $B$ layer, the distribution curve becomes linear and the distribution coefficient itself attains a maximum value. This limiting coefficient, shown in Figure 6.10b, is akin to the Henry's constants, which we have seen in gas–liquid and fluid–solid equilibria, and is sometimes denoted as such. An indication of its magnitude may be obtained from the low-concentration distribution coefficients listed in Table 6.6 for a number of systems comprising aqueous and organic solutions.

### *Illustration 6.6: The Mixture or Lever Rule in the Triangular Diagram*

With this example, we wish to draw the reader's attention to certain simple geometric constructions, which can be carried out in a triangular diagram to obtain important information in easy and rapid fashion. One such construction, sketched in Figure 6.11, leads to what is known as the *mixture rule*, or *lever rule*, for ternary liquid systems. Briefly stated, the rule asserts that the composition that results when two liquid solutions are mixed lies on a straight line connecting their compositions. Thus, if $R$ kg of a mixture represented by point $R$ is combined with $E$ kg of a solution located at point $E$, the resulting composition $F$ will lie on a straight line connecting points $R$ and $E$. Furthermore, the location of $F$ will be such that the line segments it defines stand in the ratio of the weights of the parent solutions, i.e.,

$$\frac{R}{E} = \frac{\overline{EF}}{\overline{RF}} \tag{6.8a}$$

**TABLE 6.6**

Low-Concentration Solute Distributions in Water–Organic Solvent Systems (25°C)

| Solute | Solvent | Mole% Aqueous Phase | Mole% Organic Phase | Distribution Coefficient |
|--------|---------|---------------------|---------------------|--------------------------|
| Acetic acid | Carbon tetrachloride | 5.088 | 0.916 | 5.56 |
| | Hexane | 14.810 | 1.614 | 9.18 |
| | Toluene | 7.850 | 2.440 | 3.22 |
| Acetone | Chloroform | 16.07 | 0.959 | 16.76 |
| | Diethyl ether | 1.519 | 5.446 | 0.279 |
| Methanol | Benzene | 4.067 | 0.798 | 5.10 |
| | Toluene | 5.945 | 0.286 | 20.79 |
| Ethanol | Chloroform | 4.187 | 1.784 | 2.35 |
| | Diethyl ether | 1.382 | 2.422 | 0.571 |
| | Toluene | 4.621 | 0.398 | 11.61 |
| | Hexane | 30.11 | 1.297 | 23.22 |
| | Benzene | 0.994 | 1.177 | 0.845 |

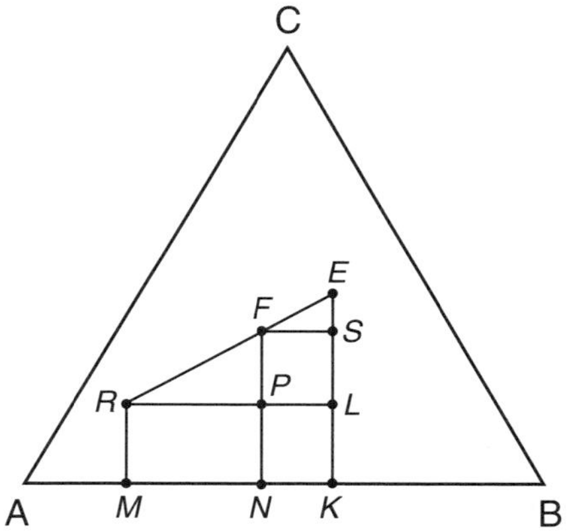

**FIGURE 6.11**
The mixture rule in a triangular diagram.

We now proceed to present a proof of these statements. This is done by composing total and component mass balances for the mixtures and relating the resulting compositional changes to the line segments of Figure 6.11. Thus, for the total mass balance

$$R + E = F \tag{6.8b}$$

where $F$ = mass of final mixture, and for the component mass balance

$$Rx_{CB} + Ex_{CE} = Fx_{CF} \tag{6.8c}$$

Eliminating $F$ from these equations yields

$$\frac{E}{R} = \frac{x_{CF} - x_{CD}}{x_{CE} - x_{CF}} \tag{6.8d}$$

where $x_{CF}$ will be located somewhere on the straight line connecting $E$ and $R$. We now show that its location is such that it subdivides the line in the ratio given by Equation 6.8a. From the diagram it follows that

$$x_{CF} = \overline{FN}, \quad x_{CD} = \overline{RM}, \quad x_{CE} = \overline{EK} \tag{6.8e}$$

and consequently

$$\frac{E}{R} = \frac{x_{CF} - x_{CD}}{x_{CE} - x_{CF}} = \frac{\overline{SL}}{\overline{ES}} = \frac{\overline{RF}}{\overline{EF}} \tag{6.8f}$$

where the last equality follows from the similarity of the two triangles *FPR* and *ESF*. The proof is thus complete. Note that the closer the parent mixture *F* is to *E*, the more *R* is formed, and vice versa.

*Comments:*

It is seen from this discussion that the mixture or lever rule can be expressed both algebraically and in geometrical form. The algebraic version, given by Equation 6.8d, is the preferred tool for numerical calculation, while the geometrical construction of Figure 6.11 serves to provide a quick visual estimate of the quantities involved. Thus, if *F* lies midway between *R* and *E*, the two parent solutions will be equal in weight, and vice versa. It also follows that if an amount *E* with a composition located at the point *E* of Figure 6.11 is removed from the mixture, the point *R* representing the residue will lie on a straight line *EF* extended through *F*, and the above relationship (Equation 6.8f) will apply.

The lever rule finds its most notable application in the use of tie-lines to establish the compositions and relative proportions of the two solvent layers in an extraction process. Suppose, for example, that two liquid solutions with overall composition represented by *M* in Figure 6.9 are contacted and allowed to settle into a two-phase equilibrium. The compositions will then be given by the tie-line end points *L* and *N*, and the relative amounts of *L* and *N* will be in the ratio $\overline{MN} / \overline{ML}$. Here again the geometrical construction serves to convey a quick visual notion of the events, which can then be followed up by actual algebraic calculations.

### 6.2.5 Equilibria Involving a Supercritical Fluid

In Section 6.1 we have drawn the reader's attention to the existence of a threshold temperature known as the critical temperature, above which a pure substance can exist only as a single phase. That phase possesses characteristics of a dual nature: It behaves, on the one hand, like a dense gas that does, however, still have a sufficiently open structure to allow rapid passage of solute molecules by diffusion. The relatively short distance between neighboring molecules, on the other hand, allows it to attract and hold substantial amounts of solute material, thus acting for practical purposes as an efficient solvent. These dual features have led to the use of supercritical fluids (SCF) as solvents in a process known as supercritical fluid extraction (SCE). Its advantages over conventional liquid extraction are twofold: It shortens the required contact time because of the higher prevailing diffusivities, and it replaces costly and potentially harmful liquid solvents with inexpensive and benign gases such as carbon dioxide and the lower hydrocarbons. A number of commercial processes have been developed that make use of this new technology, and there is an ongoing intensive quest for new applications.

To acquaint the reader with the key features of SCE, we display the supercritical region of interest (Figure 6.12) and a map of diffusivities in liquids

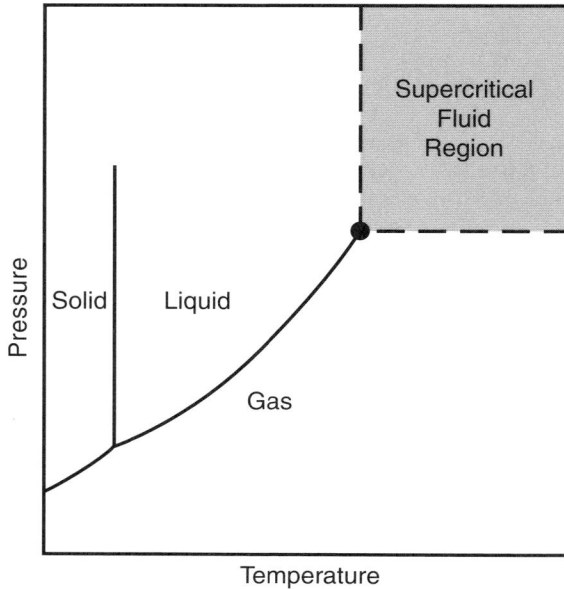

**FIGURE 6.12**
Temperature–pressure diagram for a pure substance and the region of SCE.

and supercritical carbon dioxide (Figure 6.13), which clearly shows the superior transport properties of the latter. The increased extraction power of carbon dioxide with temperature and pressure becomes evident in Figure 6.14. We note in particular that doubling the pressure from 70 bar (~70 atm) to 150 bar at temperatures slightly above critical increases the solute concentration in the extracting medium by well over an order of magnitude.

The earliest indication of this property of SCF is to be found in the work of J.B. Hannay and J. Hogarth (1879), who reported that an increase in pressure of supercritical ethanol caused increased dissolution of certain inorganic salts such as potassium iodide, while conversely a relaxation in pressure resulted in the precipitation of the salts as "a white snow."

This early work was followed intermittently by a flurry of activity that intensified during the 1970s and 1980s, partly as a result of environmental concerns over the use of conventional solvents. Major commercial processes currently in use are the SCE of caffeine from coffee and tea, the SCE of spice aromas, and the fractionation and purification of polymers. Processes under investigation include the treatment of wastewaters, activated carbon regeneration, and the SCE of edible oils and therapeutic agents from plant materials.

The scale of decaffeination processes is indicated by the extraction vessel shown in Figure 6.15, which measures about 2 m in diameter and 20 m in height. The operation is countercurrent, with $CO_2$ passing upward and coffee discharged intermittently at the rate of about 5000 kg/h. Operating pressure is typically in the vicinity of 300 atm.

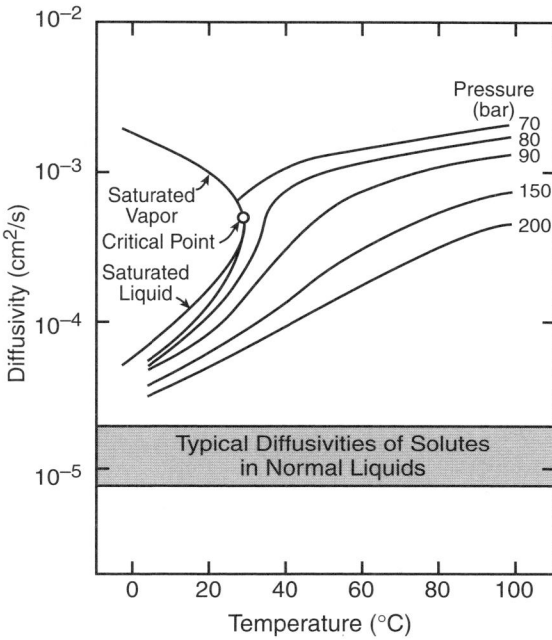

**FIGURE 6.13**
Comparison of diffusivities in supercritical carbon dioxide and normal liquids. (From McHugh, M. and Krukonis, V. *Supercritical Fluid Extraction*, 2nd ed., Butterworth-Heinemann, Boston, 1994. With permission.)

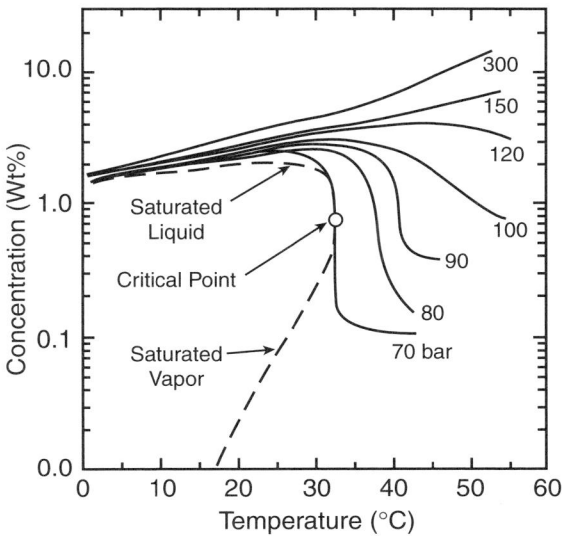

**FIGURE 6.14**
Solubility of naphthalene in supercritical carbon dioxide. (From McHugh, M. and Krukonis, V. *Supercritical Fluid Extraction*, 2nd ed., Butterworth-Heinemann, Boston, 1994. With permission.)

**FIGURE 6.15**
Extraction vessel used in the Maxwell House® Coffee Company supercritical $CO_2$ decaffeination plant in Houston, Texas. (From McHugh, M. and Krukonis, V. *Supercritical Fluid Extraction*, 2nd ed., Butterworth-Heinemann, Boston, 1994. With permission.)

## *Illustration 6.7: Decaffeination in a Single-Equilibrium Stage*

To gain an idea of the $CO_2$ requirements in decaffeination, we consider a hypothetical process in which the carbon dioxide is circulated through the extraction vessel until it is in equilibrium with its charge of coffee. We term this contact an *equilibrium stage*, a concept that is discussed more fully in Chapter 7.

The equilibrium distribution of caffeine under a particular set of conditions is shown in the log-log plot of Figure 6.16. The fitting of the data leads to the expression

$$x = 1.24y^{0.316} \tag{6.9a}$$

where $x$ and $y$ are the weight percentages of caffeine in the coffee and the carbon dioxide.

Suppose the requirement is to reduce the caffeine content from 1 to 0.05%, which is typical of commercial decaffeination processes. The task is to calculate the ratio $G/S$ kg $CO_2$/kg coffee required to achieve this reduction.

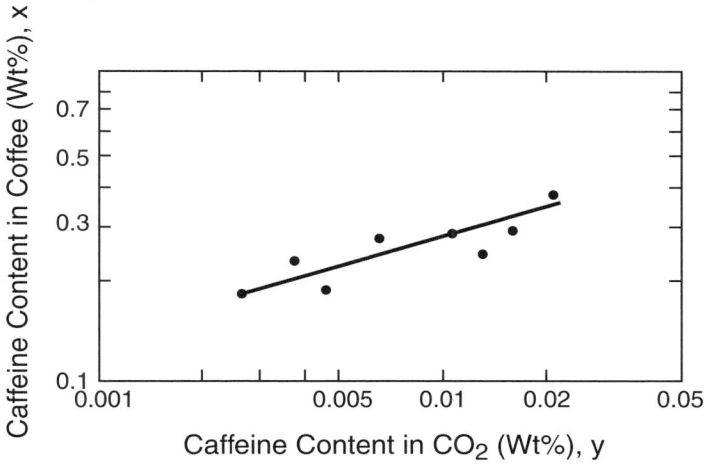

**FIGURE 6.16**
Caffeine equilibrium distribution at 60°C, 272 atm. (From McHugh, M. and Krukonis, V. *Supercritical Fluid Extraction*, 2nd ed., Butterworth-Heinemann, Boston, 1994. With permission.)

The carbon dioxide is assumed to be devoid of caffeine initially. The equilibrium content in the gas at the end of the operation is given by

$$y^* = \frac{1}{1.24} 0.05^{1/0.36} \tag{6.9b}$$

$$y^* = 1.96 \times 10^{-4}\% \tag{6.9c}$$

A caffeine mass balance for the process leads to the expression

$$\text{Rate of caffeine in} - \text{Rate of caffeine out} = 0$$
$$S \times 0.01 \quad - \quad (Gy^* + S \times 0.005) \quad = 0 \tag{6.10a}$$

and consequently

$$\frac{G}{S} = \frac{0.0095}{1.96 \times 10^{-4}} \tag{6.10b}$$

$$\frac{G}{S} = 45.9 \, kg \, CO_2 \, / \, kg \; coffee \tag{6.10c}$$

*Comments:*

This is clearly an inordinate amount of $CO_2$ required, brought about by what appears to be a rather unfavorable distribution of caffeine between the extracting gas and the coffee. That seeming drawback can be overcome by "staging" the extraction process, i.e., providing a continuous contact between the two phases as is done in commercial countercurrent operations. This type of operation is taken up in Illustration 8.4. It is shown there that the gas requirements can be considerably reduced by adopting this type of contact, a fact that has been a major contributor to the commercial success of the process.

### 6.2.6  Equilibria in Biology and the Environment: Partitioning of a Solute between Compartments

In biology and the environment, the equilibria of interest are those of a substance distributed or "partitioned" among the major compartments of the system. In biology we usually seek to establish the distribution of a substance between the circulating blood and the separate compartments of fat or lipids, muscle and bone. The substances in question can be toxic or benign (e.g., a drug).

   In the environment, the major compartments of interest are air, water, and soil, and the solute involved is usually toxic or objectionable. An interesting confluence of biology and the environment occurs in the exposure of animals to toxic chemicals contained in the air or in water. The question then becomes one of establishing equilibrium concentrations between environmental and body compartments. In fish, for example, the usual aim is to determine the distribution of the solute between water and animal fat. A description of this case is given in Illustration 6.8.

   The equilibrium concentrations between various compartments are usually taken to be in a constant ratio termed the *partition coefficient K*. Thus, for a substance distributed between blood and muscle, we have

$$\frac{C_B}{C_M} = K_{BM} \qquad (6.11a)$$

or

$$C_B = K_{BM}C_M \qquad (6.11b)$$

The relation is, in other words, a *linear* one, and the partition coefficient can be viewed as the equivalent of a Henry's constant, relating the two concentrations in question. These concentrations represent hypothetical maximum levels, which may appear briefly during the initial period of exposure. Thereafter, metabolic processes and excretion intervene to reduce the concentrations to new steady-state values. These can still be substantial, as was shown in Illustration 5.7.

### Illustration 6.8: The Octanol–Water Partition Coefficient

In environmental work, extensive use is made of a special $K$ value, the *octanol–water partition coefficient*. It describes the distribution of a solute between octanol and water; i.e., it is defined by

$$K_{OW} = \frac{C_o}{C_w} \qquad (6.11c)$$

where $K_{OW}$ is in units of m³ water/m³ octanol.

Octanol, or more properly 1-octanol, was chosen as a correlating substance because its carbon-to-oxygen ratio is similiar to that of lipids and, in general, mimics the dissolution of solutes in organic matter. $K_{OC}$ is also a direct measure of hydrophobicity, i.e., the tendency of a chemical to partition out of water, and is consequently an inverse measure of the solubility of a substance in water. The higher the $K_{OC}$, the greater the effect of a chemical on an animal. A short list of $K_{OC}$ values for various substances appears in Table 6.7.

The following serves as an example of the application of octanol–water partition coefficients: Suppose it is desired to estimate the effect on fish of the pesticide DDT dissolved in water. The lipid content of most fish clusters about a value of 4.8%. We can then define a fish–water partition coefficient $K_{FW}$ and relate it to $K_{OC}$ as follows:

$$K_{FW} = 0.048 K_{OW} \qquad (6.12a)$$

This relation expresses the assumption that a fish is composed of 4.8% by volume octanol. Using the value for DDT listed in Table 6.7, we obtain

$$K_{FW} = \frac{C_F}{C_W} = 0.0048 \times 1.6 \times 10^6 = 7.7 \times 10^4 \qquad (6.12b)$$

**TABLE 6.7**

Octanol–Water Partition Coefficients for Various Substances

| Solute | $K_{OC}$ (m³/m³) |
| --- | --- |
| *n*-Hexane | 13,000 |
| Benzene | 135 |
| Styrene | 760 |
| Chloroform | 93 |
| Carbon tetrachloride | 440 |
| Chlorobenzene | 630 |
| DDT | $1.6 \times 10^6$ |
| Range of PCBs | $10^4$ to $10^8$ |

i.e., an almost *100,000-fold* increase in concentration over that in the surrounding water. Since DDT solubility in water is $1.2 \times 10^{-3}$ mg/l, the concentration in the fish rises to $7.7 \times 10^4 \times 1.2 \times 10^{-3} = 92$ mg/l, or approximately 1/10 g in a fish of 1-l volume.

The reader is reminded that this is the maximum attainable in the absence of metabolic degradation. The steady-state bioconcentration factors (BCF), some examples of which appeared in Table 5.9, are considerably lower but still sufficiently high to cause concern.

## 6.3 Multicomponent Equilibria: Distribution of Several Components

### 6.3.1 The Phase Rule

The reader will have noted that in the equilibria considered so far, temperature was assumed to be constant or fixed. The wider question of how many such variables have to be prescribed to define a particular state of equilibrium was not addressed. It was tacitly assumed that, once a temperature and the concentration or pressure in one phase was chosen, a unique composition in the second phase was automatically assured. This approach worked without difficulty in the simple equilibria we have considered so far, but becomes somewhat tenuous when more complex systems are to be dealt with. What is required here is a formalism that will tell us exactly how many variables have to be fixed to assure a unique state of equilibrium. That formalism is provided by the Gibbsian phase rule, which states that the number of variables $F$ to be prescribed equals the difference of the number of components and phases $C - P$, plus 2. Thus,

$$F = C - P + 2 \tag{6.13a}$$

$F$, which is also referred to as the degrees of freedom of the system, is a measure of the latitude we have in arbitrarily assigning values to the independent variables of the system. Let us see how this rule is applied in practice.

### Illustration 6.9: Application of the Phase Rule

We start by examining several simple equilibria, which have been dealt with previously.

1. *Pure Component Vapor–Liquid Equilibrium*

Here the phases number 2 and the components 1. Hence

$$F = C - P + 2 = 1 - 2 + 2 = 1 \qquad (6.13b)$$

and we are allowed only one degree of freedom. Thus, if we choose for water a temperature of 100°C, a corresponding *unique* pressure of 101.3 kPa results, which is set by the system itself. Note that in the supercritical region only one phase exists and as a consequence arbitrary values can be assigned to both $T$ and $P$.

### 2. The Triple Point

Systems at the triple point contain three phases, solid, liquid, and vapor, all in equilibrium with one another. Consequently,

$$F = C - P + 2 = 1 - 3 + 2 = 0 \qquad (6.13c)$$

and the degrees of freedom vanish. Thus, we cannot, for example, prescribe a triple-point temperature and expect the system to respond with a particular triple-point equilibrium pressure. The system itself sets the values of both $T$ and $P$, neither of which is controlled or set by the observer.

### 3. Binary Vapor–Liquid Equilibria

Consider next a two-component liquid solution that is in equilibrium with its vapor. Here we have

$$F = C - P + 2 = 2 - 2 + 2 = 2 \qquad (6.13d)$$

Thus, we can fix, for example, temperature and the vapor composition, and expect the system to set its own values of liquid composition as well as total pressure. If, on the other hand, we choose to prescribe pressure and vapor composition, the system will respond with a particular temperature, i.e., its boiling point, as well as a particular liquid composition. A third possibility is to fix both temperature and total pressure, in which case the system will set its own values of both liquid and vapor composition. All three cases are encountered in practice, and are expressed in terms of appropriate phase diagrams, which are taken up below.

## 6.3.2 Binary Vapor–Liquid Equilibria

In the phase equilibria considered so far, the principal focus has rested on the distribution of a single key component, usually referred to as the solute, between the constituent phases. Thus, in the gas–liquid and liquid–solid equilibria taken up in Sections 6.2.1 and 6.2.2, our concern was with only one of the components present, while the remaining bulk components, such as liquid solvent or solid adsorbent, were left out of consideration.

In vapor–liquid equilibria, all components participate in some measure in determining equilibrium behavior. No single substance dominates the picture, and none is relegated to the status of a passive constituent. Although these considerations apply to multicomponent vapor–liquid equilibria, our discussions here are confined for illustrative purposes to binary systems. These systems possess, as shown in Illustration 6.9, two degrees of freedom, and this fact has led to the development of several distinct phase diagrams. In these diagrams, which are discussed in detail below, either temperature or pressure have prescribed constant values, along with one of the phase compositions. Once these two variables are fixed, the system is uniquely defined and all other variables fall into place. The graphical representations that result all have their own distinct features, which are exploited in different ways to suit the needs of the user.

### 6.3.2.1   *Phase Diagrams*

The principal phase diagrams used to describe binary vapor–liquid equilibria are three in number. In the first of these, total pressure and one of the compositions are the prescribed variables. This leads to the *boiling-point diagram*, shown in Figure 6.17a, and the compositional *x–y diagram* that appears in Figure 6.17c. When total pressure is replaced by temperature as the prescribed variable, a third type of diagram results: the *vapor-pressure diagram*, shown in Figure 6.17b.

The boiling-point diagram of Figure 6.17a provides the best global representation of binary vapor–liquid equilibria. It consists of a plot of temperature against both the vapor and liquid compositions, expressed as mole fractions. The lens-shaped domain encompasses the two-phase region, while the spaces above and below the lens denote single-phase vapor and liquid behavior, respectively.

Consider first a point $P$ below the lens. This point lies entirely in the liquid region. The system is below the boiling point corresponding to the prescribed total pressure $P_T$ and the vapor phase is completely absent. If we next move up the vertical axis $PQ$, i.e., raise the temperature, a point $A$ is eventually reached where the first bubble of vapor is formed. That point, which lies on the *bubble-point curve*, is in equilibrium with a vapor of composition $A'$, which is richer in the component $A$ than the bulk composition of the parent liquid. As more of the mixture is vaporized, more vapor forms at the expense of liquid, giving rise, for example, to the liquid composition corresponding to the point $K$ and a vapor composition denoted by $L$. The line connecting $K$ and $L$, and similar horizontal lines connecting points on the two curves, are termed *tie-lines*. They play the same role as the tie-lines we have seen in liquid–liquid equilibria and obey the lever rule we have derived in Illustration 6.6. Thus, the moles vapor $V$ and the moles liquid $L$ present in the two-phase mixture stand in the ratio of the corresponding line segments; i.e., we have

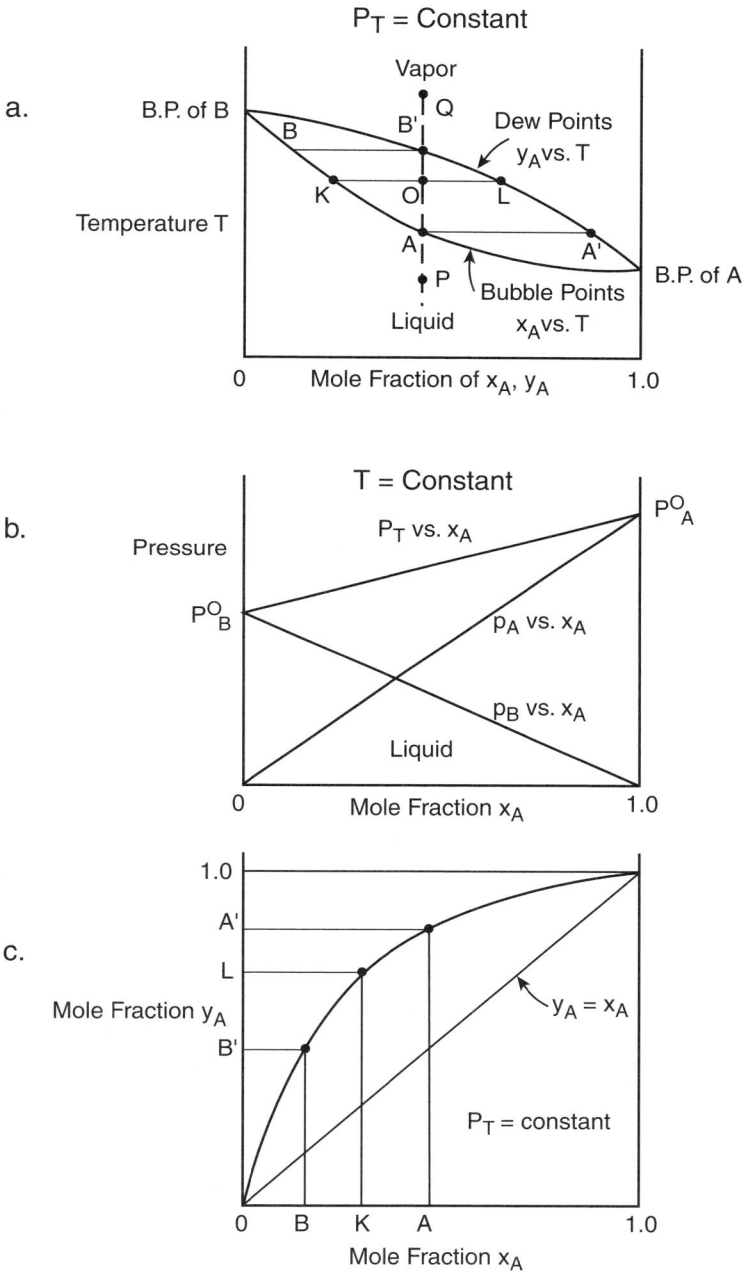

**FIGURE 6.17**
Binary vapor–liquid equilibria of ideal systems: (a) boiling-point diagram; (b) vapor-pressure diagram; (c) x–y diagram.

$$\frac{V}{L} = \frac{\overline{LO}}{\overline{KO}} \qquad\qquad (6.14)$$

Continuing our path along the $PQ$ axis, a point $B$ is eventually reached where the last drop of liquid is vaporized. Further increases in temperature result in superheated vapor. If the process is now reversed, i.e., the vapor mixture is cooled, all the phenomena reappear in reverse order. Condensation, for example, starts at point $B$, yielding the first drop of liquid "dew." The upper curve of the lens where this occurs is consequently termed the *dew-point curve*. The reader will note the similarity of this diagram to the melting-point diagram for liquid–solid systems shown in Figure 2.9a.

When temperature replaces total pressure as one of the prescribed variables, the result is the vapor-pressure diagram, examples of which are shown in Figure 6.17b and again in Figure 6.20b (see Section 6.3.3). The plot here is one of vapor pressure at constant $T$ against one of the liquid mole fractions. Both the component partial pressures $p_A$ and $p_B$ and the total pressure $P_T$ are plotted and result in straight lines or curves, depending on whether the system is ideal or nonideal. Ideal behavior, and deviations from it, are taken up in greater detail in the next section.

A third type of plot, the $x$–$y$ diagram, is shown in Figure 6.17c as well as in Figure 6.20c (see Section 6.3.3). The plotted quantities here are the vapor and liquid mole fractions of a particular component, while total pressure is held constant at a prescribed level. These diagrams play a dominant role in the analysis of distillation processes and will be encountered in considerable numbers in the succeeding chapters.

### 6.3.2.2    Ideal Solutions and Raoult's Law: Deviation from Ideality

The prediction and, failing that, the correlation of vapor–liquid equilibria are topics of considerable practical interest. In particular, we wish to address the following question: Given a prescribed liquid composition and certain standard physical properties of a system, is it possible to predict the corresponding vapor composition? Evidently, if this could be done, an immense amount of experimentation could be dispensed with. It turns out that this is accomplished most easily if the system shows what is termed *ideal behavior*. Such ideal systems, although relatively rare in practice, serve as a convenient reference for vapor–liquid equilibria in general. This is reminiscent of the concept of an ideal gas, which provides a similar yardstick against which the behavior of gases in general can be measured. These two cases, although similar in the approach used, differ in some important aspects.

For ideal gases, a total absence of intermolecular forces is assumed. Neither attractive nor repulsive forces are taken into account and collisions between molecules are taken to be entirely elastic in nature. These assumptions hold well at low pressures because of the large distances between particles and the vanishingly short time they spend in collision with each other. In liquids, the molecules are closely packed and in constant intimate contact with each other. To neglect intermolecular forces in this instance would be grossly unrealistic. What is done instead is to accept their existence and to assume that they are uniformly constant and independent of concentration. Thus, if in a binary solution a molecule $A$ is replaced by a molecule $B$, the interaction between neighboring particles remains the same. This will clearly be the case only if the two species are similar in chemical structure and in size. The important consequence of this assumption is that the partial vapor pressure of a particular component is unaffected by the presence of other species and is subject only to its own molar concentration. Thus, if the mole fraction $x_A$ is doubled, the partial pressure $p_A$ of the component is similarly doubled. Continuing this process, a linear increase of partial pressure with mole fraction results until ultimately, at $x_A = 1$, the partial pressure equals the full pure component vapor pressure $P_A^o$ of the species. We can consequently write

$$p_A = x_A P_A^o \tag{6.15a}$$

where the product $x_A P_A^o$ expresses both the linear increase in partial pressure with mole fraction and the ultimate attainment of the full vapor pressure $P_A^o$. This expression is known as Raoult's law, and systems obeying it are said to be *ideal solutions*.

Several important subsidiary laws flow from this relation. In the first instance, we can extend Raoult's law to the second component of the binary mixture and obtain

$$p_B = x_B P_B^o = (1 - x_A)P_B^o \tag{6.15b}$$

Adding the two expressions then yields

$$p_A + p_B = P_T = x_A P_A^o + (1 - x_A)P_B^o \tag{6.15c}$$

where $P_T$ is the total pressure.

All three of these expressions yield straight line plots of vapor pressure vs. liquid mole fraction, which are displayed in Figure 6.17b. Systems that obey these relations, i.e., ideal solutions, are relatively uncommon and are usually confined to neighboring substances taken from a homologous series, and to isomer and isotope mixtures. We address these in the context of the so-called separation factor of the system, which is taken up in a subsequent section.

Deviations from ideal behavior occur when there is a marked difference in the molecular structure of the participating species. Typical combinations that give rise to nonideal behavior are pairs of polar and nonpolar substances, in which the attractive and repulsive forces vary with compositional changes. When repulsive forces predominate, the vapor pressures rise above the values predicted by ideal solution theory and we speak of *positive deviations*. Most vapor–liquid equilibria fall in this category. A predominance of attractive forces, on the other hand, leads to a lowering of the vapor pressure, which is termed a *negative deviation*. The resulting nonideal vapor-pressure diagrams are displayed in Figure 6.18.

We note two interesting limiting cases that arise in systems with both positive and negative deviations: For low concentrations of a particular component, its partial pressure becomes linear in the mole fraction of the solution in question. This, in essence, is a Henry's-law-type relation; i.e., we can write

$$\left(p_A\right)_{x_A \to 0} = H_A x_A \tag{6.16a}$$

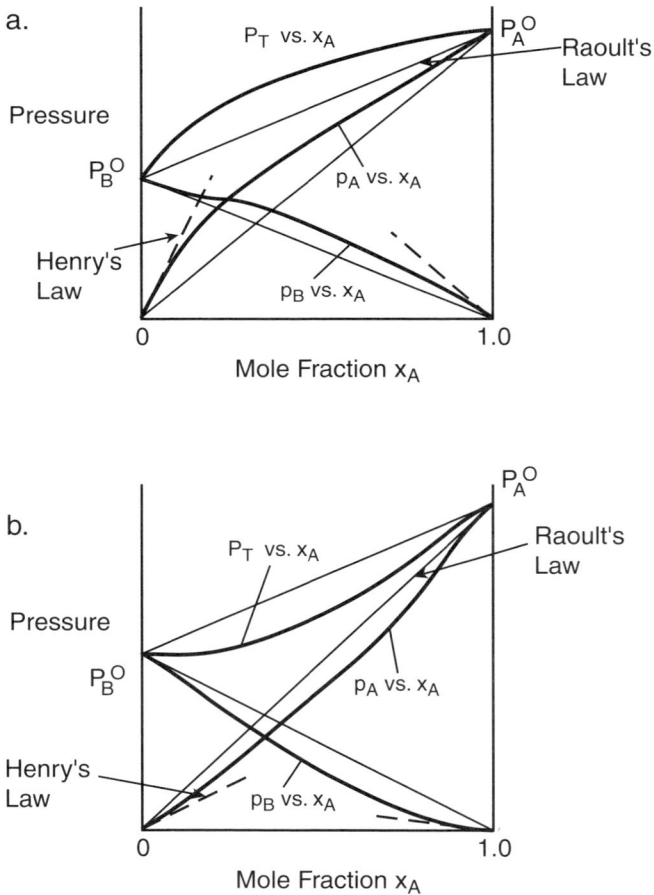

**FIGURE 6.18**
Vapor pressure diagrams for nonideal systems: (a) positive deviations from ideality; (b) negative deviations.

and

$$\left(p_B\right)_{x_B\to 0} = H_B x_B \tag{6.16b}$$

where $H_A$ and $H_B$ are the Henry's law constants for the two components in question. If, on the other hand, the mole fractions are allowed to tend to unity, the partial pressure will, in the limit, approach the values given by Raoult's law. We have in this case

$$\left(p_A\right)_{x_A\to 1} = x_A P_A{}^o \tag{6.16c}$$

and

$$\left(p_B\right)_{x_B\to 1} = x_B P_B{}^o \tag{6.16d}$$

These two cases, which have been entered in the diagrams of Figure 6.18, provide useful approximations in the limit of low-solute concentrations or high-solvent content.

### 6.3.2.3 Activity Coefficients

The reader will recall that deviations from ideal gas behavior are often expressed in terms of a correction factor referred to as a compressibility factor $z$; i.e., we write

$$PV = z(P_r, T_r)RT \tag{6.17}$$

where $P_r$ and $T_r$ are the so-called reduced pressure and temperature, defined as the ratios of $P$ and $T$ to their critical counterparts $P_C$ and $T_C$.

In much the same way we can define correction factors $\gamma$, termed *activity coefficients*, which describe deviations from Raoult's law for ideal solutions. The resulting expression, sometimes referred to as extended Raoult's law, has the form

$$p_A = \gamma_A(x_A)x_A{}^o P_A{}^o \tag{6.18a}$$

and

$$p_B = \gamma_B(x_B)x_B P_B{}^o \tag{6.18b}$$

where the correction factors are now functions of the binary mole fraction $x$. That functional relationship, expressed in the form of plots of ln $\gamma$ vs. $x_A$,

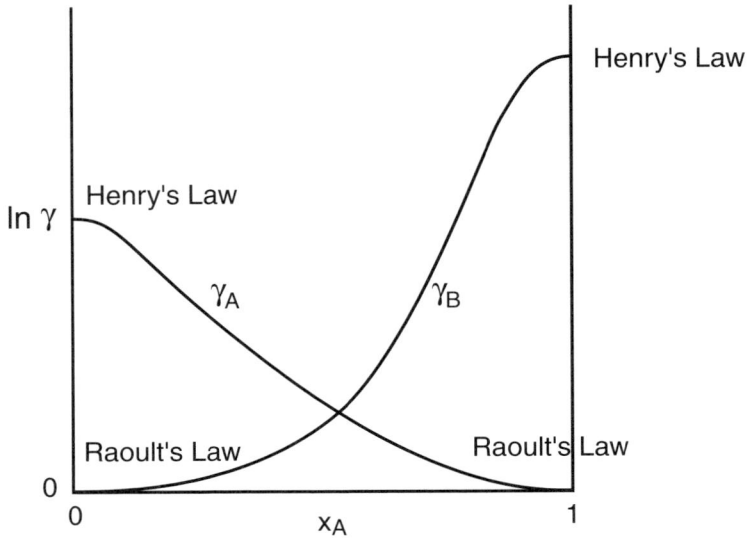

**FIGURE 6.19**
Activity coefficient for systems with positive deviations.

is shown in Figure 6.19. Note that in the limit of high-solvent content, ln $\gamma$ approaches zero, i.e., $\gamma \to 1$. At the same time, Equation 6.18a and Equation 6.18b revert to Raoult's law. At the other end of the spectrum, i.e., at low-solute content, the activity coefficients tend to constant limiting values and Henry's law results, where $H = \gamma^{po}$.

It is evidently of considerable interest to be able to predict activity coefficients without resorting to experimentation, and immense strides have in fact been made in recent decades to accomplish this goal. Among a number of promising approaches, an analytical expression known as the UNIQUAC equation (UNIversal QUAsi Chemical equation) has received the most widespread acceptance. In this model, the activity coefficient is decomposed into two constituents, one of which, termed *combinatorial* (C), accounts for molecular size and shape differences, while the other, denoted *residual* (R), expresses effects due to molecular interactions.

Thus,

$$\ln \gamma = \ln \gamma^C + \ln \gamma^R \tag{6.19}$$

It has become possible to express $\gamma^C$ and $\gamma^R$ in terms of group contributions due to molecular subunits such as $CH_3$, $CH_2$, $CH_2O$, etc. This approach has led to the successful prediction of vapor–liquid equilibria of a large number of systems. There is evidently still an ongoing need for experimentation, but this requirement is now at a much more subdued level than would otherwise have been the case.

### 6.3.3 The Separation Factor α: Azeotropes

A convenient and consistent measure of separation is provided by the so-called separation factor or relative volatility α, which is composed of the *product* of two mole-fraction ratios. These ratios are defined in such a way as to minimize changes in α with composition and are represented by the following expression

$$\alpha = \frac{y}{x} \frac{(1-x)}{(1-y)} \tag{6.20a}$$

Here $x$ and $y$ refer to the more volatile component, i.e., the constituent with the higher pure-component vapor pressure or the lower boiling point. In general, the higher the value of α above unity, the greater the degree of separation or enrichment.

The reader will have noted that any decrease in $y/x$ toward 1 as $y$ and $x$ approach unity is neatly offset by the compensating ratio $(1-x)/(1-y)$. Thus, if $y = 0.99$ and $x = 0.98$, for example, α will still be considerably above unity.

$$\alpha = \frac{0.99}{0.98} \frac{1-0.98}{1-0.99} = 2.02 \tag{6.20b}$$

Consequently, the use of α predicts, correctly, that there is still substantial separation to be obtained even when the mole fractions are near unity.

The essence of the relative volatility is best understood by examining its effect on the $x$–$y$ diagrams shown in Figure 6.17c and Figure 6.20c. It turns out that α bears a distinct and sensitive relation to the shape of these curves. The following features in particular stand out:

1. Separation factors greater than unity result in $x$–$y$ curves that lie entirely above the 45° diagonal. The higher the value of α, the greater the distance between the two lines. This implies that separation becomes easier as the $x$–$y$ curve bulges out and away from the diagonal. Systems with $x$–$y$ curves close to the diagonal are by contrast difficult to separate.

2. Separation factors that are constant, or nearly so, result in symmetrical $x$–$y$ curves. Larger variations yield asymmetrical curves, with some portions lying far above the diagonal while others are close to it.

3. When $\alpha = 1$, vapor and liquid compositions become equal, i.e., $x = y$, and the corresponding mole fraction lies on the 45° diagonal. This condition is referred to as an *azeotrope* and is shown in Figure 6.20c. The $x$–$y$ curves for such systems will inflect across the 45° line, with portions on either side of the azeotrope lying above and below the diagonal, respectively. The portion below the line yields enrichment in the less volatile component. Above the diagonal it is the more volatile component that is enriched.

Azeotropic behavior is a common occurrence in vapor–liquid equilibria (VLE). Approximately one-third of all systems listed in standard VLE handbooks exhibit azeotropes. In general, the more dissimilar the two components, the greater the likelihood that such mixtures will be formed. Combinations of polar-nonpolar substances and those with widely differing structural features are particularly prone to azeotropic behavior. Table 6.7, which lists some of the more conventional azeotropic pairs, conveys a sense of the degree of dissimilarity that leads to the formation of azeotropes.

Several additional features of these systems are to be noted. First and foremost is the fact that mixtures that lead to azeotrope formation cannot be separated into their constituent components by simple distillation. This is best seen from the boiling-point diagram shown in Figure 6.20a, where we indicate the pathway that results from a process of repeated vaporization and condensation. Starting with a liquid feed at *F*, the mixture is brought to a boil at *K* and the first vapor (*L*) collected and condensed. The cycle of vaporization and condensation is then repeated until the azeotropic composition at *A* is reached. At this point no further enrichment by vaporization is obtained, as the compositions in the two phases are the same. The mixture continues to boil at a constant temperature, yielding a mixture of constant composition, until the liquid charge is exhausted.

Azeotropic mixtures require special methods for their separation, which usually consist of adding a third component that has the ability to "break" the azeotrope. Perhaps the most famous case is that of ethanol–water, which has an azeotropic mole fraction in ethanol of 0.8943 at atmospheric pressure (Table 6.8). Here the added component is benzene and results, on distillation, in the recovery of pure ethanol and a ternary azeotrope containing benzene. That mixture, on condensation, results in two immiscible aqueous and organic layers, which are separated and further processed by distillation.

We now turn to the question of the behavior of $\alpha$ for ideal liquid solutions. Here the Raoult's law equations (Equation 6.15a and Equation 6.15b) provide a quick answer. Dividing the two expressions, we obtain in the first instance

$$\frac{p_A}{p_B} = \frac{x_A P_A^{\,o}}{(1-x_A)P_B^{\,o}} \tag{6.21a}$$

Replacing partial pressures by $yP_T$ yields

$$\frac{y_A P_T}{(1-y_A)P_T} = \frac{x_A}{(1-x_A)} \frac{P_A^{\,o}}{P_B^{\,o}} \tag{6.21b}$$

and after some rearrangement,

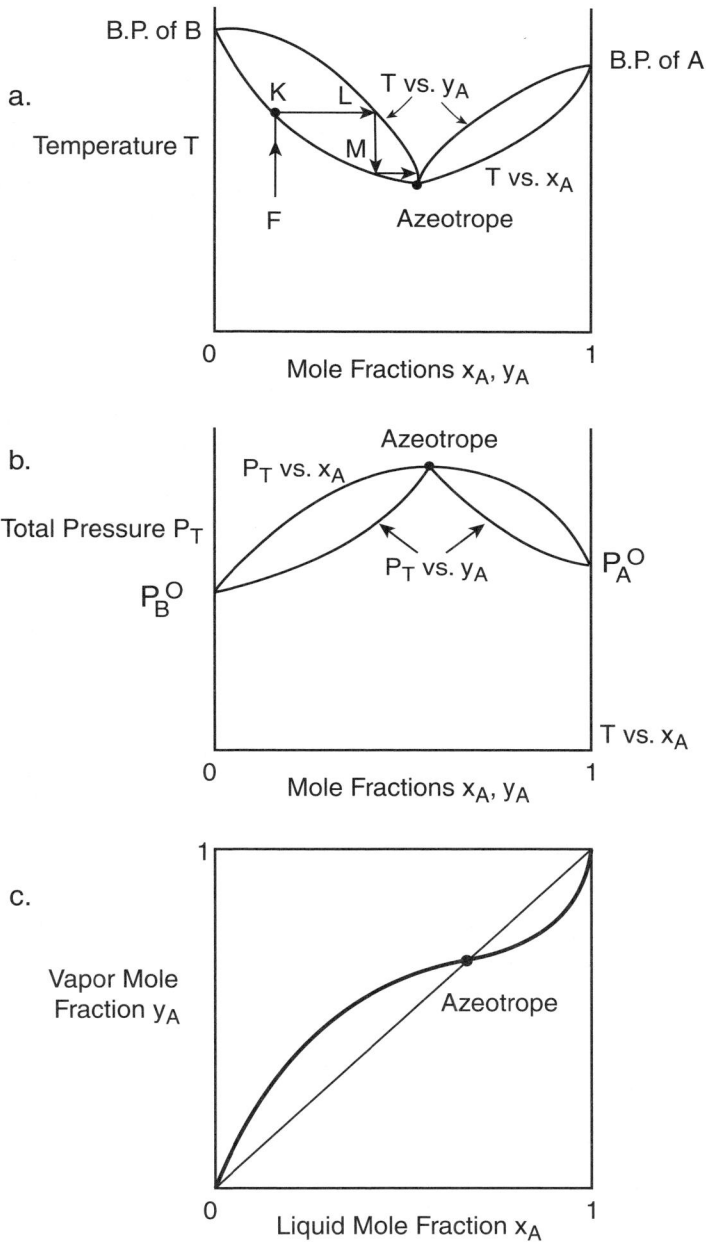

**FIGURE 6.20**
Azeotropic systems: (a) boiling-point diagram; (b) vapor-pressure diagram; (c) $x$–$y$ diagram.

**TABLE 6.8**

Binary Azeotropes at 1 atm

| System | Azeotrope Mole Fraction $x_1 = y_1$ | | Temperature $T$, °C |
|---|---|---|---|
| Water (1)–Methanol (2) | — | | — |
| Water (1)–Ethanol (2) | 0.1057 | | 78.15 |
| Water (1)–Propanol (2) | 0.5680 | | 87.80 |
| Water (1)–$i$-Propanol (2) | 0.3187 | | 80.16 |
| Water (1)–$n$-Butanol (2) | 0.75 | | 92.95 |
| Water (1)–Formic acid (2) | 0.415 | | 107.60 |
| Water (1)–Acetic acid (2) | — | | — |
| Water (1)–Dioxane (2) | 0.525 | | 87.75 |
| Water (1)–Tetrahydrofurane (2) | 0.18 | | 63.5 |
| Water (1)–Pyridine (2) | 0.77 | | 94.40 |
| Methanol (1)–Acetone (2) | 0.1980 | | 54.40 |
| Methanol (1)–Methyl acetate (2) | 0.3280 | | 53.80 |
| Methanol (1)–Ethyl acetate (2) | 0.287 | | 62.1 |
| Methanol (1)–Chloroform (2) | 0.35 | | 53.5 |
| Methanol (1)–Carbon tetrachloride (2) | 0.55 | | 55.7 |
| Methanol (1)–Hexane (2) | 0.498 | | 45 |
| Methanol (1)–Heptane (2) | 0.747 | | 58.8 |
| Methanol (1)–Benzene (2) | 0.60 | | 57.8 |
| Methanol (1)–Toluene (2) | 0.90 | | 63.6 |
| Ethanol (1)–Ethyl acetate (2) | 0.48 | | 72.1 |
| Ethanol (1)–Chloroform (2) | | Yes | |
| Ethanol (1)–Carbon tetrachloride (2) | | Yes | |
| Ethanol (1)–Hexane (2) | 0.345 | | 58.0 |
| Ethanol (1)–Heptane (2) | | Yes | |
| Ethanol (1)–Benzene (2) | 0.445 | | 67.8 |
| Ethanol (1)–Toluene (2) | | Yes | |
| Acetone (1)–Chloroform (2) | 0.3600 | | 64.4 |
| Acetone (1)–Carbon tetrachloride (2) | 0.93 | | 56.0 |
| Acetone (1)–Carbon disulfide (2) | 0.34 | | 39.10 |
| Acetone (1)–Pentane (2) | 0.25 | | 32 |
| Acetic acid (1)–Octane (2) | 0.675 | | 105.1 |
| Benzene (1)–Cyclohexane (2) | 0.55 | | 77.5 |

$$\alpha = \frac{y_A}{x_A} \frac{(1 - x_A)}{(1 - y_A)} = \frac{P_A^{\,o}}{P_B^{\,o}} \qquad (6.21c)$$

Thus, for ideal solutions, $\alpha$ stands in the simple ratio of the vapor pressure of the more volatile component to that of its less-volatile counterpart. This ratio varies somewhat with temperature but the variation is not severe and is often accounted for by composing the arithmetic or geometric average of the two end values at $x = 0$ and $x = 1$.

To provide the reader with a sense of the validity of Equation 6.21c, we have composed in Table 6.9 a comparison of calculated values with the experimental range of $\alpha$ values taken over the span of measured mole frac-

**TABLE 6.9**

Separation Factors for Some Near-Ideal Systems at 1 atm

| System | Experimental Range of α | Range of $P_1^\circ/P_2^\circ$ |
|---|---|---|
| *Homologues* | | |
| Methanol (1)–Ethanol (2) | 2.20–1.88 | 1.67–1.76 |
| Methanol (1)–Propanol (2) | 3.55–3.30 | 3.20–3.97 |
| Methanol (1)–*i*-Propanol (2) | 2.34–1.64 | 1.91–2.17 |
| Ethanol (1)–*i*-Propanol (2) | 1.15–1.21 | 1.17–1.27 |
| Ethanol (1)–*n*-Butanol (2) | 4.21–5.84 | 3.95–5.92 |
| Ethanol (1)–sec-Butanol (2) | 2.48–2.54 | 2.20–2.37 |
| Ethanol (1)–1-Pentanol (2) | 8.96–7.45 | 7.15–10.9 |
| Formic acid (1)–Acetic acid (2) | 1.65–1.41 | 1.62–1.72 |
| Chloroform (1)–Carbon tetrachloride (2) | 1.41–2.11 | 1.62–1.72 |
| Acetone (1)–Methyl ethyl ketone (2) | 1.56–2.23 | 1.71–2.22 |
| Benzene (1)–Toluene (2) | 2.23–2.26 | 2.36–2.60 |
| *Isotopes* | *T, K* | $P_1^\circ/P_2^\circ$ |
| $H_2$–HD | 20.4 | 1.73 |
| $C^{12}O$–$C^{13}O$ | 68.3 | 1.0109 |
| $O^{16}O^{16}$–$O^{16}O^{18}$ | 69.5 | 1.01 |
| $CH_4$–$CH_3D$ | 90 | 1.0025 |
| $N^{14}O$–$N^{15}O$ | 109.5 | 1.032 |
| $H_2O$–HDO | 373.1 | 1.026 |

tions. The systems involved are those that can be expected to behave ideally, i.e., members of a homologous series, isotope pairs, and the like. For isotope mixtures, experimental α values are rare and we have instead reported the vapor-pressure ratio, which for these substances is nearly constant and equal to the relative volatility.

## Illustration 6.10: The Effect of Total Pressure on α

In principle, distillation can be carried out over a wide range of pressures, spanning the extremes of near-total vacuum and the critical point. While atmospheric distillation is the preferred mode of operation, reduced pressures offer the advantage of lower temperatures and energy requirements. They are often employed to avoid thermal degradation of heat-sensitive materials (ethylbenzene/styrene distillation) or are forced on the operator by the low vapor pressure of the charge (metal distillation). The drawbacks of this mode of operation are the obvious cost and inconvenience of a vacuum process and the lowering of material throughput. Distillation above atmospheric pressures is often employed with volatile charges containing dissolved gases and is routinely used in refinery operations. It shows superior throughput but has higher operating temperatures and energy requirements than atmospheric distillation.

An important factor to be taken into account in any overall assessment is the effect of total pressure on the separation factor α. We can explore this dependence for the simple case of ideal solutions by first drawing on Equation 6.21c

$$\alpha = \frac{y_1(1-x_1)}{x_1(1-y_1)} = \frac{P_1^0}{P_2^0} \qquad (6.21c)$$

and then using the Clausius–Clapeyron equation (Equation 6.1g) to relate $P^o$ to $T$:

$$\log \frac{P_1^0}{P_2^0} = (B_2 - B_1)/T + (A_1 - A_2) \qquad (6.22)$$

It is clearly seen from the latter that the vapor-pressure ratio and hence the separation factor both increase with a decrease in boiling point or total pressure. This result has sometimes been cast into a sweeping rule-of-thumb that low-pressure distillation leads to improved separation. Many nonideal systems follow this rule, but there are also numerous exceptions, particularly in the azeotropic category. Nevertheless, low-pressure distillation is a worthwhile alternative to explore, provided some low-pressure equilibrium data are available to confirm the expected results.

*Comments:*

Even ideal systems have been known to defy the rules. The isotope pair $CH_4$–$CH_3D$, for example, has a separation factor of 1.001 at approximately 100 mmHg, which *rises* to 1.0035 at 1 atm. This anomaly led to the suggestion that heavy water ($D_2O$) might be produced economically by high-pressure distillation of liquefied natural gas, which contains the isotope $CH_3D$. Because the capital cost of such a plant varies inversely with $\alpha - 1$ (see Illustration 7.9), even small increases in $\alpha$ can be hugely beneficial. Some exploratory work took place in the 1960s, and although the results appeared promising, market forces intervened to bring the proceedings to a halt. The large-scale use of heavy water is nowadays confined to the CANDU nuclear reactor, which faces an uncertain future.

### Illustration 6.11: Activity Coefficients from Solubilities

Activity coefficients are generally either determined from Equation 6.18a and Equation 6.18b using measured vapor- and liquid-equilibrium compositions or are estimated from the UNIQUAC equation mentioned earlier. A third method arises when the two components in question have low mutual solubilities. $\gamma$ values are then confined to the Henry's law regions near $x = 0$ and $x = 1$ (Figure 6.19) and cease to exist outside those regions. This provides a way of calculating activity coefficients using classical chemical thermodynamics.

Suppose an organic solvent with a low solubility is equilibrated with water. The aqueous phase will then contain small amounts of the solvent, while water will appear in similar amounts in the organic phase. Because the concentrations involved are vanishingly small, we can approximate the chemical potential of each major constituent by its *pure component* chemical potential $\mu^o$ (T.P). We can write:

$$\mu_{H_2O} \cong \mu^o_{H_2O}(T.P) = \mu^o_{H_2O}(T.P) + RT(\ln \gamma \, x_s)_{H_2O} \tag{6.23a}$$

and

$$\mu_{org} \cong \mu_{org}(T.P) \cong \mu^o_{org}(T.P) + RT(\ln \gamma \, x_s)_{org} \tag{6.23b}$$

This leads to the result

$$\gamma_{org} \cong 1/(x_s)_{org} \tag{6.23c}$$

and

$$\gamma_{H_2O} \cong 1/(x_s)_{H_2O} \tag{6.23d}$$

where $x_s$ represents the saturation solubility, expressed as a mole fraction.

We illustrate these results by examining the system carbon tetrachloride–water. The solubility of $CCl_4$ in water is listed in Table 6.3 as 0.8 g/l $\cong$ 800 ppm. The solubility of water in carbon tetrachloride is almost 10 times lower at 84 ppm. Converting to mole fraction, we obtain

$$(x_s)_{CCl_4} \cong 800(18/154)10^{-6} = 9.4 \times 10^{-5}$$

and

$$(x_s)_{H_2O} \cong 804(154/18)10^{-6} = 7.2 \times 10^{-4}$$

Taking the inverse of these results, we obtain for the activity coefficients:

In the aqueous phase: $\qquad \gamma_{CCl_4} \cong 11,000$

In the organic phase: $\qquad \gamma_{H_2O} \cong 1,400$

These are enormously high values, which indicate that the equilibrium vapor mole fraction will by far exceed that of the liquid (see Practice Problem 6.13).

There is another simplification to be made: Since $\gamma x_s$ is very nearly unity, partial pressures will be closely approximated by the pure component vapor pressures, which can be computed from the Antoine constants of Table 6.1. We have

For the aqueous phase:

$$p_{CCl_4} = (\gamma\, x_s\, P^o)_{CCl_4} \cong P^o{}_{CCl_4} = 86.8\,\text{mmHg}$$

For the organic phase:

$$p_{H_2O} = (\gamma\, x_s\, P^o)_{H_2O} \cong P^o{}_{H_2O} = 23.6\,\text{mmHg}$$

The results for $CCl_4$ are particularly noteworthy in an environmental context. The high partial pressure indicates that a water basin contaminated with carbon tetrachloride will quickly lose most of the contaminant to the atmosphere by evaporation. We return to this topic in Illustration 7.4.

## Practice Problems

6.1. *The Vapor Pressure of Ice and Snow*

Ice has a substantial vapor pressure, which causes it to evaporate at surprisingly high rates. Most of the disappearance of snow during a lull in precipitation is due to evaporation rather than melting.

a.  Suppose that a shallow puddle of water with a surface temperature of 20°C evaporates in 2 h during a dry windy day. How long would it take the same amount of ice at −5°C to evaporate under identical wind and humidity conditions? The vapor pressures of the water and the ice at their respective temperature are 17.5 and 3.0 mmHg.

b.  Estimate the rate of evaporation of ice.

*Answer:* 11.7 h, 0.032 g/m² s

6.2. *Oxygen Content of Lake Waters*

Calculate the maximum total oxygen content in a lake containing $10^7$ m³ water.

*Answer:* 85.1 tons

6.3. *Water Intake of Fish*
The oxygen required by a 2.5-kg fish amounts to approximately 50 g/day. Calculate the minimum water intake by the fish to supply the necessary amount of oxygen.

*Answer*: 4.1 l/min

6.4. *Sea Salt by Evaporation*
Sea salt contains 454 mmol/kg NaCl and 9.6 mmol/kg KCl. What percentage of the water has to be evaporated for the first crystals of KCl to precipitate? The total solids content of sea water is approximately 2%. (*Hint*: Consult Table 6.3.)

*Answer*: 57.1%

6.5. *More about Adsorption: The Toth Equation Applied to Air Purification*
The Langmuir isotherm, while adequate for many purposes, has since its inception been superseded by more refined equations with a broader range of applications. One such equation is that due to Toth, which is given by

$$X = \frac{ap}{(b + p^C)^{1/C}} \tag{6.24a}$$

Consider a factory air space of $10^5$ m³, which is regularly contaminated with 10 ppm by volume of hydrogen sulfide, a toxic and foul-smelling gas. It is proposed to use adsorption to purify the air. Activated carbon does not lend itself to this purpose because of its low capacity, but the zeolite *mordenite* has a suitably high loading factor represented by the following Toth isotherm:

$$X = \frac{4.5675\,p}{[0.3252 + p^{0.2425}]^{1/0.2425}} \tag{6.24b}$$

where $p$ is in units of kPa, and $q$ in mol/kg adsorbed. Calculate the minimum weight of mordenite required to purify the air.

*Answer*: 561 kg

6.6. *Ion-Exchange Resins*
The process of ion exchange relies on the use of synthetic organic resins, which carry ionic groups capable of exchanging cations or anions with similar species in an aqueous solution. The resins come in the form of small beads (2 to 3 mm) made up of a polymeric skeleton into which the ionic groups are introduced by suitable

chemical reactions. In one version of this process, styrene is polymerized in emulsion to provide the resin matrix, which is then sulfonated, thereby introducing sulfonic acid groups $-SO_3^-H^+$ into the polymer skeleton. We can describe this process in the following form:

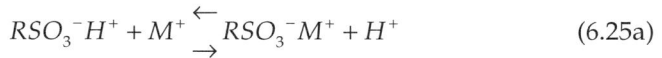

$$RSO_3^-H^+ + M^+ \underset{\rightarrow}{\overset{\leftarrow}{\phantom{x}}} RSO_3^-M^+ + H^+ \qquad (6.25a)$$

The hydrogen ion of this group can be readily released and its place taken by another cation from a neighboring aqueous solution. Anion exchangers use amine groups $R_3N^+OH^-$ to affect a similar exchange of anions. Thus,

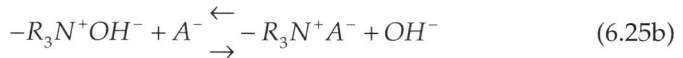

$$-R_3N^+OH^- + A^- \underset{\rightarrow}{\overset{\leftarrow}{\phantom{x}}} -R_3N^+A^- + OH^- \qquad (6.25b)$$

Ion-exchange resins find extensive use in the deionization of water and in the recovery of valuable ions from aqueous solutions. In water-softening processes, for example, calcium ions contained in the hard water are exchanged for hydrogen ions released by the sulfonic acid groups, while the anions are replaced by hydroxyl ions provided by the amine groups. The net result is the replacement of the calcium salts by water molecules and an attendant softening of the process water. Ion-exchange resins behave in much the same way as adsorbents, with ion uptake increasing in proportion to the concentration in the contacting solution and ultimately leveling off at a saturation value when all ions in the resin have been exchanged. The phase equilibrium is consequently well-described by a Langmuir-type isotherm. To account for the ionic nature of the species, it is customary to express both resin loadings and aqueous concentrations as milliequivalents rather than as the actual weight of ions. Suppose that the uptake of copper ions $Cu^{2+}$ by a particular resin is described by the Langmuir form

$$X = \frac{11.2\,C}{1 + 3.0\,C}$$

where $C$ is in meq $Cu^{2+}/l$ and $X$ in meq $Cu^{2+}/g$ resin. Use this equation to calculate the resin loading $X$ for an aqueous solution containing 0.01 by weight of copper ions.

*Answer*: 0.322 meq/g

6.7. *Special Types of Liquid–Liquid Equilibria*
Consider the following two cases:

a.  A solute distributes itself between two partially soluble solvents and is itself only partially soluble in one of these solvents. Draw the triangular diagram for this case and locate the plait point.

b.  A solute distributed between two partially soluble solvents reverses selectivity at some intermediate concentration level. Draw the distribution curve for this case and identify the corresponding behavior in vapor–liquid equilibria.

6.8. *Decaffeination by Supercritical Extraction: The Freundlich Isotherm*
The data of Figure 6.16 showing the equilibrium distribution of caffeine between coffee beans and supercritical carbon dioxide can be fitted by a power relation termed the *Freundlich isotherm*:

$$x = ay^m \qquad\qquad (6.9a)$$

where $x$ = kg caffeine/kg beans and $y$ = kg caffeine/kg $CO_2$. This relation, an empirical one, is often used to fit adsorption equilibria in the intermediate concentration range. Use the data of Figure 6.16 to derive the values for $a$ and $m$.

6.9. *Repartitioning of a Solute between Compartments*
A solute is partitioned among three compartments with volumes of $V_1 = 100$, $V_2 = 50$, $V_3 = 10$ (arbitrary units) and partition coefficients of $K_{12} = 10^{-2}$ and $K_{13} = 10^{-3}$. Degradation is extremely slow and the system is considered to be at a quasi-equilibrium. If the concentration in compartment 1 undergoes a sudden increase from $10^{-3}$ to $10^{-2}$, what will be the ultimate new equilibrium concentrations, assuming no degradation takes place?

*Answer*: $C_1 = 2.13 \times 10^{-3}$

6.10. *Vapor–Liquid Separation Factors for Light and Heavy Water*
Monodeuterated water HDO has an abundance in natural water of approximately 600 ppm. One of the methods used in early attempts to separate the two isotopes was fractional distillation, for which the separation factor at 1 atm was found to be 1.026. Although this procedure was ultimately superseded by the more efficient chemical-exchange processes for larger-scale production of heavy water, distillation remains the separation method of choice for upgrading small amounts of heavy water that have been contaminated by atmospheric water vapor. Distillation is in this case carried out at reduced pressure to take advantage of the higher separation factor. Suppose we wish to carry out the distillation of $H_2O$–HDO at ambient temperatures, i.e., at subatmospheric

pressures. Given that the separation factor at 40°C is 1.056, calculate its value at a temperature of 25°C.

*Answer*: 1.069

*Note:* The actual distillation process at high $D_2O$ concentrations involves the species $HDO-D_2O$. Separation factors of this binary pair are nearly identical to those for $H_2O-HDO$.

6.11. *More on Nonideal Systems*
Consider the case of a nonideal liquid solution in which attractive forces outweigh the repulsive. Describe how this changes the diagrams shown in Figure 6.18 and Figure 6.20.

6.12. *Steam Distillation*
It often happens, particularly in the food and pharmaceutical industries, that a heat-sensitive liquid has to be purified by distillation, or a dissolved substance has to be concentrated by boiling off solvent without risking thermal degradation. Vacuum distillation is one way of lowering the boiling point but suffers from low production rates and other disadvantages (see Illustration 6.10). A second method is steam distillation, which involves passing live steam through the solution and using it as a carrier gas to sweep off the evaporating liquid or solvent. The charge consists of partially condensed steam and a separate organic phase. The latent heat of vaporization is provided by external heating.

a. Show that the normal boiling point is always less than 100°C and increases with time.

b. Consider the steam distillation of a dilute solution of a heat-sensitive substance in toluene. Estimate its initial boiling point and the initial steam consumption in kg steam/kg toluene. (*Hint:* Use Table 6.1.)

*Answer*: a. 84.3°C, b. 0.24

6.13. *Transfer of Pollutants from a Water Basin to the Atmosphere*
To obtain a sense of the magnification in concentration, which occurs when a sparingly soluble substance evaporates into the atmosphere, calculate the so-called *K*-factor, $K = y/x$, for carbon tetrachloride in water using the data given in Illustration 6.11.

*Answer*: 8300

# 7

## Staged Operations: The Equilibrium Stage

Chapter 6 has laid the groundwork for the topic to be taken up here by examining in considerable detail the various phase equilibria that enter into the formulation of an *equilibrium stage*. The equilibrium stage, also termed an *ideal stage* or *theoretical stage,* plays a central role in the analysis of an important class of mass transfer operations termed *staged processes.* In these operations, two phases are brought into intimate contact in a stirred tank or its equivalent and the desired mass transfer process is allowed to take place. The two phases are then separated, and the process is either repeated (multistage contact) or brought to a halt (single-stage contact). While there is only one mode of single-stage contact, multistage operations can be arranged in a variety of geometrical configurations. In co-current operations, the two phases move through the stages and cascade parallel to each other and in the same direction. When the phases move in opposite directions but still run parallel, we speak of a countercurrent operation. Finally, when the two phases move at right angles to each other, the process is termed a crosscurrent operation. These three modes are sketched in Figure 7.1.

Analysis of these processes by the equilibrium-stage model proceeds in two steps. In the first step, the two phases are visualized as entering a stirred tank or its equivalent where they are allowed to come to equilibrium. The two phases are then conceptually withdrawn and separated, and the concentrations in each phase are calculated by means of appropriate mass balances and equilibrium data. We practice this step extensively in this chapter using both analytical expressions and specially designed geometrical constructions termed *operating diagrams.*

In the second step of the process, the two phases are contacted only for a finite time interval leading to *incomplete* equilibrium. This step, which has to be carried out experimentally, reflects conditions that prevail in an actual operation. The resulting concentration changes, which differ from those attained at equilibrium, are cast in the form of a fractional approach to equilibrium or stage efficiency $E$. The value of this efficiency is then grafted onto the results of the first step to arrive at an estimate of the actual prevailing concentrations.

The concepts we have just described are illustrated in Figure 7.2. Figure 7.2a represents an equilibrium stage in which $G$ kg of a carrier gas containing

a.

b.

c.

d.

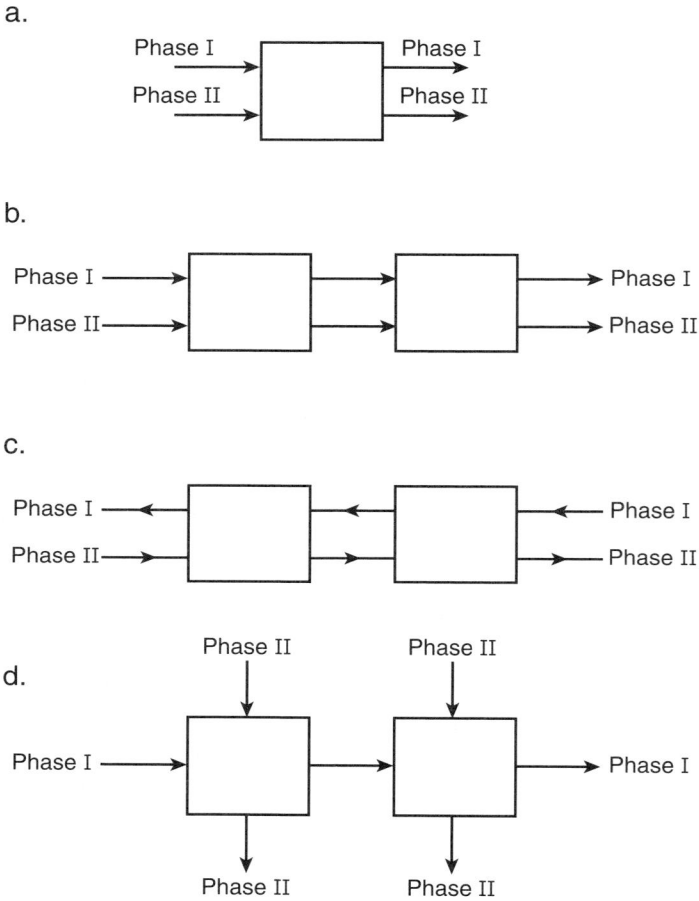

**FIGURE 7.1**
Staged operations: (a) single stage; (b) co-current; (c) countercurrent; (d) crosscurrent.

$Y_F$ kg solute/kg carrier is contacted with $L$ kg of a solvent carrying $X_F$ kg of the solute/kg solvent. The solute transfers from the gas to the liquid phase until equilibrium is attained. The phases are then withdrawn and separated, with their concentrations at their respective equilibrium values $X$ and $Y^*$. In Figure 7.2b, the same feed enters the stage but attains only partial equilibrium, with the exiting liquid concentration $X'$ falling short of the equilibrium value $X$. We express this through the stage efficiency $E$ by writing

$$X' = EX \tag{7.1}$$

Thus, when $E = 1$, the fractional approach to equilibrium is 1, and equilibration is consequently complete. When $E$ is zero, no transfer takes place and the concentration remains frozen at the level $X_F$ of the incoming feed.

a.

b.

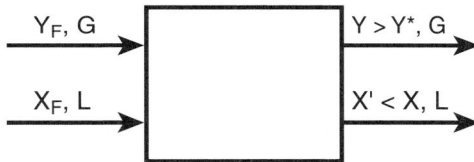

**FIGURE 7.2**
Staged contact of two phases: (a) equilibrium stage; (b) non-equilibrium or actual stage.

This chapter is divided into two parts: In the first, we take up the topic of equilibrium stages in their various configurations and apply them to a number of different mass transfer operations. The second part, which is less extensive, considers the effect of mass transfer resistance expressed through an appropriate stage efficiency.

## 7.1 Equilibrium Stages

### 7.1.1 Single-Stage Processes

The single equilibrium stage is the key unit on which the more complex configurations such as the crosscurrent and countercurrent cascades are based. We use it here to introduce the reader to some basic notions of equilibrium stage processes and to make a first presentation of a key tool, the operating diagram.

Figure 7.3 shows the flow diagrams for two single-stage processes. In Figure 7.3a, we display the streams entering and leaving a liquid phase adsorption stage. $S$ and $L$ represent the mass of solute-free adsorbent and solvent, and $X$ and $Y$ are the corresponding mass ratios in units of kg solute/ kg adsorbent and kg solute/kg solvent. Figure 7.3b shows a similar process of single-stage liquid extraction. The principal difference here lies in the

a.

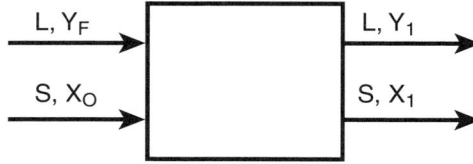

b.

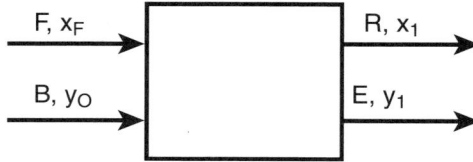

**FIGURE 7.3**
Single-stage equilibrium processes: (a) adsorption from a liquid solution; (b) extraction.

choice of units, with concentrations expressed as mass fractions $x$ and $y$, and the capitalized symbols expressing the total mass of each phase entering and leaving. The symbol $R$ denotes the *raffinate*, i.e., the solution being extracted, and $E$ stands for the *extract* containing the bulk of the solvent and the extracted solute. Our considerations here are confined to *finite* amounts of both phases. The case where differential amounts of one phase are continuously withdrawn is taken up in Section 7.1.2.

For *single-stage* operations, the following questions are to be answered:

1. With the amount and concentration of feed fixed, what is the amount of adsorbent or solvent needed to achieve a prescribed reduction in solute content?

2. Conversely, and with feed conditions again fixed, what is the concentration in the effluent if we use a given amount of adsorbent or solvent, with a known impurity level?

When dealing with *multistage* operations, the number of questions escalates. We may then ask for the number of stages required to achieve a prescribed concentration change, for example, or the most economic use of adsorbent or solvent. These will be taken up at the appropriate time.

Let us now consider the tools needed to answer the single-stage questions. In the first instance, we require a solute mass balance over the stage. For the adsorption process shown in Figure 7.3a, it takes the form

Amount of solute in − Amount of solute out = 0

$$(LY_F + SX_0) \quad - \quad (LY_1 + SX_1) \quad = 0 \qquad (7.2a)$$

Each term here is made up of the product of kg solvent $L$ or kg adsorbent $S$, which do not change during their passage, and the concentrations $X$ (kg solute/kg adsorbent) and $Y$ (kg solute/kg solvent), which do. For the extraction process of Figure 7.3b, additional balances are required because we do not have a convenient constant "carrier" mass available. This is taken up in Illustration 7.2.

The second tool required is a statement that the concentrations $X_1$, $Y_1$ leaving the stage are in equilibrium with each other. These values are obtained experimentally and are formally expressed by the relation

$$Y^*_i = f(X_i) \qquad (7.2b)$$

where the asterisk denotes equilibrium conditions.

Laboratory data of $X_i$, $Y_i$ are not always easily expressed in simple analytical form, which leads to complications in attempting to solve Equation 7.2a and Equation 7.2b. This has led to the development of a graphical construction termed an operating diagram, which is shown in Figure 7.4. This diagram is in essence a graphical representation of both the material balance and the equilibrium relation (Equation 7.2a and Equation 7.2b). It displays in vivid form the interrelation of the operational variables and enables the analyst to make various calculations of interest in rapid fashion.

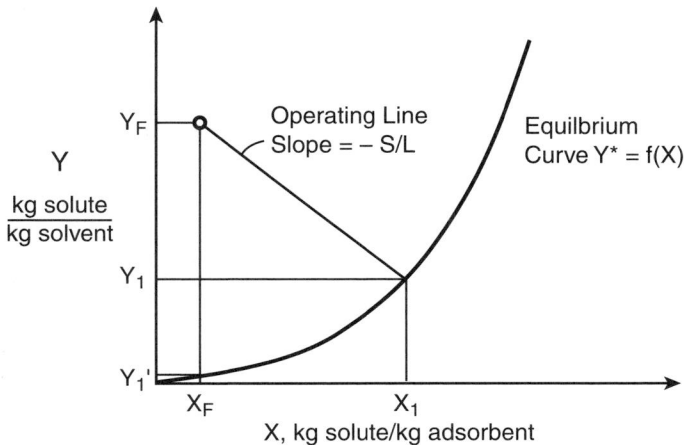

**FIGURE 7.4**
Operating diagram for single-stage equilibrium adsorption.

### *Illustration 7.1: Single-Stage Adsorption: The Rectangular Operating Diagram*

The first step in constructing the operating diagram is to plot the experimental equilibrium data obtained in the laboratory in the form of an equilibrium curve. This curve is shown in Figure 7.4 and is of the Langmuir form, where solid-phase concentration now appears on the abscissa instead of the ordinate as previously shown in Figure 6.5. This is done for greater clarity of construction.

The second step is to represent the mass balance, Equation 7.2a, which is best achieved by rearranging it in the form

$$-\frac{S}{L} = \frac{\Delta Y}{\Delta X} = \frac{Y_F - Y_1}{X_F - X_1} \qquad (7.3)$$

This is the equation of a straight line termed the *operating line*, which has a slope of $-S/L$. In this equation, the amount of solvent $L$ and the initial or feed concentrations $X_F$, $Y_F$ are known variables; the others are either prescribed or unknown.

There are several features that the diagram conveys at a glance. For example, the steeper the slope of the operating line, the greater the amount of adsorbent used and hence the higher the fractional recovery of solute, here given by the ratio $(Y_F - Y_1)/Y_F$. Note that when the amount of adsorbent becomes infinite, the recovery is not complete but instead stabilizes at the value $(Y_F - Y_1')/Y_F$. This is because of the residual impurity level $X_0$, which the incoming adsorbent carried with it. When an infinite amount of *clean* adsorbent is used, the operating line drops to the origin and recovery becomes complete.

Suppose now that, with the feed condition $Y_F$, $X_F$, and $L$ known and the amount $S$ prescribed, we wish to establish the resulting degree of recovery. This becomes a simple matter of locating the point $(Y_F, X_F)$ on the diagram and drawing a line of slope $-S/L$ through it. Its intersection with the equilibrium curve determines the concentrations of the exiting streams, $Y_1$ and $X_1$, from which we can compute the recovery $(Y_F - Y_1)/Y_F$. If, on the other hand, recovery is *prescribed* and the unknown is the amount of adsorbent required, the procedure is reversed: A line is drawn through the two known points $(Y_F, X_F)$ and $(Y_1, X_1)$ and the unknown $S$ calculated from its slope — $S/L$.

Let us turn next to the extraction process shown in Figure 7.3b. We start by noting that in the case of mutually insoluble solvents, single-stage extraction can be analyzed in exactly the same fashion as the single-stage adsorption process described above. With $A$ and $B$ denoting the mass of raffinate and extract solvent $x$, the operating diagram becomes identical to that shown in Figure7.4, with the slope of the operating line now given by $-B/A$.

In most practical cases, the raffinate and extract solvents will show some degree of mutual solubility, and the use of the $X–Y$ diagram becomes less appropriate. We must turn instead to the triangular diagrams for a full description of the extraction process. The initial tools are again two in number, i.e., a solute mass balance and a statement of solute equilibrium between the two liquids. Here, however, we encounter an unexpected difficulty: Both the amounts of raffinate and extract are new unknowns and require us to introduce two additional equations. One such equation is the total mass balance, which we did not require or have occasion to use in the adsorption process. A second component balance would complete our requirements, but the resulting escalation in equations is not welcome news. This complication is avoided by returning to our graphical tools and making use of some of the properties of triangular diagrams. One such property, stated in Illustration 6.6, was that when two solutions are combined, the resulting composition ($F$ in Figure 6.11) is located on the straight line connecting the concentrations of the parent solutions (points $R$ and $E$). This rule, which was not explicitly proved, springs from the requirement that the lever rule, Equation 6.8f, must hold for all three component substances. In other words, the extract-to-raffinate ratio $E/R$ must be the same, irrespective of which component mass fraction is used in Equation 6.8f. The construction of Figure 6.11 shows that this is the case, and only the case, if $F$ lies on a straight line connecting $R$ and $E$. In other words, all three component balances are silent partners in the construction of Figure 6.11. The graphical construction consequently provides the additional component balance to satisfy the algebraic requirement we had confronted earlier. We amplify these points in the illustration that follows.

### Illustration 7.2: Single-Stage Liquid Extraction: The Triangular Operating Diagram

Consider the case of a feed with a composition located at $F$ in Figure 7.5 being contacted with $B$ kg of pure solvent, located at point $B$. $M$ is the location of the mixture that results when $F$ kg feed are contacted with the solvent, and $\overline{RE}$ represents the equilibrium tie-line. Composing total and component solute mass balances as previously stipulated, we obtain

Rate in      Rate out

$$F + B = M = R + E \tag{7.4a}$$

$$x_F F = x_M M = x_R R + y_E E \tag{7.4b}$$

It is desired to calculate the equilibrium compositions that result, as well as the amounts of raffinate $R$ and extract $E$ produced. We first locate the composition $x_M$ of the mixing point $M$ by eliminating total mass $M$ from the left side of the two balances. Thus,

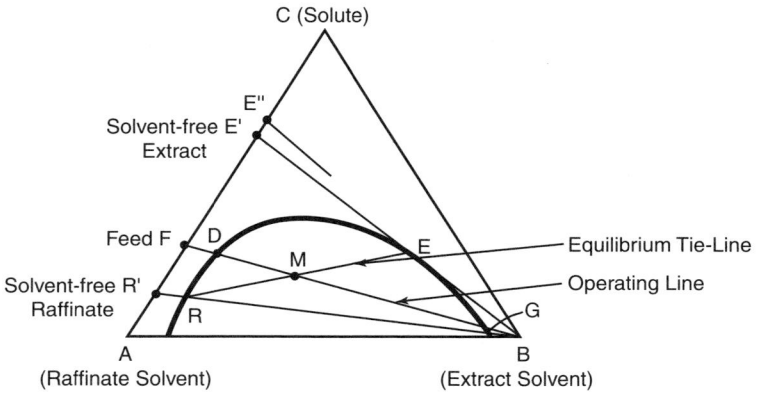

**FIGURE 7.5**
Operating diagram for single-stage equilibrium extraction.

$$x_M = \frac{x_F F}{F + B} \tag{7.4c}$$

This composition is entered on the operating line connecting $F$ and $B$.

Next a tie-line is drawn through $M$ to establish the raffinate and extract compositions at $R$ and $E$. This requires some visual interpolation between neighboring tie-lines obtained from experimental equilibrium data and completes the first part of the task. The second part is accomplished by again eliminating $M$ from the two balances, this time from the right side of Equations 7.4a and 7.4b. Thus,

$$E = \frac{(F + B)(x_M - x_R)}{Y_E - x_R} \tag{7.4d}$$

and

$$R = \frac{(F + B)(y_E - x_M)}{y_E - x_R} = F + B - E \tag{7.4e}$$

Thus, the operating diagram, together with the two algebraic mass balances, has provided us with a complete answer to the problem. The reverse task, that of calculating the solvent requirement for a prescribed solute recovery, is solved by drawing both the tie-line (which is now known) and the companion operating line, and, from their intersection, establishing the value of $x_M$. The solvent requirement is then obtained from the left side of the two mass balances as

$$B = \frac{F(x_F - x_M)}{x_M} \tag{7.4f}$$

*Comments:*

Several additional interesting features may be deduced from the diagram:

1. Removal of the solvent from the raffinate and extract, which can (for example) be accomplished by distillation, results in the solvent-free compositions located at $R'$ and $E'$. These points lie on the straight-line extrapolations of $\overline{RB}$ and $\overline{EB}$, as required by the mixing rule. We note in addition that a maximum in the solute concentration results when the product extract is located on the tangent drawn through $B$ to the solubility curve, leading to point $E''$.

2. An *increase* in the amount of solvent used results in a movement of the mixing point $M$ toward $G$, while a *reduction* in solvent causes it to approach the point $D$. When $M$ coincides with $D$, the amount of solvent is at a minimum and the amount of extract is infinitesimally small. This follows from the lever rule, Equation 6.8f. Under these conditions, any solvent present in the system resides entirely in the raffinate phase. Conversely, the point $G$ represents the maximum amount of solvent we can use and is attended by an infinitesimally small amount of raffinate. Evidently, the actual amount of solvent will lie somewhere between the two extremes. Note that these two limiting values are established from the intersection of the operating lines with the solubility curve and do not require any tie-line data.

### 7.1.2 Single-Stage Differential Operation

A special type of single-stage contact arises when one phase, which is at all times fully equilibrated with its partner, is slowly withdrawn in infinitesimally small amounts while the second phase remains within the stage and undergoes a correspondingly slow change in concentration. This process is best visualized, and most easily implemented, in the case of distillation, which is then referred to as simple distillation. The equations to be used again involve integral mass balances, but they now contain an unsteady term to reflect the slow changes in the contents of the stage.

Figure 7.6 shows the variables involved in a differential distillation process. For a binary system, they are four in number: the moles liquid in the still or boiler at any instant $W$ and its mole fraction $x_B$, the rate of vapor withdrawal $D$ (mol/s), and the instantaneous vapor composition $y_D$. The mass balances and the equilibrium relation $y_0^* = f(x_W)$ provide only three of the required equations. For the fourth we must draw on an energy balance. This stands to reason since the rate of vapor production $D$ will evidently depend on the energy input to the system. The temperature of the still will then come into

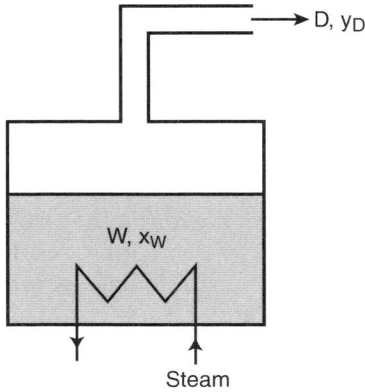

**FIGURE 7.6**
Differential or simple distillation.

play as a fifth variable, and must be accounted for by an appropriate equilibrium relation. This set of equations, which has to be solved numerically, yields the changes with time of the five variables in question.

A reduced solution to the problem can be obtained if we eliminate time as a variable and establish the instantaneous relation between the state variables, for example, $W = f(x_W)$. This is still a highly useful result because it can tell us how much of a given initial charge must be distilled to obtain a desired enrichment $x_B$. It also provides us with other items of interest and falls short only by failing to establish the full time dependence of the variables.

### Illustration 7.3: Differential Distillation: The Rayleigh Equation

We start the derivation of the model by composing the unsteady integral mass balances for the system:

$$\text{Rate of moles in} - \text{Rate of moles out} = \begin{array}{c} \text{Rate of change} \\ \text{of contents} \end{array}$$

$$-D = \frac{d}{dt}(W) \tag{7.5a}$$

and

$$-y_D D = \frac{d}{dt}(x_W W) \tag{7.5b}$$

This is supplemented by the equilibrium relation

$$y_D{}^* = f(x_W) \tag{7.5c}$$

To eliminate the time variable, we resort to a favorite trick of ours, one that is frequently used and should be part of the analyst's tool box: The two mass balances are divided, thereby eliminating not only time as a variable but also the unknown distillation rate $D$. Thus,

$$y_D = \frac{d(x_W W)}{dW} \tag{7.5d}$$

which can be combined with the equilibrium relation and expanded to yield

$$f(x_W) = x_W + W \frac{dx_W}{dW} \tag{7.5e}$$

and consequently

$$\frac{dx_W}{f(x_W) - x_W} = \frac{dW}{W} \tag{7.5f}$$

Formal integration of this expression then leads to

$$\ln \frac{W}{W^o} = \int_{x_W{}^o}^{x_W} \frac{dx_W}{f(x_W) - x_W} \tag{7.5g}$$

where the superscript $o$ denotes the initial conditions.

This expression is known as the Rayleigh equation and relates the amount $W$ left in the still at any instant to its composition $x_W$.

*Comments:*

We start by noting, as we have done on other occasions, that any of the variables appearing in Equation 7.5g may be regarded as an unknown. Thus, for a given boiler content $W$, the equation will yield the corresponding composition $x_W$ and conversely we can calculate the fraction to be distilled $1 - W/W^o$, by prescribing a desired enrichment $(x_W - x_W{}^o)/x_W{}^o$. What is often overlooked is that the equilibrium relation $f(x_W)$ can also be established from experimental $x_W$ vs. $W$ data. A particularly simple case arises when the $f(x_W)$

is linear in $x_W$, which holds at low values of the mole fraction $x$ (Henry's law region). We then have

$$\ln \frac{W}{W^o} = \int_{x_W^o}^{x_W} \frac{dx_W}{(H-1)x_W} \tag{7.6}$$

and consequently

$$\frac{W}{W^o} = \left(\frac{x_W}{x_W^o}\right)^{1/H-1} \tag{7.7}$$

This expression can be exploited to derive isotope or isomer separation factors, which here equal the Henry constant. Suppose, for example, that 95% of the charge has been boiled off, i.e., $W/W^o = 0.05$, and the isotope has been depleted by 2%, i.e., $x_W/x_W^o = 0.98$. Then $H - 1 = \alpha - 1 = \ln 0.98/\ln 0.05$ and $\alpha = 1.0067$.

We can also use Equation 7.7 as an adjunct to the *cumulative* balances for the process, which were derived in Illustration 2.5 and which we repeat here. They are

$$W^o = W + D' \tag{2.15a}$$

and

$$x_W^o W^o = x_W W + x_D' D' \tag{2.15b}$$

where $D'$ = moles in the distillate receiver. These can be used to derive the composite distillate concentration $x_D'$, which prevails at any given instant. With $W$ prescribed and the initial conditions known, we can calculate $x_W$ from Equation 7.7 and by substitution into Equation 2.15b arrive at a value for $x_D'$. Thus, with the exception of temperature and distillation rate $D$, both of which require an energy balance, we have managed to establish all pertinent variables of the system. The reader will find all these equations applied in Illustration 7.10 dealing with batch-column distillation.

### Illustration 7.4: Rayleigh's Equation in the Environment: Attenuation of Mercury Pollution in a Water Basin

We have shown in Illustration 6.10 that substances with low solubility in water, such as hydrocarbons, chlorinated organics, and mercury, can still have high partial pressures $p$ because of their exceptionally high activity coefficients $\gamma$. The relevant relation is given by the extended Raoult's law, Equation 6.18a:

$$p = \gamma x P^o \qquad (6.18a)$$

which can be recast in the form

$$y = \gamma \frac{P^o}{P_T} x \qquad (7.8a)$$

The group $\gamma P^o / P_T$ can be viewed as a partition coefficient $K$, which for the substances mentioned is generally quite high. The loss of even a small amount of solution by evaporation could thus be expected to result in a marked drop of pollutant concentration. This is small comfort for the environment as a whole since the toxic substances are merely transferred from one medium to another. Their fate nevertheless needs to be tracked.

We consider here a body of water containing dissolved mercury at the saturation level and set ourselves the task of calculating the reduction in mercury content that occurs when 0.01% of the contents is evaporated. It is assumed that the vapor in question is at all times in equilibrium with a well-mixed liquid, i.e., that Rayleigh's equation applies.

The equilibrium relation (Equation 7.8a), which will be needed in the model, contains the activity coefficient $\gamma$, which is not usually known or easily measured. The difficulty was circumvented in Illustration 6.10 by relating $\gamma$ to solubility $x_s$, using the chemical potentials of the respective phases. We obtain

$$\mu^o(T_1 P) = \mu^o(T_1 P) + RT \ln \gamma \, x_s \qquad (7.8b)$$

Mercury      Saturated solution

and hence

$$\gamma = 1/x_s \qquad (7.8c)$$

Substitution into Equation 7.8a yields

$$y_{Hg} = \frac{x_{Hg}}{x_s} \frac{P^o_{Hg}}{P_T} \qquad (7.8d)$$

Here we must be careful to exclude the partial pressure of air from $P_T$ since the equilibrium involved is one of the binary $Hg$–$H_2O$ vapor in equilibrium with the liquid phase. We must therefore write

$$y_{Hg} = \frac{x_{Hg}}{x_s} \frac{P^o{}_{Jg}}{P_{Hg} + P^o{}_{H_2O}} \cong \frac{x_{Hg}}{x_s} \frac{P^o{}_{Hg}}{P^o{}_{H_2O}} \tag{7.8e}$$

where we have omitted mercury partial pressure since it is small compared to the full vapor pressure of water $P^o{}_{H_2O}$.

Substituting Equation 7.8e into the Rayleigh equation (Equation 7.5g) and integrating yields the following result:

$$\left[\frac{P^o{}_{Hg}}{x_s P^o{}_{Hw}} - 1\right]\ln\frac{W}{W_o} = \ln\frac{x}{x_o} \tag{7.8f}$$

The data to be introduced at this stage are as follows:

Mercury solubility (Table 6.3): $3 \times 10^{-2}$ mg/l = $2.7 \times 10^{-7}$ mole fraction
Mercury vapor pressure (25°C): 0.173 Pa
Water vapor pressure (25°C): $3.17 \times 10^3$ Pa

Noting that $W/W_o = 1 -$ (fraction evaporated) we obtain by substitution into Equation 7.8f

$$\left[\frac{0.173}{2.7 \times 10^{-9} \times 3170} - 1\right]\ln(1 - 10^{-4}) = \ln\frac{x_{Hg}}{x_s} \tag{7.8g}$$

Taylor series expansion of the logarithmic term then yields

$$\ln(1 - 10^{-4}) \cong 10^{-4} \tag{7.8h}$$

so that

$$\ln\frac{x_{Hg}}{x_s} = -2.02 \tag{7.8i}$$

and

$$x_{Hg}/x_s = 0.133 \tag{7.8j}$$

Thus, nearly 87% of the original mercury has been transferred to the atmosphere, a phenomenal amount considering only 1/10,000th of the solution has evaporated.

*Comments:*

It takes astuteness to overcome the difficulties and avoid the pitfalls of this problem. The fact that $\gamma$ is unknown and could not be located in the literature on vapor–liquid equilibria could have brought the proceedings to a halt. Instead, it had to be realized that $\gamma$ could be related to solubility (which is tabulated) and the relationship established by means of elementary thermodynamics. These are considerable leaps of thought.

A second point concerns the calculation of $y_{Hg}$. The first instinct would be to use the conventional relation $y = p/P_T$ and set total pressure equal to 1 atm. This would have led to the wrong result since the partial pressure of air would have been included. It required some thought to realize that $y_{Hg}$ refers to the system $H_2O$–Hg, not $H_2O$–Hg-air.

It is of some interest to calculate the ratio or partition coefficient $K = y/x$, which is a measure of enrichment in the vapor phase. Drawing on Equation 7.8a, we have

$$K = y / x = \frac{1}{x_s} \frac{P^o_{Hg}}{P^o_{Hg} + P^o_{H_2O}} = \frac{1}{2.7 \times 10^{-9}} \frac{0.173}{0.173 + 3.17 \times 10^3} = 2 \times 10^4$$

This value explains and confirms the astoundingly high depletion of mercury in the water basin.

### 7.1.3 Crosscurrent Cascades

The linkage of several single-stage contacts leads to a composite array termed a *cascade*. As mentioned earlier, the mode of contact in these configurations results in three distinct types of cascades: crosscurrent, co-current, and countercurrent. In the crosscurrent cascade, which is taken up here, the two phases enter and leave in directions that are, at least symbolically, at right angles to each other. This can take place in two different ways: In one mode of contact, the extracting agent, i.e., the solvent or adsorbent, is used in only one stage and is then removed for processing before being reused. This mode of contact, shown in Figure 7.7a for the case of adsorption, is sometimes and not very accurately referred to as co-current and has a high extraction efficiency, but also a high inventory. In the second mode of contact, the extracting agent is used *repeatedly* within the cascade before being withdrawn for regeneration. This type of configuration is shown in Figure 7.7b and has the advantage of low adsorbent or solvent inventory.

The model equations for the case shown in Figure 7.7a are now three in number: one solute mass balance each for the two stages and a companion equilibrium relation. We have

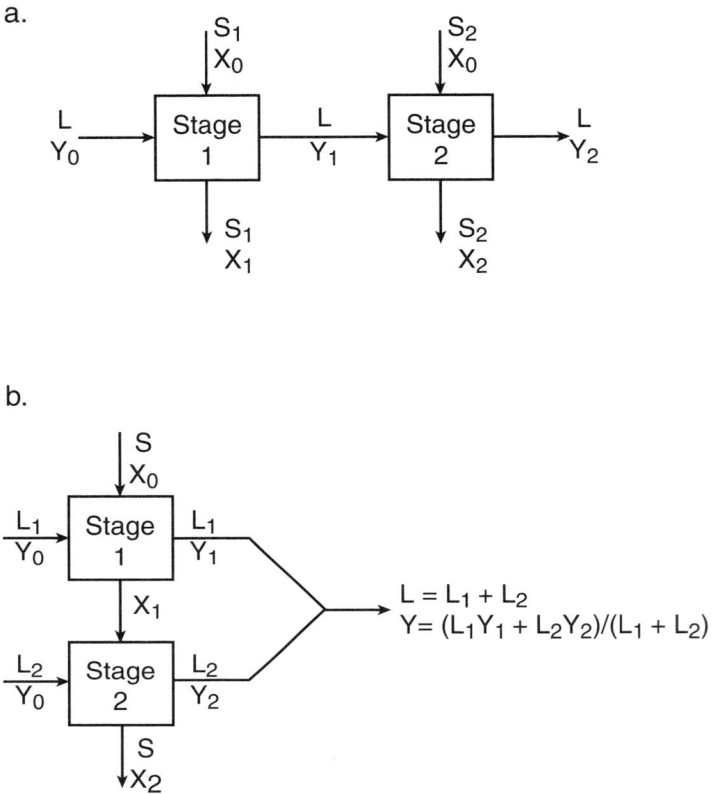

**FIGURE 7.7**
Two modes of crosscurrent adsorption processes: (a) use of fresh adsorbent in each stage; (b) repeated use of the same adsorbent.

Balance on Stage 1

$$\text{Mass of solute in} - \text{Mass of solute out} = 0$$

$$(LY_0 + S_1X_0) \quad - \quad (LY_1 + S_1X_1) \quad = 0 \qquad (7.9a)$$

Balance on Stage 2

$$\text{Mass of solute in} - \text{Mass of solute out} = 0$$

$$(LY_1 + S_2X_0) \quad - \quad (LY_2 + S_2X_2) \quad = 0 \qquad (7.9b)$$

Equilibrium Relation

$$Y^* = f(X) \qquad (7.9c)$$

Here again it is convenient to represent the equations in graphical form by means of an operating diagram in much the same way as was done in the case of single-stage operations. This is best done by rearranging the two mass balances into the following form:

$$-\frac{S_1}{L} = \frac{\Delta Y}{\Delta X} = \frac{Y_0 - Y_1}{X_0 - X_1} \tag{7.9d}$$

and

$$-\frac{S_2}{L} = \frac{\Delta Y}{\Delta X} = \frac{Y_1 - Y_2}{X_0 - X_2} \tag{7.9e}$$

Suppose now that the amounts of adsorbent $S_1$ and $S_2$ entering each stage have been prescribed and it is desired to determine the final effluent concentration $Y_2$. The feed condition, comprising the impurity level in the incoming adsorbent $X_0$, the feed concentration $Y_0$, and the solvent content $L$, is known. The procedure for constructing the operating diagram, shown in Figure 7.8, is then as follows:

1. Locate the feed point $(X_0, Y_0)$ in the diagram.
2. Draw a line of slope $-S_1/L$ through the feed point. Its intersection with the equilibrium curve fixes the concentrations $(X_1, Y_1)$ leaving stage 1.
3. Draw a second operating line of slope $-S_2/L$ through the point $(X_0, Y_1)$. Its intersection with the equilibrium curve defines the final impurity level $Y_2$.

It is clear from this construction that there exists an infinite number of combinations of $S_1$ and $S_2$ that will reduce the feed-solute level $Y_0$ to $Y_2$. It is also seen that if either $S_1$ or $S_2$ is set equal to zero, i.e., a single-stage contact is used, the resulting operating line, which extends from $(X_0, Y_0)$ to $(X_2, Y_2)$, will always have a slope $-S/L$ greater than the combined slopes $(S_1/L + S_2/L)$ so that

$$S_{\text{single stage}} > (S_1 + S_2)_{\text{2-stage cascade}}$$

i.e., the single stage always consumes more adsorbent for a given reduction in solute content than a two-stage arrangement.

It follows that between the two extremes $S_1 = 0$ and $S_2 = 0$ there exists a combination $S_1 + S_2$, which represents a *minimum* total adsorbent inventory $(S_T)_{\text{Min}}$. This minimum amount can be determined by simple calculus for the case of linear equilibrium. We show this in the following illustration.

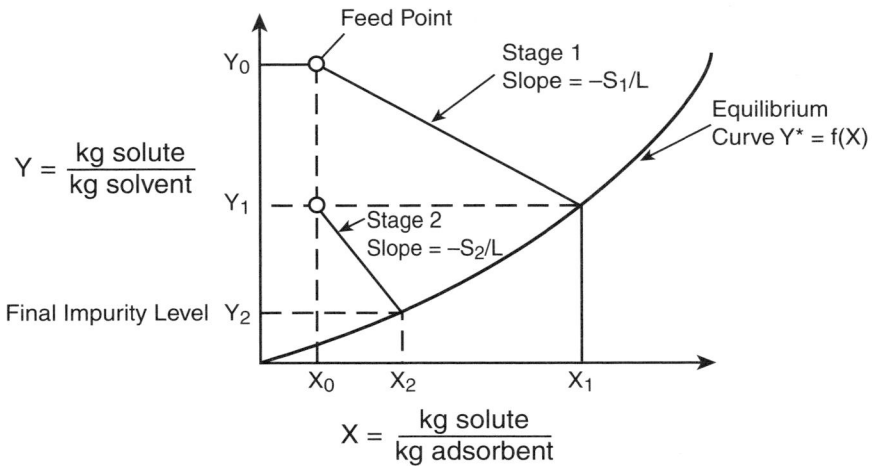

**FIGURE 7.8**
Operating diagram for a two-stage crosscurrent adsorption process using fresh adsorbent.

### Illustration 7.5: Optimum Use of Adsorbent or Solvent in Crosscurrent Cascades

Consider a crosscurrent adsorption process in which the equilibrium is linear, a condition that holds at low solute level. We then have

$$X_i = HY_i \quad H > 1 \tag{7.10a}$$

where $H$ is the Henry's constant.

Assume further that the adsorbent is initially clean, i.e., $X_0 = 0$ (this is frequently the case). Substitution of Equation 7.10a into the two mass balances (Equation 7.9a and Equation 7.9b) then yields

$$S_T = S_1 + S_2 = HL\left(\frac{Y_0}{Y_1} + \frac{Y_1}{Y_2} - 2\right) \tag{7.10b}$$

The condition $dS_T/dY_1 = 0$ is now used to establish the optimum intermediate concentration $(Y_1)_{opt}$, which will yield the minimum value of $S_T$. We obtain

$$dS_T/dY_1 = Y_0 Y_1^{-2} + Y_2^{-1} = 0 \tag{7.10c}$$

and hence

$$(Y_1)_{opt} = (Y_0 Y_2)^{1/2} \tag{7.10d}$$

In other words, the optimum solute level from the first stage that will minimize $S_T$ is the geometric mean of the concentrations entering and leaving the cascade. Let us see how this affects the amount of adsorbent to be used.

Backsubstitution into the material balances (Equation 7.9a and Equation 7.9b) yields

$$\frac{S_1}{L} = H\left[\frac{Y_0}{(Y_0 Y_2)^{1/2}} - 1\right] \tag{7.10e}$$

and

$$\frac{S_2}{L} = H\left[\frac{(Y_0 Y_2)^{1/2}}{Y_2} - 1\right] \tag{7.10f}$$

But

$$\frac{Y_0}{(Y_0 Y_2)^{1/2}} = \frac{Y_0 (Y_0 Y_2)^{1/2}}{Y_0 Y_2} = \frac{(Y_0 Y_2)^{1/2}}{Y_2} \tag{7.10g}$$

so that the bracketed terms in Equation 7.10e and Equation 7.10f are identical and consequently $S_1 = S_2$. Hence it is seen that in this case of a two-stage cascade with linear equilibrium, optimum operation calls for the use of *equal amounts of adsorbent* given by either Equation 7.10e or Equation 7.10f.

It has been shown that this principle applies to *any number of stages*; i.e., the optimum use of adsorbent requires an *equal division* of that adsorbent among the stages of a crosscurrent cascade. The same principle applies to crosscurrent extraction of systems with mutually insoluble solvents. The solute recovery or removal that results in such cascades is depicted graphically in Figure 7.9. In this plot, $m$ represents the distribution coefficient for extraction or Henry's constant $H$ for adsorption, $E$ is the so-called extraction factor $mB/A$ or $HS/L$, and $Y_n$ or $x_n$ is the effluent concentration from the $n$th stage of the solution being treated.

Let us apply this plot to a specific extraction process. We consider 100 kg of a feed of 1% nicotine in water that is to be extracted in a three-stage crosscurrent cascade employing 50 kg kerosene in each stage. The distribution has a slight curvature, with $m$ varying over the range 0.80 to 0.90. We choose a mean value of 0.85 and obtain the extraction factor:

$$E = mB / A = 0.85\frac{50}{99} = 0.43 \tag{7.11a}$$

Using Figure 7.9, this yields for solute-free kerosene ($y_0 = 0$) the extraction ratio

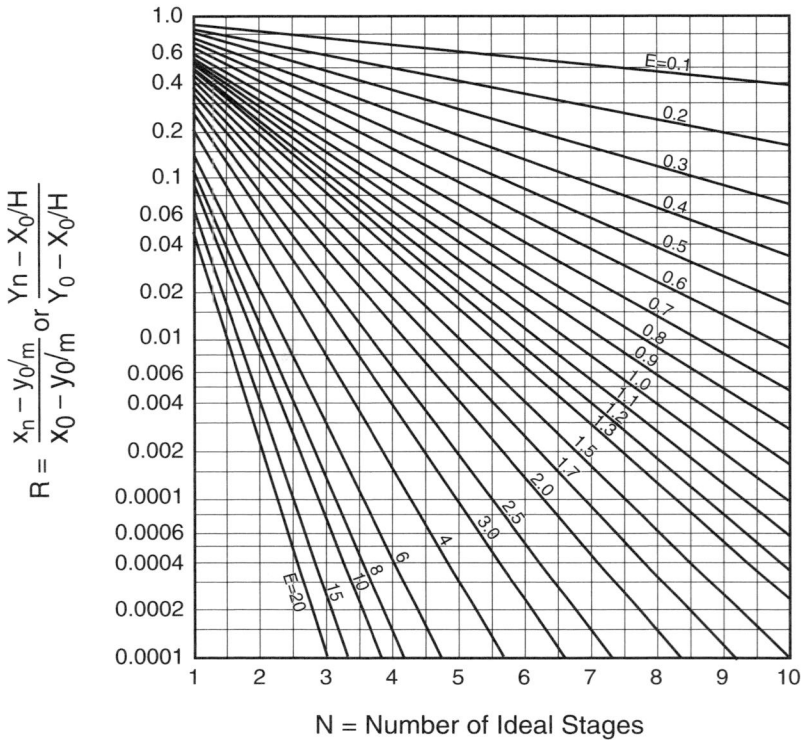

**FIGURE 7.9**
Crosscurrent cascades using equal amounts of adsorbent or solvent.

$$\frac{x_n}{x_0} = 0.34 \tag{7.11b}$$

and consequently

$$x_n = 0.34 \times 0.010 = 0.0034 \tag{7.11c}$$

The same value is obtained using the graphical construction of the operating diagram.

*Comments:*

The optimization problem considered indicates that the best way to operate a crosscurrent cascade is by equal subdivision of solvent or adsorbent. The reader should be aware, however, that the complete optimization of a plant must also consider the cost of the stages, the cost of solvent recovery, and the value of the extracted solute itself. Thus, in addition to optimum solvent use, we also need to determine the optimum number of stages, and this requires bringing in all of the above-mentioned factors. Still, equal subdivision of solvent or adsorbent is a good policy to pursue and comes close to meeting the requirements of a global optimum for systems with linear distributions. When the equilibrium relation becomes nonlinear, the optimum policy begins to deviate from this norm and has to be determined anew, usually by numerical means.

### Illustration 7.6: A Crosscurrent Extraction Cascade in Triangular Coordinates

We consider 1000 kg of a feed containing 50% by weight of acetone in water, which is to be reduced to 10% by extraction with 1,1,2-trichloroethane in a crosscurrent cascade; 250 kg solvent are to be used in each stage. Representative tie-line data are shown in Figure 7.10.

Solution of this problem calls for the repeated application of the methods established for single-stage extraction in Illustration 7.2. We start by drawing a line connecting the solvent at $B$ to the feed located at $F$, and follow this by locating the mixing point $M$, using Equation 7.4c. We obtain

$$x_{M_1} = \frac{x_F F}{F + B} = \frac{0.5 \times 1000}{1000 + 250} = 0.40 \tag{7.12a}$$

The tie-line through this coordinate is shown in Figure 7.10 and yields raffinate and extract concentrations corresponding to the end points of the tie lines, i.e., $x_{R_1} = 0.35$ and $y_{E_1} = 0.475$. These values can then be used to calculate the amounts of raffinate and extract using Equation 7.4d and Equation 7.4e.

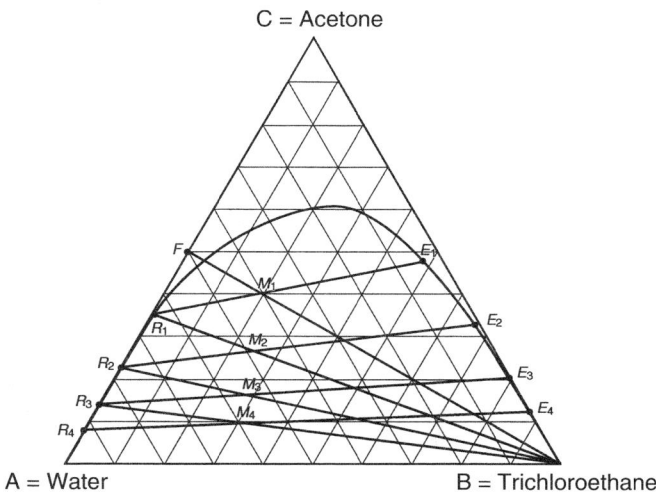

FIGURE 7.10
Operating diagram for Illustration 7.6.

$$E_1 = \frac{(F+B)(x_{M_1} - x_{R_1})}{y_{E_1} - x_R} = \frac{(1000+250)(0.40-0.35)}{0.475-0.35} = 500\,\text{kg}$$

$$R_1 = F + B - E_1 = 1000 + 250 - 500 = 750 \text{ kg}$$

The results of these calculations for four stages are given below:

| Stage | $x_M$ | $x_R$ | $y_E$ | E | R |
|-------|-------|-------|-------|-----|-----|
| 1 | 0.400 | 0.35 | 0.475 | 500 | 750 |
| 2 | 0.262 | 0.223 | 0.325 | 382 | 618 |
| 3 | 0.159 | 0.134 | 0.204 | 310 | 558 |
| 4 | 0.0925 | 0.075 | 0.120 | 305 | 503 |

From this tabulation it emerges that three stages are too few and four stages too many to reduce acetone content in the raffinate to $x_R = 0.10$. There are three alternatives we can pursue:

1. Use three stages and slightly more solvent per stage, determining the quantity to be used by trial and error until $x_{R_3} = 0.10$.
2. Use four stages and slightly less solvent, again with a trial-and-error procedure.
3. Use four stages with the same amount of solvent and accept a lower acetone content of $x_{R_4} = 0.075$ in the final raffinate.

The last alternative is the most convenient, as well as the most practical, since the stage inefficiency will inevitably consume the margin provided by the fourth stage.

## 7.1.4 Countercurrent Cascades

The countercurrent cascade is the most popular among the various existing stage configurations. It combines economy of adsorbent or solvent consumption with a high recovery of solute, but pays for it in part with a greater number of stages than would be required in a comparable crosscurrent cascade. Perhaps its largest application apart from distillation operations occurs in the field of gas absorption, the flow sheet for which appears in Figure 7.11a. A corresponding countercurrent adsorption cascade is sketched in Figure 7.11b. The two assemblies are identical in concept but differ in the implementation of the countercurrent contact. Gas absorbers usually consist of vertical columns containing trays on which the two phases come into intimate contact. The gas enters from below through openings in the plate or tray and bubbles through the liquid, which flows across and down to the next tray. Displays of various types of trays are shown in Figure 7.12. Adsorption cascades in contrast consist of a series of stirred tanks from which the

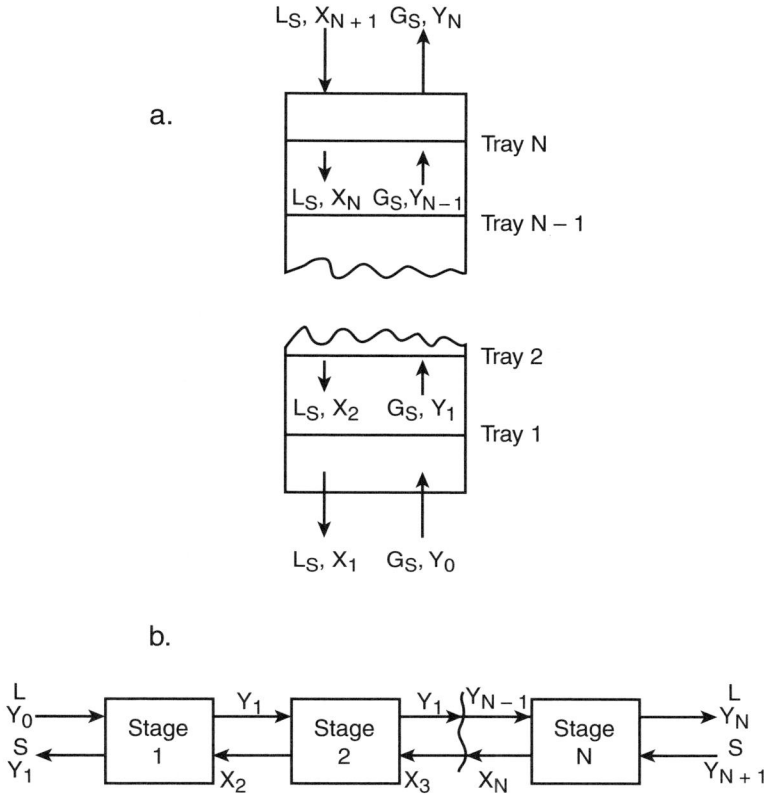

**FIGURE 7.11**
(a) Countercurrent gas adsorber and (b) countercurrent adsorption cascade.

two streams are withdrawn after allowing the two phases to separate by settling.

In gas absorption, gas and liquid streams enter the bottom tray 1 and the top tray $N$ at opposite ends of the cascade, carrying with them the solute concentrations $Y_0$ and $X_{N+1}$. They leave the same trays with concentration $Y_1$ and $X_N$. Note that the subscripts refer to the tray from which a particular stream exits, with the subscripts 0 and $N + 1$ representing imaginary stages numbered 0 and $N + 1$. Concentrations are expressed in mass ratios, and flow rates $G_S$ and $L_S$ in kg/s of solute-free carrier gas and solvent. Similar considerations apply to the adsorption cascade.

Our tools again comprise mass balances and equilibrium relations, which are expressed graphically in the operating diagrams shown in Figure 7.13. They contain an operating line that represents the solute material balances and a staircase construction, which spans the interval between operating line and equilibrium curve and represents the various stages or trays.

A material balance over the entire gas absorber leads to the equation

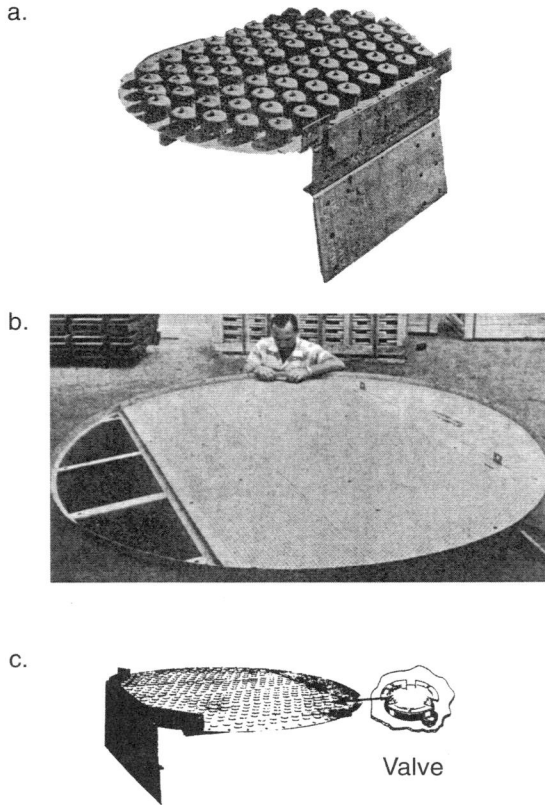

a.

b.

c.

Valve

**FIGURE 7.12**
Types of trays: (a) bubble-cap tray: vapor rises through openings in the plate, reverses direction, and escapes through the slots of an inverted cup; (b) sieve tray; (c) valve tray: tray openings are adjusted by means of floating disks that rise and fall with the vapor flow rate.

$$\text{Rate of solute in} - \text{Rate of solute out} = 0$$

$$(L_S X_{N+1} + G_S Y_0) - (L_S X_1 + G_S Y_N) = 0 \tag{7.13a}$$

Rearrangement then gives

$$L_S / G_S = \frac{Y_N - Y_0}{X_{N+1} - X_1} \tag{7.13b}$$

Usually in the design of countercurrent absorbers, the solvent rate $L_S$ and purity $X_{N+1}$ as well as $G_S$ and $Y_0$ are known, and the desired exit gas concentration $Y_N$ is specified. Equation 7.13b can then be plotted in the operating diagram by drawing a line of slope $L_S/G_S$ through the point $(X_N, X_{N+1})$.

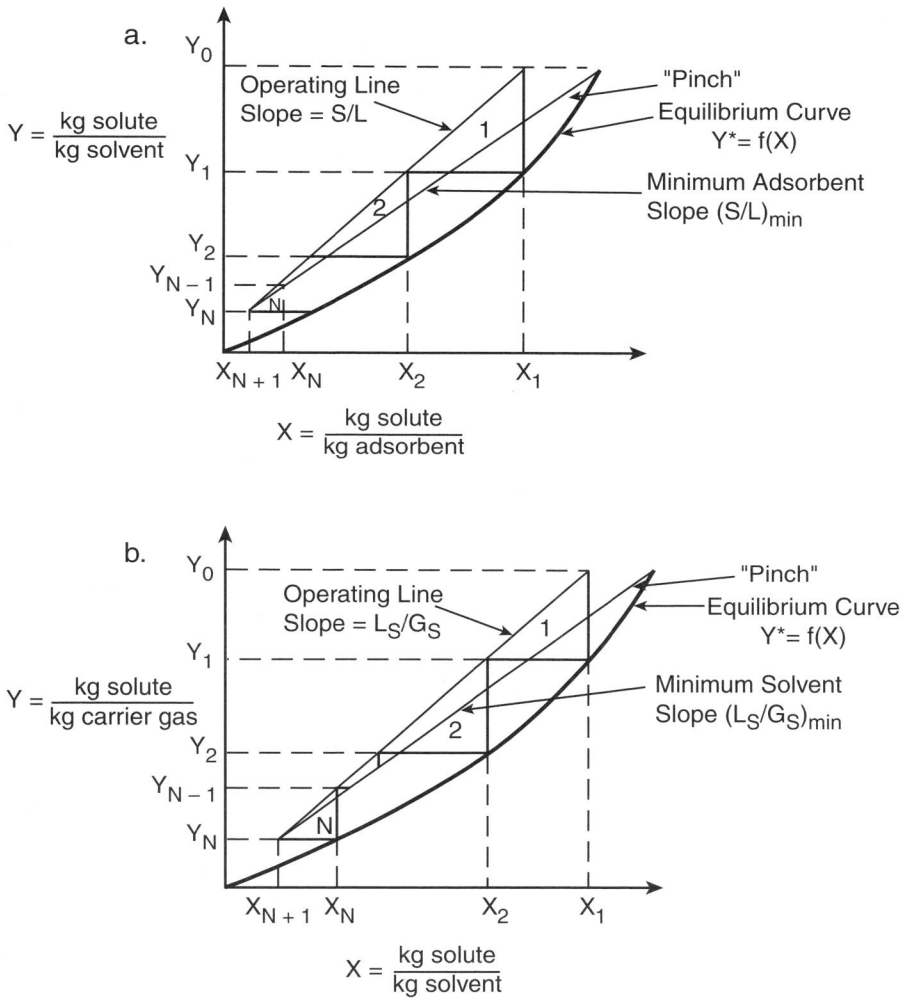

**FIGURE 7.13**
Operating diagram for countercurrent cascades: (a) adsorption; (b) absorption.

Consider next a solute balance on a *single* stage, for example, tray 1. We obtain

$$\text{Rate of solute in} - \text{Rate of solute out} = 0$$

$$(L_S X_2 + G_S Y_0) \quad - \quad (L_S X_1 + G_S Y_1) = 0 \qquad (7.13c)$$

This equation can be represented in the operating diagram as follows: We locate the point $(Y_0, X_1)$, where the abscissa value $X_1$ is obtained from the intersection of the operating line with the horizontal ordinate value $Y = Y_0$. We next drop a *vertical* line from this point to the equilibrium curve $Y^* = f(X)$. The point of intersection will have the coordinates $(Y_1, X_1)$, since these values represent the concentration levels leaving tray 1 and are known to be in equilibrium with each other. This is followed by drawing a *horizontal* line through $(Y_1, X_1)$ all the way to the operating line. The point of intersection will have the coordinates $(Y_1, X_2)$. This is shown by rearranging Equation 7.13c to read

$$L_S / G_S = \frac{Y_1 - Y_0}{X_2 - X_1} \qquad (7.13\text{d})$$

which is the expression of a line of slope $L_S/G_S$ that passes through the points $(Y_1, X_2)$ and $(Y_0, X_1)$. But Equation 7.13b, which has the same slope, also passes through the point $(Y_0, X_1)$. The two lines (Equation 7.13b and Equation 7.3d) must therefore coincide, and we have established the first step of the staircase construction seen in Figure 7.13. This construction of alternating vertical and horizontal lines between operating line and equilibrium curve is continued until the known concentrations at the top of the column $(Y_N, X_{N+1})$ are reached. A count is then made of the number of trays between gas inlet and outlet concentrations. That number represents the number of stage contacts required to reduce the feed concentration $Y_0$ to the prescribed value of $Y_N$ using a fixed solvent flow rate $L_S$. A fractional step at the outlet end of the staircase is usually rounded off to one stage.

We now note a number of features of this construction, which are of use in both engineering calculations and analysis.

- The operating line in essence represents solute mass balances around a single stage or an aggregate of stages. Its coordinate points $(Y_j, X_{j+1})$ establish the relation between concentrations *entering* a stage, while the equilibrium curve relates concentrations *leaving* a stage.

- Reducing the amount of solvent $L_S$ (or adsorbent) *lowers* the slope of the operating line and *simultaneously increases* the number of stages required for a prescribed reduction in solute content. This process can be continued until the operating line intersects the equilibrium curve. At that point a "pinch" results at the high concentration end (Figure 7.13) yielding an *infinite* number of stages and a corresponding *minimum* flow rate of solvent or adsorbent. This is the lowest flow rate that will achieve the desired solute removal and is a useful limiting value to establish. Below that value, the prescribed effluent concentration $Y_N$ can no longer be attained, even if an infinite number of stages were used.

- An increase in solvent or adsorbent flow will reduce the number of contact stages and consequently lower the capital cost of the plant. This advantage is earned at the expense of an increase in operating costs occasioned by the higher flow of solvent or adsorbent. There consequently exists an optimum flow rate that will *minimize* the combined operating and capital costs. That optimum can only be established with precision by a detailed economic analysis of the process. It has been found in such studies, however, that the optimum flow usually lies in the range 1.5 to 2 times the minimum value. That range is commonly used to establish the flow of solvent or adsorbent to be used, and hence the operating line, in preliminary designs of the process.

- The reverse task to the design problem considered above, i.e., the prediction of the effluent concentration of an existing or hypothetical plant (number of stages $N$ and feed conditions known) cannot be achieved in the same direct fashion. We must resort instead to a trial-and-error procedure by drawing a series of parallel operating lines of known slope $L_S/G_S$ or $S/L$ until a staircase construction accommodating $N$ stages reaches the known impurity level of the solvent or adsorbent. This presents no undue difficulties and can be accomplished quite rapidly.

- The operating diagrams we have discussed are not only elegant in their simplicity but are also capable of conveying important information in rapid fashion. Thus, the effect of a change in flow rate, concentrations, or the number of stages can be quickly assessed, at least qualitatively, by visual inspection. Although in practice much of this work is now dealt with using appropriate computer packages, particularly in the case of multicomponent systems, the operating diagram remains unsurpassed in conveying the essence of staged operations and in providing a valuable educational tool.

### Illustration 7.7: Comparison of Various Stage Configurations: The Kremser–Souders–Brown Equation

This example brings together the various stage arrangements discussed in the preceding sections and compares their performances. We consider an extraction process that is to be carried out, first in a single stage, and then for comparison in two-stage crosscurrent and countercurrent cascades. The two solvents involved are taken to be mutually insoluble, and the task is to determine the quantity of extraction solvent required per unit mass of raffinate solvent to reduce the solute content from a mass ratio of $X_F = 0.1$ to $X_1 = 0.01$. The distribution coefficient is constant at 3.0; i.e., the equilibrium is linear. We obtain the following results.

### 1. Single Stage

The solute balance for this case is given by

$$\text{Mass of solute in} - \text{Mass of solute out} = 0$$
$$AX_F - (BY_1 + AX_1) = 0 \tag{7.14a}$$

with $Y_1$ given by the distribution coefficient $m'$:

$$Y_1 = m'X_1 \qquad (7.14b)$$

Combining the two equations and rearranging yields

$$B/A = \frac{X_F - X_1}{mX_1} = \frac{0.1 - 0.01}{3 \times 0.01} = 3.3 \qquad (7.14c)$$

and hence $B$ = 3.3 kg solvent/kg raffinate solvent.

2. *Two-Stage Crosscurrent Cascade*

Here we make use of the plot of Figure 7.9 to compute the solvent quantity, which is assumed to be divided equally between the two stages for optimum operation. We have for the ordinate value, assuming pure solvent $y_0 = 0$,

$$\frac{x_n - y_0/m}{x_o - y_0/m} = \frac{X_1 - Y_0/m}{X_0 - Y_0/m} = \frac{0.01 - 0}{0.1 - 0} = 0.1 \qquad (7.14d)$$

For $N$ = 2, this yields the parameter value

$$E = mB/A = 3.3 \qquad (7.14e)$$

and hence from Figure 7.9

$$B/A = 3.3/3.0 = 1.1 \qquad (7.14f)$$

which is the quantity of solvent used in each stage. Multiplying by 2 we obtain for the solvent total

$$B_{Tot} = 2 \times 1.1 = 2.2 \text{ kg solvent/kg raffinate solvent} \qquad (7.14g)$$

3. *Two-Stage Countercurrent Cascade*

Here we make use of the fact that for systems with linear equilibria, an analytical expression can be derived that relates the number of stages to the operating parameters. That expression is known as the Kremser–Souders–Brown equation (or Kremser equation for short), and is given by

$$R = \frac{E^{n+1} - E}{E^{n+1} - 1} \qquad (7.15a)$$

or in rearranged form

$$N = \frac{\log[R(1-1/E)+1/E]}{\log E} \tag{7.15b}$$

A plot of Equation 7.15b appears in Figure 7.14. By suitably defining the parameters $R$ and $E$ the Kremser equation can be used for any countercurrent cascade involving linear-phase equilibria. These parameter definitions are listed in Table 7.1 and provide a convenient dictionary for use in a number of important operations.

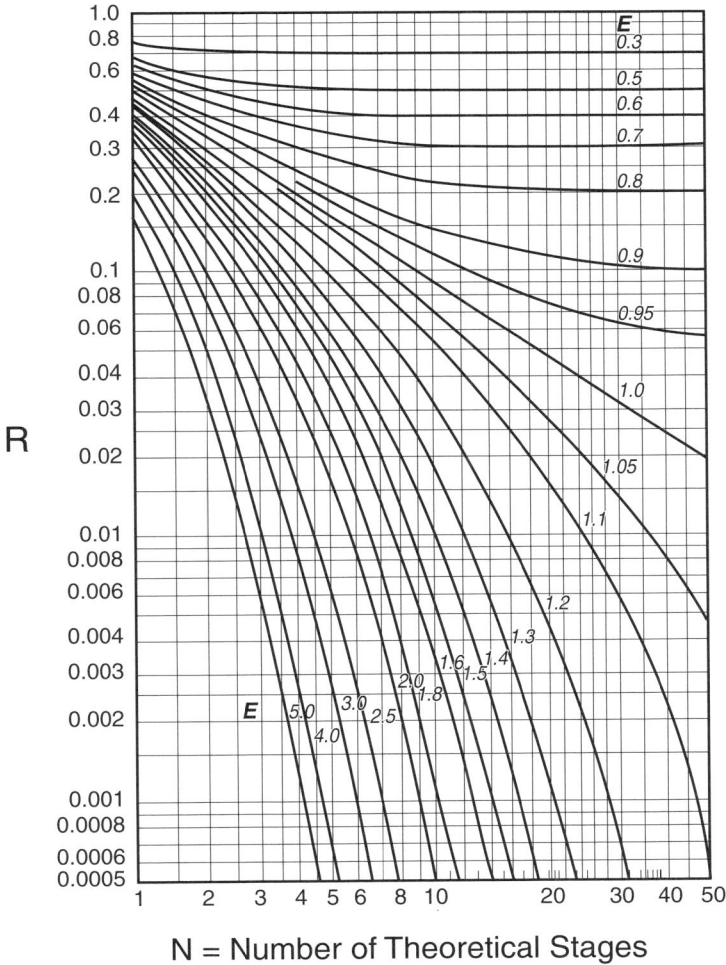

FIGURE 7.14

Countercurrent cascade with linear equilibrium; graphical representation of the Kremser equation; for definition of the residue factor $R$ and extraction ratio $E$, see Table 7.1.

**TABLE 7.1**

Operating Parameters for Use in the Kremser Plot

| Operation | R | E | Equilibrium |
|---|---|---|---|
| Gas absorption | $\dfrac{Y_N - HX_{N+1}}{Y_0 - HX_{N+1}}$ | $\dfrac{L}{HG}$ | $Y$ (gas) $= HX$ (liquid) |
| Liquid extraction | $\dfrac{X_n - Y_0/m}{X_0 - Y_0/m}$ | $\dfrac{mB}{A}$ | $Y$ (extract) $= mX$ (raffinate) |
| Liquid adsorption | $\dfrac{Y_N - X_{N+1}/H}{Y_0 - X_{N+1}/H}$ | $\dfrac{HS}{L}$ | $X$ (solid) $= HY$ (liquid) |
| Solids leaching | $\dfrac{Y_N - mX_{N+1}}{Y_0 - mX_{N+1}}$ | $\dfrac{S}{mL}$ | $Y$ (liquid) $= mX$ (solid) |

The parameter $R$, which we term the *residue factor,* is a direct measure of the amount of residual solute leaving the cascade. The smaller the value of $R$, the lower the effluent concentrations $Y_N$ or $X_N$, and hence the higher the degree of solute recovery. The parameter $E$, on the other hand, is identical to the extraction ratio we have defined for crosscurrent extraction (Equation 7.11a) and varies directly with the amount of solvent or adsorbent used and its capacity. Large values of $E$ lead to good recoveries and low effluent concentrations, both desirable features.

With these definitions in place, we can calculate the solvent requirement for the two-stage countercurrent process. We have for the residue factor $R$

$$R = \frac{X_2 - Y_0/m}{X_0 - Y_0/m} = \frac{0.01 - 0}{0.1 - 0} = 0.10 \qquad (7.15c)$$

and hence a solute recovery $r = 1 - 0.10 = 0.9$ or 90%.

The corresponding value of $E$, read from Figure 7.14, is 2.6. Hence

$$E = mB/A = 2.6 \qquad (7.15d)$$

and

$$B/A = 2.6/3 = 0.87$$

Therefore,

$$B = 0.87 \text{ kg solvent/kg raffinate solvent}$$

This example shows the distinct advantage that the countercurrent process holds, not only over the single-stage operation, which was expected, but also over its crosscurrent counterpart: Consumption by the latter is higher by a factor of $2.2/0.87 \cong 2.5$. The principal advantage of the crosscurrent cascade is that it is more easily adaptable to *batch* processing, whereas the countercurrent cascade is by necessity continuous.

The Kremser plot, Figure 7.14, serves several additional useful purposes. We illustrate this with the following examples:

Suppose that an existing gas absorber using clean solvent has its feed rate $G$ doubled over the previous design value. A quick scan of the Kremser plot shows that to maintain the same effluent concentration $Y_N$ or solute recovery as before, the solvent flow rate $L$ would likewise have to be doubled. Similar considerations apply to adsorption, extraction, and leaching.

Consider next the same unit subjected to a twofold increase in feed concentration $Y_o$. If no remedial action is taken, the residue ratio $Y_N/Y_o$ will remain the same, but the effluent concentration $Y_N$ will double. To bring it down to its previous value, solvent flow rate and hence $E$ would have to be increased. The amount of adjustment needed can once again be determined quickly through the use of Figure 7.14.

Finally, a closer examination of Figure 7.14 shows that at $E$ values below unity, the plots veer off and asymptotically approach a constant value of $R$. This implies that we cannot, under these conditions, attain arbitrarily low effluent concentrations, no matter how high we set the level of solvent or adsorbent flow rate or indeed the number of stages. The value of $E = 1$ consequently represents an important watershed point, below which it becomes impractical or impossible to attain a desired goal. The reasons for this behavior are addressed more fully in Illustration 8.2.

### 7.1.5 Fractional Distillation: The McCabe–Thiele Diagram

The operation of a conventional tray fractionation column, and the associated variables, is sketched in Figure 7.15. Although the process has most of the standard properties of a staged countercurrent operation we have seen before, it does display a number of new features that require special mention:

- The feed to be fractionated, which can be a liquid, a vapor, or even a combination of both, enters the column at some central location rather than at one of the ends of the cascade as had previously been the case. This results in a division of the column into two parts, the *rectification* or *enriching section* above the feed tray, and the *stripping* or *exhausting section* below it. The upper section serves to enrich the vapor in the more volatile components, a portion of which is ultimately withdrawn as liquid "overhead product" or "distillate." In the lower section, residual volatile components are progressively stripped off the liquid and conveyed upward as vapor, while the downward flow of liquid becomes enriched in the heavier or less-volatile components.

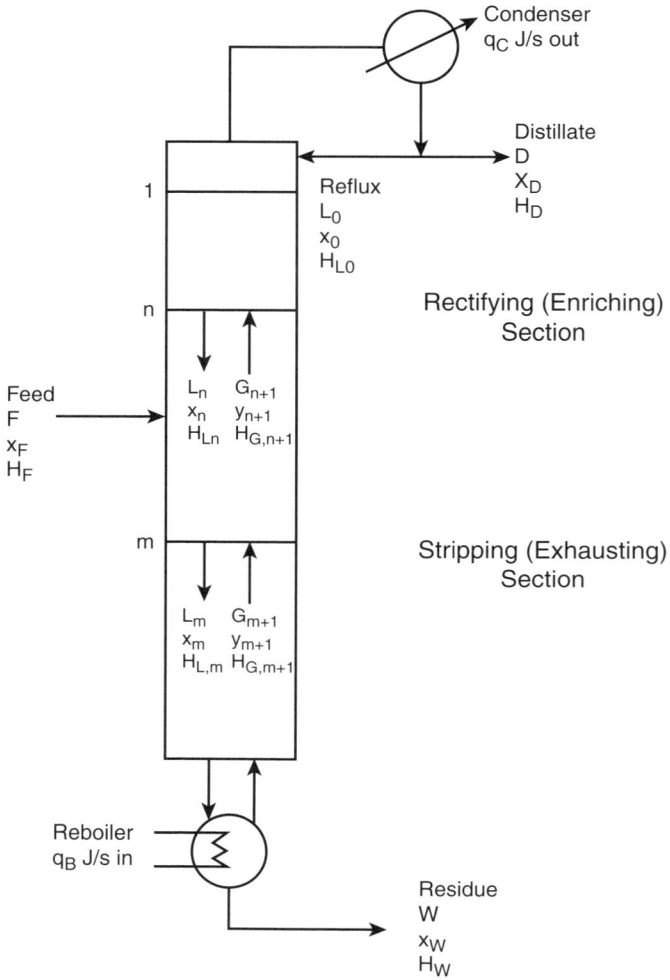

**FIGURE 7.15**
The fractionation column.

- The two streams entering the ends of the column are generated in separate vaporizers or "reboilers" and in condensers located at the two outlets. At the top, vapor leaving the column is condensed and returned in part to the column as "reflux," while the remainder is withdrawn as distillate. At the bottom, the exiting liquid is partly revaporized in a reboiler and the vapor is diverted back into the column. The remainder of the stream is removed as "bottom product" or residue.

- In contrast to the processes we encountered previously, which were largely or entirely isothermal in nature, distillation has substantial heat effects associated with it. Consequently, we expect heat balances to be involved in modeling the process, as well as the usual mass balances and equilibrium relations. These balances are formulated entirely in molar units because the underlying equilibrium relations, such as Raoult's law and its extension, or the separation factor $\alpha$, are all described in terms of mole fractions. Thus, the flow rates $L$ and $G$, which appear in Figure 7.16, are both in units of mol/s, enthalpies $H$ in units of J/mol, and the liquid and vapor compositions are expressed as mole fractions $x$ and $y$ of a binary system.

- Column operation is usually taken to be isobaric, so that temperatures within the column will vary and lie in the range between the boiling points of the overhead and bottom products. The appropriate equilibrium diagrams for this case are the boiling-point diagram (Figure 6.17a) and the $x$–$y$ diagram (Figure 6.17c). The latter is used in the construction of the operating diagram that is taken up further on.

### 7.1.5.1 Mass and Energy Balances: Equimolar Overflow and Vaporization

There are three balances for the binary system considered here: the total mole balance, the component mole balance, and the heat balance. It is customary to apply these in unison and in turn to three separate regions of the fractionation column. One set each is used to describe conditions above and below the feed tray, i.e., the rectifying and stripping sections. These balances are taken over the entire upper and lower portions and include the reboiler and condenser as well as the product stream. A third set of balances is applied to an isolated stage, the feed tray, and includes the flow rates and thermal condition of the incoming feed. Let us see how this works out in practice. For the rectifying section, we have

Rate of total moles in – Rate of total moles out = 0

$$G_{n+1} \quad - \quad (L_n + D) \quad = 0 \qquad (7.16a)$$

Rate of component moles in – Rate of component moles out = 0

$$y_{n+1}G_{n+1} \quad - \quad (x_n L_n + x_D D) \quad = 0 \qquad (7.16b)$$

Rate of heat in – Rate of heat out = 0

$$G_{n+1}H_{G,n+1} - (L_n H_{L,n} + DH_D + q_C) \quad = 0 \qquad (7.16c)$$

where $q_C$ is the heat removed in the condenser.

Elimination of $D$ from the mass balances leads, after some rearrangement, to the following expression:

a.

b.

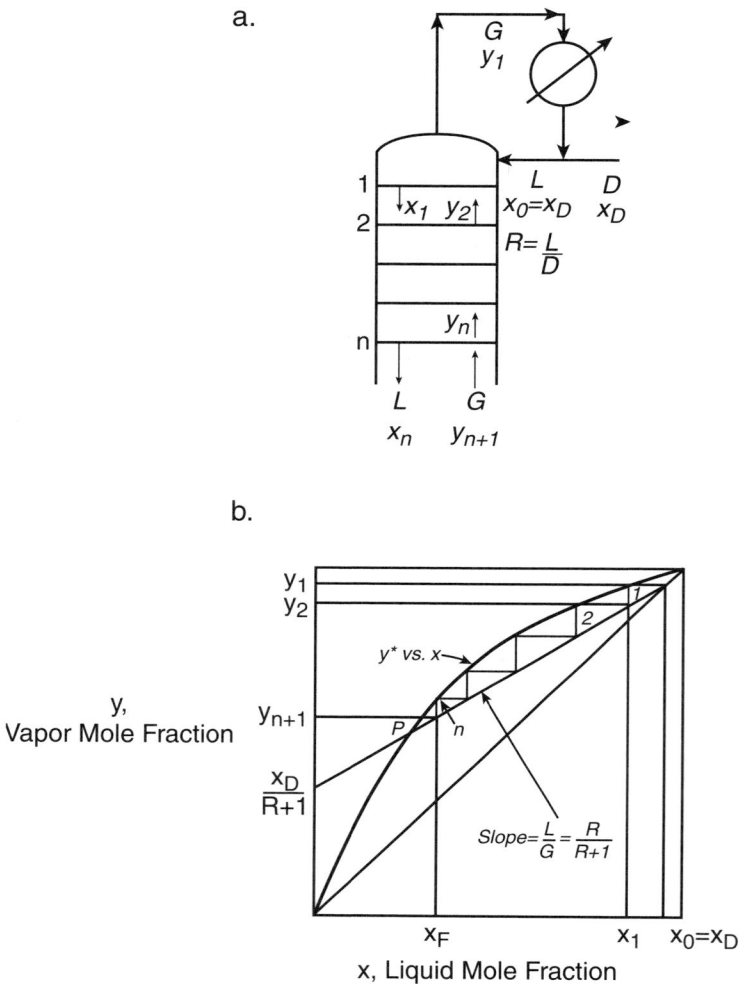

**FIGURE 7.16**
Details of the rectifying (enriching) section: (a) flow diagram; (b) operating diagram.

$$\frac{L_n}{G_{n+1}} = \frac{y_{n+1} - x_D}{x_n - x_D} \qquad (7.16d)$$

This is the operating line for the enriching section. It relates the vapor composition $y_{n+1}$ entering a tray and that of the liquid leaving, $x_n$, and has a slope equal to the ratio of liquid-to-vapor flow rates $L_n/G_{n+1}$.

The corresponding operating line for the stripping section is given by

$$\frac{\overline{L}_n}{\overline{G}_{n+1}} = \frac{\overline{y}_m - x_w}{\overline{y}_{n+1} - x_w} \tag{7.16e}$$

where the bar is used to indicate the position *below* the feed tray to distinguish these quantities from their counterparts in the rectifying section.

To this point it has been assumed that all quantities vary from tray to tray and that consequently $3(N + 2)$ balances will be required to describe the operation. Here $N$ denotes the number of stages, with an additional $3 \times 2$ equation needed to balance the flow of mass and heat about the condenser and reboiler. These have to be further supplemented by expressions relating each enthalpy to the key state variables $x$, $y$, and $T$. Evidently, we are dealing with a model of considerable complexity and dimensionality, which would require a numerical solution.

We now introduce the reader to a concept that avoids this complication and drastically reduces the complexity of the model. We draw for this purpose on the heat balance, Equation 7.16c, and rearrange it to read

$$\frac{L_n}{G_{n+1}} = 1 - \frac{H_{G,n+1} - H_{L,n}}{q_C / D + H_D - H_{L,n}} \tag{7.16f}$$

where $q_C$ is the condenser heat load in J/s. A first simplification comes about by noting that the sensible heat of the liquid on tray $n$, $H_{L,n}$, is much smaller than the term $q_C/D + H_D$, which involves the latent heat of condensation. $H_{L,n}$ can, therefore, to a good approximation, be neglected in comparison to this term. If we make the further assumption that the substances have very similar molar latent heats of vaporization $\Delta H_v$, we can approximate the numerator by the relation

$$H_{G,n+1} - H_{L,n} \doteq \Delta H_v \tag{7.16g}$$

so that the ratio of flow rates becomes a *constant* for *all* trays of the enriching section. Similar arguments can be applied to the exhausting section, with the result that

$$\frac{L_n}{G_{n+1}} = K_1 \tag{7.16h}$$

$$\frac{\overline{L}_m}{\overline{G}_{n+1}} = K_2 \tag{7.16i}$$

Consider next a total mole balance about the $n$th tray:

$$(L_{n-1} + G_{n+1}) - (L_n + G_n) = 0 \qquad (7.16j)$$

If we now substitute Equation 7.16h into this relation, there results

$$K_1 G_n + G_{n+1} - K_1 G_{n+1} + G_n = 0 \qquad (7.16k)$$

and consequently

$$G_{n+1} = G_n \qquad (7.17a)$$

and

$$L_{n-1} = L_n \qquad (7.17b)$$

We have, in other words, shown that if we assume latent heat to be the predominant thermal quantity in distillation and that this property comes close to being identical for the two components, the vapor and liquid flow can, for practical purposes, be considered constant in the rectifying section. A similar procedure applied to the stripping section leads to the conclusion that here also the flow rates will remain practically constant although different in value from those of the enriching section because of the intervening flow of feed.

What we have just derived is referred to as the principle of equimolal overflow and vaporization.

### 7.1.5.2    The McCabe–Thiele Diagram

One immediate consequence of the principle of equimolal overflow and vaporization is a drastic simplification of the operating lines (Equation 7.16d and Equation 7.16e). Whereas previously these expressions had to be plotted laboriously step by step from tray to tray, we are now dealing with a straight line that can be drawn easily, knowing only its slope and the location of one point or, alternatively, two points. The slope $L/G$ of the enriching operating line is referred to as the *internal reflux ratio*. It is not a convenient quantity to deal with since it is usually neither known nor specified, nor can it be easily manipulated in actual column operations. A more suitable parameter, and one that is easily controlled by appropriate valve settings, is the *external reflux ratio*, or reflux ratio $R$ for short. This quantity represents the ratio of liquid flow $L$ returned to the column as reflux to the flow of distillate $D$ withdrawn as product, and is given by

$$R = L/D \qquad (7.18a)$$

Thus, when $R = 3$, 3 moles of liquid product are returned to the column for each mole withdrawn.

We can now recast the operating line (Equation 7.16d) in terms of this new parameter $R$ by using a set of revised mole balances around the envelope shown in Figure 7.16a. We have in the first instance

$$\text{Rate of total moles in} - \text{Rate of total moles out} = 0$$

$$G - (L + D) = 0 \tag{7.18b}$$

and obtain after combining this expression with Equation 7.18a:

$$G = D(R + 1) \tag{7.18c}$$

The corresponding component mole balance is given by

$$Gy_{n+1} - (Lx_n + Dx_D) = 0 \tag{7.18d}$$

or

$$y_{n+1} = \frac{L}{G}x_n + \frac{D}{G}x_D \tag{7.18e}$$

Consequently, using Equation 7.18a and Equation 7.18c yields

$$y_{n+1} = \frac{R}{R+1}x_n + \frac{x_D}{R+1} \tag{7.18f}$$

This is the equation of a straight line with slope $R/(R + 1)$, which has an intercept on the ordinate of $x_D/(R + 1)$ and passes as well through the point $y = x_D$ on the 45° diagonal. This point and the $y$-intercept, which are usually known or prescribed, permit easy construction of the operating line, as shown in Figure 7.16b. The concentration associated with the various trays can then be derived using the "staircase" construction we employed for the countercurrent gas absorber described in Section 7.1.4 and Figure 7.13a. Each point on the curve $y^* = f(x)$ represents the concentrations leaving a particular tray, $(y_n, x_n)$, which are in equilibrium with each other, while those on the operating line relate the entering vapor composition to that of the exiting liquid.

Let us next consider the corresponding balances around the exhausting section, shown in Figure 7.17a. Here there is no simple relation to the reflux ratio $R$, but the operating line (Equation 7.16e) is nevertheless considerably simplified because the flow rates are now constant throughout the entire section, although they differ in magnitude from those of the rectifying portion of the column. We have for the total and component mass balances, respectively,

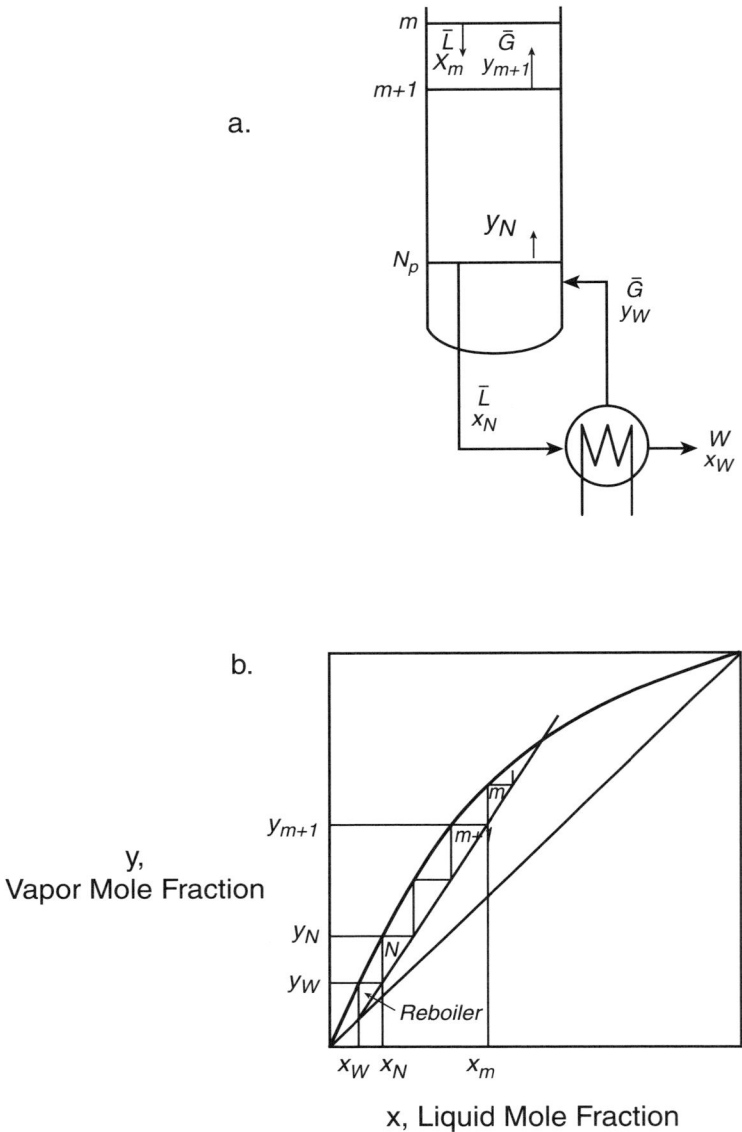

**FIGURE 7.17**
Details of the exhausting (stripping) section: (a) flow diagram; (b) operating diagram.

Rate of moles in − Rate of moles out = 0

$$\bar{L} - (\bar{G} + W) = 0 \tag{7.19a}$$

and

$$\bar{L}x_m - (\bar{G}y_{m+1} + Wx_w) = 0 \qquad (7.19b)$$

so that

$$y_{m+1} = \frac{\bar{L}}{\bar{G}}x_m - \frac{W}{\bar{G}}x_w \qquad (7.19c)$$

or alternatively

$$y_{m+1} = \frac{\bar{L}}{\bar{L} - W}x_m - \frac{W}{\bar{L} - W}x_w \qquad (7.19d)$$

This is the equation for the straight operating line shown in Figure 7.17b, which has a slope of $\bar{L}/\bar{L} - W$ and passes through the point $y = x = x_w$ on the 45° diagonal. If the vapor entering the column $y_w$ is in equilibrium with the residue composition xw, the reboiler can be taken to represent an additional equilibrium stage, and this is so indicated in the operating diagram of Figure 7.17b. Once the operating line is established, the usual staircase construction can be used to step off the number of stages, as shown in Figure 7.17b. It is not clear, however, how this is to be accomplished, because only one point, $y = x_w$ on the diagonal, is known or prescribed and we have no *a priori* knowledge of either a second point or the slope of the line. In other words, the equation for the stripping operating line, as it stands, contains too many undefined variables. Some reflection will show that this must indeed be so. This follows from the fact that the liquid flow will be influenced by what comes down from the feed tray, and we must consequently draw on an additional balance, performed around that tray.

We first compose the total mole balance, and follow this with an energy balance around the feed tray. We obtain

Rate in − Rate out = 0

$$(F + L + \bar{G}) - (G + \bar{L}) = 0 \qquad (7.20a)$$

and

$$FH_F + LH_L + \bar{G}H_G = GH_G + \bar{L}H_L \qquad (7.20b)$$

These two equations can be combined and rearranged to yield the expression

$$\frac{\overline{L} - L}{F} = \frac{H_G - H_F}{H_G - H_L} = q \qquad (7.20c)$$

The ratio of enthalpy differences, which appears in this equation and which we have denoted $q$, represents the molar heat of vaporization of the feed divided by the molar latent heat of vaporization of the binary system, assumed to be constant. Suppose, for example, that the feed consists of either saturated vapor or saturated liquid. Then the value of $q$ will be 0 or 1, respectively, and for a partially vaporized feed it will lie somewhere between these two limits. The quantity $q$ is therefore a dimensionless measure of the thermal quality of the feed. It turns out that $q$ also enters into the construction of the *locus of the points of intersection* of the two operating lines of the enriching and stripping sections. That locus is represented by the expression

$$y = \frac{q}{q-1}x - \frac{x_F}{q-1} \qquad (7.20d)$$

which for a given $q$ is the equation of a straight line of slope $q/(q-1)$ passing through the point $y = x = x_F$. A series of such lines for various thermal conditions of the feed are shown in Figure 7.18. Proof of these relations is somewhat lengthy and can be found in standard texts.

### 7.1.5.3   Minimum Reflux Ratio and Number of Plates

The reader will recall that in the discussion of the countercurrent gas scrubber, Section 7.1.4, we mention the limiting case that arises when the slope of the operating line and the associated solvent flow is progressively reduced until it intersects the equilibrium curve. This results in a condition termed a *pinch* and corresponds to a cascade with an infinite number of stages and a minimum solvent requirement. Any solvent flow rate below this value causes a rise in the effluent concentration and can therefore no longer meet the specified solute recovery.

In distillation, the liquid reflux returned to the top of the column plays, in a sense, very much the role of a solvent. In the course of its downward flow, it dissolves residues of the heavy component contained in the vapor phase and thereby contributes to its enrichment in the lighter, more volatile component. A reduction of reflux, or of the reflux ratio, may therefore be expected to result in an *increase* in the required number of stages in much the same way as happens in the case of the gas scrubber. Ultimately, when the operating lines simultaneously intersect the equilibrium curve, a pinch results and the number of stages goes to infinity. This is shown in Figure 7.19a. The reflux ratio at which this occurs can be read from the intercept $x_D/(R+1)$ and represents the *minimum required to achieve the desired separation*.

Let us now proceed in the opposite direction and progressively *increase* the reflux ratio. Both the logic of the preceding argument and Figure 7.19

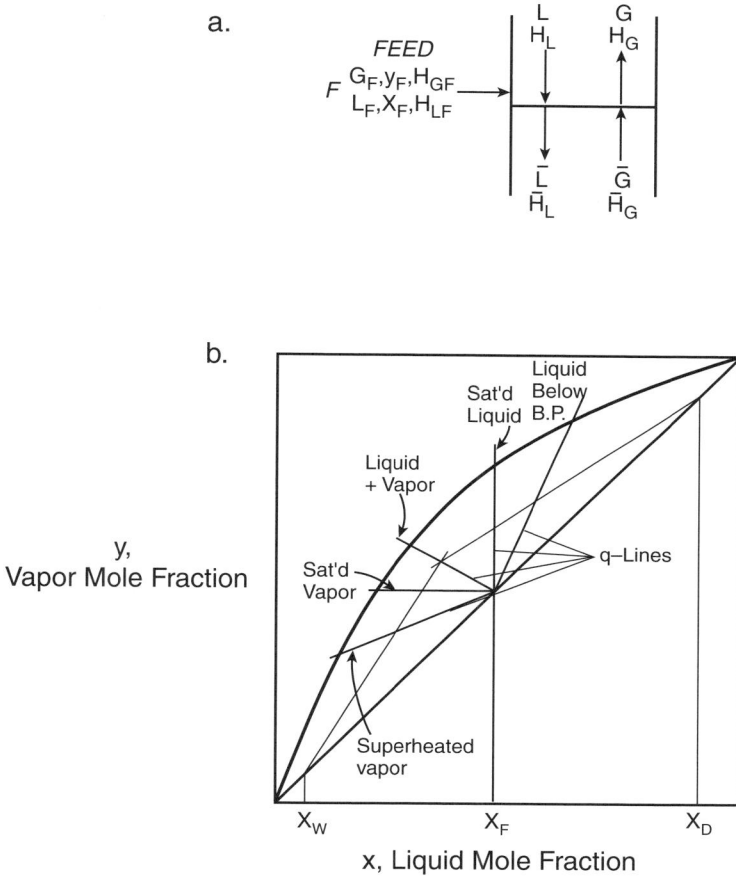

**FIGURE 7.18**
The feed plate: (a) flow diagram; (b) feed quality and the *q*-line.

indicate that this will lead to a *decrease* in the number of theoretical trays required. A limit is reached when no distillate is withdrawn and the entire overhead product is returned to the column as reflux. The operation is then said to be at total reflux. The reflux ratio becomes infinity, $R = L/D = L/0 = \infty$, and the operating line assumes a slope of 1; i.e., it coincides with the 45° diagonal. This is shown in Figure 7.19b.

These two asymptotic cases are immensely useful in conveying to the analyst the lower limits of $R$ and $N$, below which the desired separation will no longer proceed. In an actual operation, these values evidently must be exceeded and both $R$ and $N$ will assume finite values. The best or optimum value of $R$ and $N$ to be used will be determined by economic considerations, i.e., when the total cost consisting of fixed and operating costs is at a minimum. Let us see how this minimum comes about.

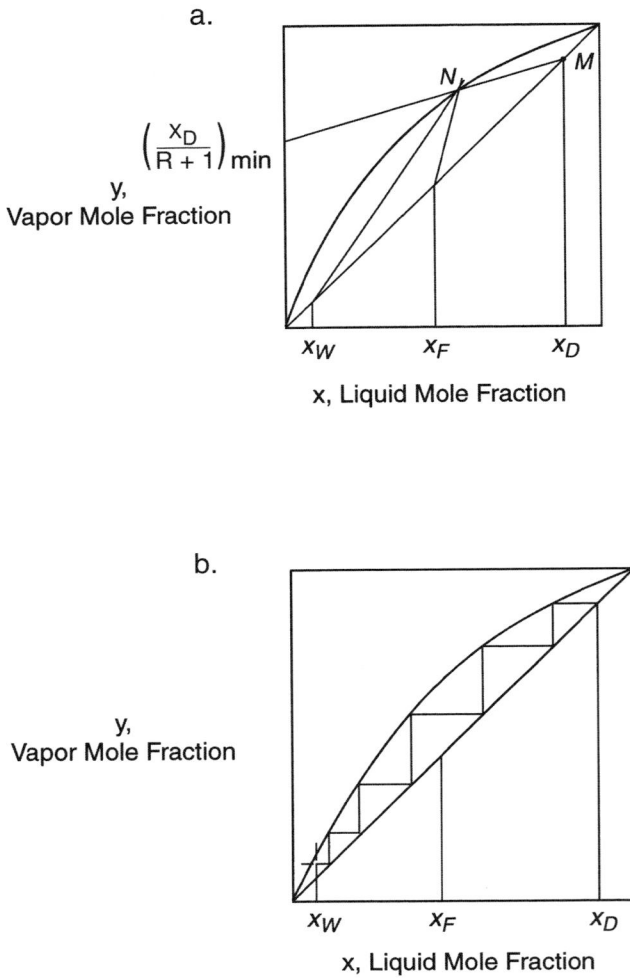

a.

$\left(\dfrac{x_D}{R+1}\right)_{min}$

y,
Vapor Mole Fraction

$x_W$       $x_F$       $x_D$

x, Liquid Mole Fraction

b.

y,
Vapor Mole Fraction

$x_W$       $x_F$       $x_D$

x, Liquid Mole Fraction

**FIGURE 7.19**
Two limiting conditions: (a) minimum reflux; (b) minimum number of trays.

Operating costs, consisting of condenser and reboiler loads as well as pumping costs, are at their lowest value when $R$ is at a minimum. As $R$ is gradually increased, these costs begin to rise in almost direct proportion and ultimately tend to infinity as $R \to \infty$. There is consequently no minimum in this item. The fixed costs, on the other hand, are at first infinite at $R_{Min}$ because of the infinite number of trays required. They then sharply drop as $R$ is slowly increased above its minimum value because the number of trays has now become finite. As this process is continued, however, the size and cost of the condenser, reboiler, and reflux pump begin to creep up, eventually overtaking the cost reduction occasioned by the decrease in the number of plates. The fixed costs consequently, and in contrast to the operating costs, pass through a minimum and the sum of the two results in an optimum reflux ratio, $R_{Opt}$. It has been found in practice that this optimum lies close to the minimum reflux ratio and is frequently, but not always, found in the range 1.2 to 1.5 times $R_{Min}$. This set of values is invariably used in the preliminary design of fractionation columns.

The reader is referred to Figure 8.2 for a similar optimization problem, which arises in the operation of gas scrubbers.

### Illustration 7.8: Design of a Distillation Column in the McCabe–Thiele Diagram

The intent of this example is to acquaint the reader with the principal steps involved in designing a distillation column and to demonstrate their implementation by means of the McCabe–Thiele diagram. We start by listing the parameters, which in the usual course of the design are either prescribed or known *a priori*:

1. Feed rate $F$ and composition $x_F$
2. Thermal condition of feed
3. Distillate composition $x_D$ and bottoms composition $x_w$
4. Reflux ratio, usually specified as a multiple of the minimum reflux ratio
5. Thermal condition of the overhead and bottoms product

The parameters or quantities to be calculated are as follows:

1. Number of theoretical stages
2. Heat load for the condenser and reboiler
3. Recovery of overhead product

We consider a hypothetical system with an $x$–$y$ equilibrium relation as shown in Figure 7.20. The feed, entering at the rate of 10 mol/s, is known to have a composition $x_F = 0.38$ and consists entirely of saturated liquid at its boiling point. Overhead and bottoms compositions are specified at $x_D = 0.92$ and $x_w = 0.01$, respectively. A value 1.5 times the minimum reflux ratio is to be used.

The first step in the procedure is to establish the $q$-line. This is a straightforward matter because the feed consists of saturated liquid and the $q$-line is consequently vertical.

In the second step, an operating line is drawn from distillate composition $y = x = x_D$ through the point of intersection of the $q$-line with the equilibrium curve, which results in a pinch and represents minimum reflux conditions. The intercept of this line on the ordinate is given by $x_D/(R_{Min} + 1)$ and establishes the minimum reflux ratio. We have

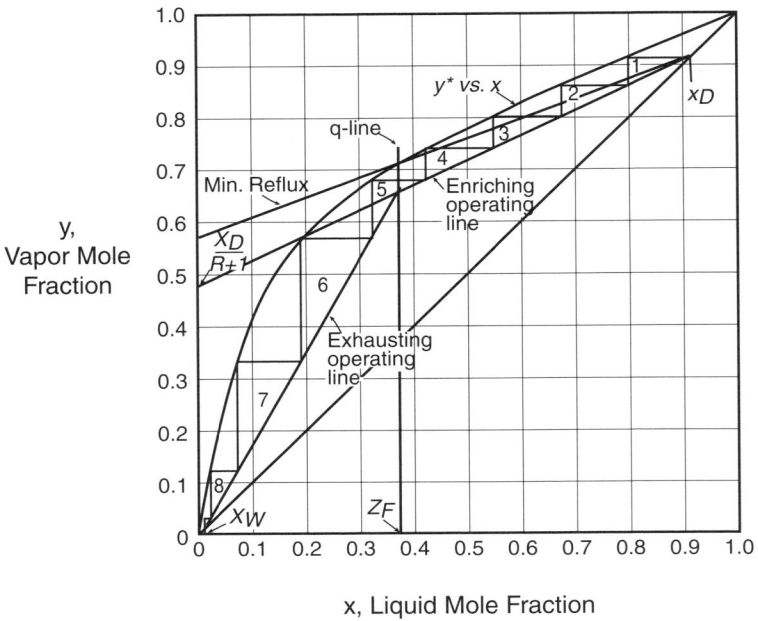

**FIGURE 7.20**
Operating diagram for Illustration 7.6.

$$\frac{x_D}{R_{Min}+1}=0.575 \tag{7.21a}$$

and consequently

$$R_{Min}=\frac{0.92}{0.575}-1=0.6 \tag{7.21b}$$

The actual reflux ratio to be used is 1.5 times this minimum; i.e., $R = 1.5 \times 0.6 = 0.9$.

We now repeat the second step, but this time draw the line through the new intercept, i.e., $x_D/(R + 1) = 0.92/(0.9 + 1) = 0.48$. This is the actual operating line to be used in stepping off the number of theoretical trays.

In step four, we start the staircase construction at the distillate composition $y = x = x_D$ and proceed downward, alternating between equilibrium curve and the enriching operating line. When the feed composition is reached, a crossover is made to the exhausting operating line, with the feed tray 5 straddling the $q$-line. The construction is then continued until we reach the bottoms composition $y = x = x_w = 0.01$. This occurs after eight stages have been stepped off. Since the liquid bottoms product and the vapor returning to the column are usually at or near equilibrium, it is customary to consider the reboiler an additional theoretical stage. The column will consequently comprise eight theoretical stages plus the reboiler.

Let us next turn to the computation of condenser and reboiler heat loads. We assume that the relevant enthalpies have been computed and are given by

$$H_G = 50 \text{ kJ/mol}, \ H_{L,D} = 10 \text{ kJ/mol}, \ H_{L,W} = 12 \text{ kJ/mol}$$

To compute the heat loads, we require the vapor flow rates into the condenser and out of the reboiler. These are obtained by a series of total and component balances, starting with those taken over the entire column. Thus,

$$\text{Rate of moles in} - \text{Rate of moles out} = 0$$

$$F - (W + D) = 0 \tag{7.21c}$$

and

$$x_F F - (x_W W + x_D D) = 0 \tag{7.21d}$$

or

$$10 - (W + D) = 0 \tag{7.21e}$$

and

$$0.38 \times 10 - (0.01 \ W + 0.92 \ D) = 0 \tag{7.21f}$$

from which

$$W = 5.93 \text{ mol/s and } D = 4.07 \text{ mol/s} \tag{7.21g}$$

We then obtain from Equation 7.18c

$$G = D(R + 1) = 4.07(0.9 + 1) = 7.73 \text{ mol/s} \tag{7.21h}$$

Because the feed enters entirely as a liquid, the vapor flow rate will remain constant over the entire column, so that $\overline{G} = G$.

We consequently have for the heat loads

$$q_C = G(H_G - H_{L,D}) = 7.73 \ (50 - 10) \tag{7.21i}$$

$$q_C = 309 \text{ kJ/s} \tag{7.21j}$$

and

$$q_B = \overline{G}(H_G - H_{L,w}) = 7.73(50 - 12) \qquad (7.21\text{k})$$

$$q_B = 294 \text{ kJ/s} \qquad (7.21\text{l})$$

The recovery $r$ of overhead product is given by

$$r = \frac{x_D D}{x_F F} = \frac{0.92 \times 4.07}{0.38 \times 10} = 0.985 \text{ or } 98.5\% \qquad (7.21\text{m})$$

*Comments:*

This example was used to illustrate to the reader the use of the McCabe–Thiele diagram and the elegant and simple way in which it conveys the design information. In practice, the systems tend to be more complex than the simple binary example used here, and the computations are done using appropriate computer packages. Today, these packages are quite powerful and are able to handle mixtures of many components without recourse to the simplifying assumption of equimolal overflow and vaporization. The McCabe–Thiele diagram nevertheless remains a valuable tool for visualizing the principal features of the fractionation process and for providing the student an entry into the treatment of more complex systems.

### Illustration 7.9: Isotope Distillation: The Fenske Equation

Vapor–liquid equilibria of isomeric or isotopic mixtures perhaps come closest to showing perfect ideal behavior and a separation factor, which for all practical purposes remains constant over the entire range of compositions. Distillation is the most commonly used technique for the separation or enrichment of several important isotopes. Both $C^{13}$ and $O^{16}O^{18}$ are produced by the low-temperature distillation of carbon monoxide and oxygen in small commercial installations. Although these processes are generally carried out in packed columns, all preliminary design questions are settled by deducing the number of theoretical plates for a required separation, starting with the minimum number needed to achieve the desired result. These values are then translated into packing heights using the concept of the height equivalent to a theoretical plate (HETP), which has been taken up in Chapter 5, Section 5.4.

Suppose we wish to gain an idea of the requirements for distilling the isotopic pair $C^{12}O$–$C^{13}O$, for which the value of $\alpha$ is 1.01 (see Table 6.9) with the equilibrium curve very close to the diagonal, which represents total reflux conditions. Here, a graphical construction on the McCabe–Thiele diagram does not recommend itself, since the operating and equilibrium curves are too close together to allow a precise determination of the number of theoretical plates. Fortunately, for systems with a constant separation factor and operating at total reflux, a simple analytical treatment is possible, which leads to a relation between the minimum number of plates, the separation factor $\alpha$, and the product compositions. This expression, known as the Fenske equation, is derived as follows.

We start by applying the definition of the separation factor, Equation 6.22a, to the reboiler and obtain

$$\frac{y_w}{1-y_w} = \alpha \frac{x_w}{1-x_w} \tag{7.22a}$$

At total reflux the operating line coincides with the 45° diagonal so that $y_w = x_N$. Equation 7.22a becomes

$$\frac{x_N}{1-x_N} = \alpha \frac{x_w}{1-x_w} \tag{7.22b}$$

A similar scheme can be applied to the Nth plate, yielding

$$\frac{y_N}{1-y_N} = \alpha \frac{x_N}{1-x_N} = \alpha^2 \frac{x_w}{1-x_w} \tag{7.22c}$$

Continuing the procedure up the column we ultimately obtain

$$\frac{x_D}{1-x_D} = \alpha^{N+1} \frac{x_w}{1-x_w} \tag{7.22d}$$

or equivalently

$$N+1 = \frac{\ln \dfrac{x_D}{1-x_D} \dfrac{1-x_w}{x_w}}{\ln \alpha} \tag{7.22e}$$

This is the expression due to Fenske.

A further simplification results for separation factors close to 1, for then we have by a Taylor-expansion of the denominator

$$\ln \alpha \cong \alpha - 1 \quad \alpha \approx 1 \tag{7.23a}$$

and consequently

$$N+1=\frac{1}{\alpha-1}\ln\left[\frac{x_D}{1-x_D}\frac{1-x_w}{x_w}\right]\qquad(7.23b)$$

Let us apply this expression to the distillation of the carbon monoxide isotopes, with the aim of obtaining products of $x_D = 0.9$ and $x_w = 0.1$. We have

$$N+1=\frac{1}{1.01-1}\ln\frac{0.9}{0.1}\frac{1-0.1}{0.1}\qquad(7.24a)$$

and therefore

$$N+1=439\qquad(7.24b)$$

or

$$N=438\qquad(7.24c)$$

Thus, a *minimum* of more than 400 theoretical plates is required to achieve the desired separation; this is an enormous number. Fortunately, high-efficiency packings are now available with equivalent heights HETP of the order of 1 cm. This limits the size of the column to heights that are not excessive and can be implemented in practice.

*Comments:*

One feature of isotope distillation that may have been noted by the reader is the extreme sensitivity of the number of plates $N$ to the value of $\alpha$. Suppose that we manage by some means — for example, by a change in operating pressure — to increase the separation factor of the CO isotopes from 1.01 to 1.02. This represents a rather modest increase of only 1% in the value of $\alpha$, which would normally have no more than a marginal effect on $N$. For separation factors close to 1, however, the effect is dramatically enhanced as a result of the appearance of the term $\alpha - 1$ in the denominator of Equation 7.23b. The seemingly insignificant change in $\alpha$ of only 1% translates here into a reduction of the number of plates from 438 to fully one half that value. This will evidently result in a considerable reduction in both the size and cost of the fractionation column. Any increase in $\alpha$, no matter how small in appearance, is therefore welcome in isotope or isomer distillation, and methods for achieving this increase should be fully explored before settling on a particular process.

### Illustration 7.10: Batch-Column Distillation: Model Equations and Some Simple Algebraic Calculations

Frequently, in the production of rare and valuable substances, the material flow is too small to make continuous distillation for the purpose of separation or purification practicable. The intermediate or raw product is then accumulated and held in storage before being fed to the fractionation column in intermittent batches.

The distillation process, which is now an unsteady one, can be carried out in two modes. In the first mode, we allow distillation to proceed without outside intervention and at constant reflux until a prescribed fraction of the charge has been boiled off or the still contents have been concentrated to some desired value. In the course of the process, both the instantaneous and cumulative distillation composition $x_D$ and $x_D'$, as well as the contents of the still, undergo a slow and continuous change.

In the second mode of operation, which is more common, the reflux ratio is continually and automatically adjusted to maintain a constant $x_D$. The boiler contents still change, as does the reflux ratio, but both the instantaneous and cumulative overhead compositions remain invariant with time.

We now introduce two assumptions that allow us to draw on the principles and diagrams we have previously established and used to describe *continuous* distillation. The first assumption presumes that the liquid content of the trays, or *hold-up* as it is termed, is negligible compared to the still contents. This allows us, in any material balance performed over the column, to ignore the contribution due to hold-up. In the second assumption, we stipulate that the process is sufficiently slow that the system has time to adjust to a quasi-steady state at any instant of the operation. This highly important assumption allows us to represent both modes of operation on a McCabe–Thiele diagram, which was originally derived for steady-state operations. The difference here is that instead of dealing with a single operating line, we now must deal with a continuous spectrum of such lines, which can, however, be accommodated on a single diagram. This has been done in Figure 7.21, which displays the two cases of constant and varying reflux we have described. The first case is displayed as follows:

*Case 1: Constant $x_D$, Variable R*

1. *Minimum Reflux Ratio.* Here we are dealing with a single operating line that extends from the desired overhead composition $x_D$ to the batch-feed composition denoted $x_F$. The line is drawn through the points $x = y = x_D$ and $x_W^o, y_W^*$, as shown in Figure 7.21b. The minimum reflux ratio is then determined from the intercept $x_D/(R_{Min} + 1)$.

2. *Number of Theoretical Plates.* An actual reflux ratio is set next, which leads to an operating line with intercept $x_D/(R + 1)$, as shown in Figure 7.21b. The number of theoretical plates is then stepped off between operating and equilibrium curves, as was done in the case of continuous fractionation, and yields a total of four stages between the composition $x_D$ and $x_W^o$. As the distillation proceeds, the reflux ratio is gradually increased to maintain a constant overhead composition, causing the bottoms mole fraction to diminish. The process can be

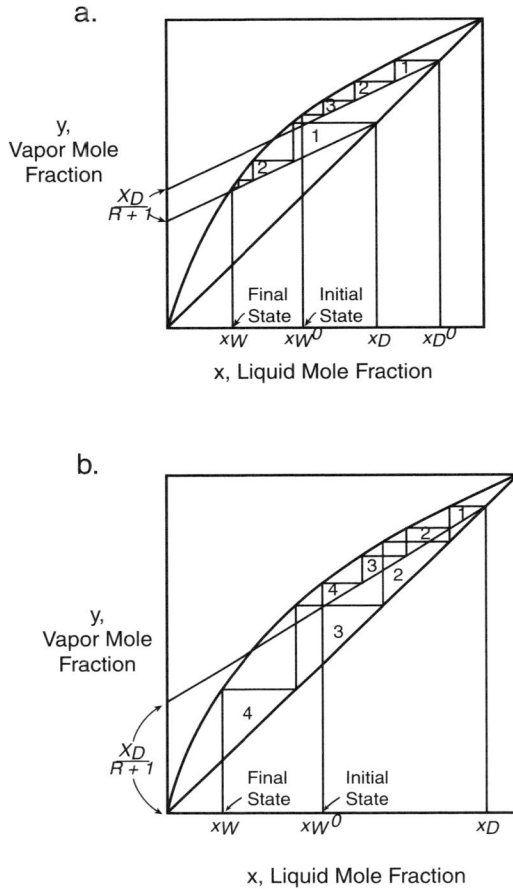

a.

y,
Vapor Mole
Fraction

$\dfrac{x_D}{R+1}$

| Final | Initial |
| State | State |

$x_W$   $x_W{}^0$      $x_D$   $x_D{}^0$

x, Liquid Mole Fraction

b.

y,
Vapor Mole
Fraction

$\dfrac{x_D}{R+1}$

Final        Initial
State        State

$x_W$        $x_W{}^0$              $x_D$

x, Liquid Mole Fraction

**FIGURE 7.21**
The two types of batch distillation processes: (a) varying distillate composition with constant reflux ratio; (b) varying reflux ratio with constant distillate composition.

stopped when a prescribed composition or recovery is reached, or it can be continued until the column is at total reflux. $x_W$ is then at its lowest point and fractional recovery at its maximum. We have reached the final state shown in Figure 7.21b.

3. *Fraction Distilled and Recovery.* Much of the time in actual operations, we wish to determine the fraction $f = 1 - W/W^o$ that needs to be distilled to achieve a prescribed bottoms composition $x_W$ or recovery $r$ defined as $r \equiv 1 - x_W W / x_W{}^o W^o = 1 - (1 - f)(x_W/x_W{}^o)$. Although it is not immediately clear how we should proceed to obtain these quantities, a good way to start is by composing total and component material balances about the entire column. We have

$$\text{Rate of moles in} - \text{Rate of moles out} = \frac{d}{dt}\text{contents}$$

$$0 - D = \frac{d}{dt}W \tag{7.25a}$$

and

$$0 - x_D D = \frac{d}{dt}x_w W \tag{7.25b}$$

These are the same equations, in form at least, as those we have seen in the Rayleigh distillation of Illustration 7.3. The difference here is that $x_D$ is *not* in equilibrium with $x_W$ but represents an independent entity set by the operator. Proceeding as in Illustration 7.3, the two equations are divided to eliminate $D$ and $dt$, and we obtain, after some manipulation, the result

$$\int_{W^o}^{W} \frac{dW}{W} = \ln\frac{W}{W^o} = \int_{x_W^o}^{x_W} \frac{dx_W}{x_D - x_W} \tag{7.25c}$$

Because $x_D$ is the prescribed and constant overhead composition, Equation 7.25c can be immediately integrated to yield

$$\frac{W}{W^o} = \frac{x_D - x_W^{\,o}}{x_D - x_W} \tag{7.25d}$$

After some rearrangement this becomes, for the volatile component,

$$W^o x_W^o \quad = \quad (W_o - W)x_D \quad + \quad W x_W \tag{7.25e}$$

| Initial | Moles | Moles |
| Moles | Distilled | Left Over |

In other words, what we have derived here is nothing but a cumulative balance to time $t$ of the volatile component, which could have been obtained directly by algebraic means. It can be rearranged to obtain the desired relation between fraction distilled $f$ and recovery $r$ in terms of the known values of $x_D$ and $x_W^o$ This results in the following simple expressions:

$$r = 1 - x_W W / x_W{}^o W^o = f x_D / x_w{}^o$$

or

$$r = 1 - (1-f)(x_W / x_W{}^o) = f x_D / x_w{}^o \qquad (7.25f)$$

Note that the first part of each expression represents the *definition* of $r$, and the second part comes from the cumulative balance (Equation 7.25e). Of the host of equations available, Expression 7.25f is by far the most fruitful and also the simplest to apply. It is quite generally valid, subject only to the condition that $x_W$ cannot fall below the minimum attained at total reflux. We demonstrate its use by returning to Figure 7.20 of Illustration 7.8 and use the same prescribed overhead composition $x_D = 0.92$ and reflux ratio $R = 0.9$, but raise $x_W{}^o$ to 0.55 so that exactly three plates are accommodated on the operating diagram. To obtain the maximum recovery possible under these conditions, we carry the process to total reflux and determine the corresponding $x_W$. The reader will verify that if we step off three plates, starting at $x_D = 0.92$ and using the diagonal as operating line, we obtain $x_W = 0.175$. To calculate $r_{Max}$, we first solve Equation 7.25f for fraction distilled $f$ and then by backsubstitution for recovery $r$. This yields

$$f = \frac{x_W{}^o - x_W}{x_D - x_W} = \frac{0.55 - 0.175}{0.92 - 0.175} = 0.503 \qquad (7.25g)$$

and by backsubstitution

$$r_{Max} = f \frac{x_D}{x_W{}^o} = 0.503 \frac{0.92}{0.55} \qquad (7.25h)$$

$$r_{Max} = 0.84 \qquad (7.25i)$$

Thus the maximum recovery we can achieve with three plates and a reflux ratio of 0.9 is 84%. Let us next turn to the second case shown in Figure 7.21a.

*Case 2: Constant R, Variable $x_D$*

This case, which is less frequently used, differs from the previous process in several respects. We are here no longer dealing with a *design* problem, since both $x_D$ and $x_W$ vary continuously and cannot be prescribed by the analyst as fixed design parameters. Rather than designing a column, we usually use the model equations to *predict* the performance of an existing or hypothetical unit with a given number of plates $N_p$ and a prescribed value of $R$.

We start by establishing the operating line that will accommodate the existing number of plates and initial boiler composition $x_W°$. This involves a trial-and-error procedure consisting of adjusting the operating line of constant slope $R/(R + 1)$ until it accommodates exactly $N_p$ plates. $x_D°$ of this line is the initial composition that emerges from the column (Figure 7.21a). The operation continues, with overhead and bottoms compositions decreasing steadily, until a final prescribed bottoms composition $x_W$ is reached. The operational parameters for the entire process are then established as follows:

1. *Fraction Distilled* f. This item is obtained from Equation 7.25c, written in the form

$$\ln \frac{W}{W°} = \ln(1-f) = \int_{w_W°}^{x_W} \frac{dx_W}{x_D - x_W} \qquad (7.26a)$$

   Here we can no longer evaluate the integral analytically because both $x_D$ and $x_W$ vary in unrelated ways. Instead, values of $x_D$ and $x_W$ must be read off the operating diagrams in pairs and the integral determined graphically or numerically. This is tedious, but not overwhelmingly so.

2. *Average Distillate Composition* $x_D'$. This is a new item that must be addressed because the overhead composition is no longer constant. It is obtained from the same cumulative balance equation (Equation 7.25e) that was used to describe the previous case of constant $x_D$. This equation applies here as well with $x_D'$ now taking the place of $x_D$. Solving Equation 7.25e for $x_D'$, we have

$$x_D' = \frac{x_W W° - x_W W}{W° - W} = \frac{x_W° - x_W(1-f)}{f} \qquad (7.26b)$$

   where $x_W$ is prescribed, and $f$ drawn from Equation 7.26a. Note that Equation 7.26b is identical to the cumulative composition we have derived for the simple batch still in Illustration 2.9.

With both $f$ and $x_W$ in hand, recovery $r$ can then be calculated from Equation 7.25f.

Let us return to the same example considered above, with $x_D$ and $x_W°$ set at 0.92 and 0.55, and $N_p$ at 3. Reflux ratio $R$ is the same as before at 0.9 but is now kept constant. The aim is to calculate $f$ and $r$ for the same final bottoms compositions $x_W = 0.175$ used previously.

We start by drawing a series of parallel operating lines and reading off values of $x_D$ and $x_W$, which are used in the graphical evaluation of the integral in Equation 7.26a. The reader will verify that this leads to a value of −0.913 and hence a fraction distilled $f$ of 0.60.

The next step is to evaluate the cumulative distillate composition $x_D'$ at the end of the process. We have from Equation 7.26b

$$x_D' = \frac{x_W^o - x_W(1-f)}{f} = \frac{0.55 - 0.15(1-0.6)}{0.6} = 0.817 \qquad (7.26c)$$

Using this value in Equation 7.25f yields the final result

$$r = f\frac{x_D'}{x_W^o} = 0.6\frac{0.827438}{0.55} \qquad (7.26d)$$

$$r = 0.89 \text{ or } 89\% \qquad (7.26e)$$

Thus, although recovery has increased by 5%, the distillate composition is lower by more than 10% over the previous case. Evidently, these results can be manipulated in any number of ways by adjusting $N_p$ and $R$ to achieve more desirable results.

## 7.1.6   Percolation Processes

Percolation processes refer to operations in which a fluid stream is passed through a bed of granular porous material and a transfer of mass takes place between the two phases. Such processes are seen in the purification of gases and liquids by adsorption and ion exchange, in the transfer of toxic substances from aqueous streams to surrounding soils or river beds, and in general whenever a fluid percolating through or over a mass of stationary porous solids exchanges material with it.

In Illustration 6.4, we introduced the reader to the notion of applying equilibrium stage concepts to operations of this type. This is at first glance a startling approach because none of the processes that fall in this category remotely resembles a stirred tank or its equivalent. They are, in fact, distributed in both time and distance and generally require PDEs for a rigorous description of the events. It has been shown, however, that if flow is sufficiently slow, we can assume local equilibrium to be established, or at least closely approached, at any point of the system. The process can then be viewed as the composite of a continuous spectrum of equilibrium stages and, therefore, that an algebraic representation becomes possible. This was shown in Illustration 6.4 for the restricted case of a linear equilibrium described by an appropriate Henry's constant.

We now extend this treatment to the more general case of a nonlinear equilibrium case of the Langmuir type. Composing a cumulative solute balance to time $t$, as was done in Illustration 6.4, we obtain

$$Y_F G_b A_c t \quad = \quad X_F \rho_b A_c z \quad + \quad \varepsilon Y_F \rho_f A_c z \tag{7.27a}$$

$$\underset{\text{Introduced}}{\underset{\text{Amount}}{}} \qquad \underset{\text{Retained by Solid}}{\underset{\text{Amount}}{}} \qquad \underset{\text{Left in Fluid}}{\underset{\text{Amount}}{}}$$

which is identical in form to Equation 6.6a, but assumes a general nonlinear relation between $q_F$ and $Y_F$. Neglecting the last term, as before, we obtain after some rearrangement

$$z/t = \frac{G_b}{\rho_b X_F / Y_F} \tag{7.27b}$$

where $G_b$ is the mass velocity of the carrier fluid in $kg/m^2 \, s$. When the solid phase already contains some solute at a concentration level $q_0$, $Y_0$, we can recast Equation 7.27b into the more general form

$$z/t = \frac{G_b}{\rho_b \Delta X / \Delta Y} = \frac{v \rho_f}{\rho_b \Delta X / \Delta Y} \tag{7.27c}$$

where $\Delta X / \Delta Y = (X_F - X_0)/(Y_F - Y_0)$ and $v$ = superficial velocity of the fluid.

An alternative form results if we solve Equation 7.27c for $t$ and set $z = L$, where $L$ is a particular position downstream from the inlet to the system:

$$t = \frac{\rho_b \Delta X / \Delta Y}{v \rho_f} L \tag{7.27d}$$

This confirms the intuitive notion that the greater the distance $L$ and the greater the adsorptive loading $\Delta X$, the longer it will take the solute front to reach position $L$. Conversely, the faster the fluid flow $v$, the shorter the time required for the solute to "break through" at position $L$.

We can now construct a diagram that contains the equilibrium relation $X^* = f(Y)$ as well as an operating line, the slope of which equals the ratio $\Delta X / \Delta Y$. This is shown in Figure 7.22 and leads to the two equivalent expressions:

$$t = \frac{\rho_b \Delta X / \Delta Y}{\rho_f v} L = \frac{\rho_b L}{\rho_f v} \times (\text{Slope of operating line}) \tag{7.27e}$$

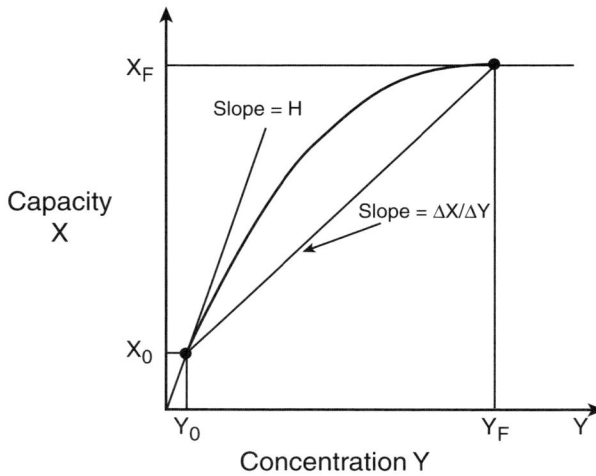

**FIGURE 7.22**
Operating diagram for Langmuir-type isotherm.

The construction has the same attractive property as previous operating diagrams in conveying at a glance the manner in which the system reacts to changes in certain operating parameters. Thus, if the feed concentration is increased, $\Delta Y$ will likewise increase, lowering the slope of the operating line, and thus leading to faster breakthrough. If, on the other hand, $\Delta X$ is increased, for example, through the use of a more efficient sorbent, the movement of the solute front will slow and breakthrough will occur much later. In typical water purification processes, $\rho_b/\rho_f$ is of the order 1, $v$ of the order 1 cm/s, and $L \times$ slope of the order $10^7$. The on-stream time is then of the order $10^7$ s $\cong$ 100 days under equilibrium conditions.

So far in our discussion, the movement of solute was assumed to be entirely in one direction, i.e., from the flowing fluid to the stationary solid. This is the case in adsorption processes or in the uptake step of ion-exchange operations. When the direction of transfer is reversed, we speak of *desorption*, *regeneration*, or, in the case of environmental systems, of *clearance*. The theoretical treatment here becomes more complex, and requires a more profound approach based on PDEs. We do not address this problem here and instead present the final result that emerges from that analysis for the case of complete desorption from a Langmuir-type isotherm under equilibrium conditions. The relevant equation is completely analogous in form to that for the adsorption step (Equation 7.27e), and reads

$$t = \frac{\rho_b H}{\rho_f v} L = \frac{\rho_b L}{\rho_f v}\left(\begin{array}{c}\text{slope of equilibrium}\\\text{curve at the origin}\end{array}\right) \qquad (7.27f)$$

Thus, the only change that has occurred in passing from adsorption to desorption is the replacement of the slope of the operating line $\Delta X/\Delta Y$ by the slope of the equilibrium curve at the origin, i.e., the Henry constant. Let us now apply these two expressions to an environmental problem of interest.

### Illustration 7.11: Contamination and Clearance of Soils and River Beds

When soils or sediments are exposed to contaminants contained in ground-water or in the river flow, an important question arises: What is the length of the recovery period required to restore the system to its original state once contamination has ceased? Equation 7.27e and Equation 7.27f provide some important guidelines that can be used to address this question.

We start by noting that in the general case of a nonlinear equilibrium relation, the adsorption and desorption periods will always differ because of the different values of $\Delta q/\Delta Y$ and $H$ (Figure 7.22). Because the latter quantity is invariably the greater of the two, we conclude that for nonlinear isotherms the adsorption step will proceed at a faster pace than the corresponding desorption step. This difference becomes more pronounced, the steeper the equilibrium curve at the origin.

Let us now turn to the case where the equilibrium is linear; i.e., the operation is entirely in the Henry's law region. This is a common, although not exclusive, occurrence in environmental systems. Inspection of Figure 7.22 shows that in this case the two relevant slopes become identical; i.e., we have

$$(\Delta X/\Delta Y)_{ads} = (\Delta X/\Delta Y)_{des} = H \tag{7.28}$$

This leads to the surprising but also reassuring conclusion that, for linear systems, the clearance period will always equal that for contamination. Long recovery times arise, but only as a consequence of long exposure times. The shorter the period of accidental contamination, the greater the prospects for a fast recovery. This agrees, of course, with our physical understanding of the process, but it requires the use of Equation 7.27e and Equation 7.27f to establish that the contamination and recovery periods are in fact identical.

---

## 7.2 Stage Efficiencies

We turn now to the consideration of the second aspect of the equilibrium stage, that of its *efficiency*. This is an area of much greater uncertainty than we have previously seen. Evidently, the efficiency of a stage will be affected in a complex manner by an array of variables whose precise influence on the operation is difficult to quantify. In the case of trays used in gas absorption and distillation, we must consider the vigor of gas–liquid contact, the

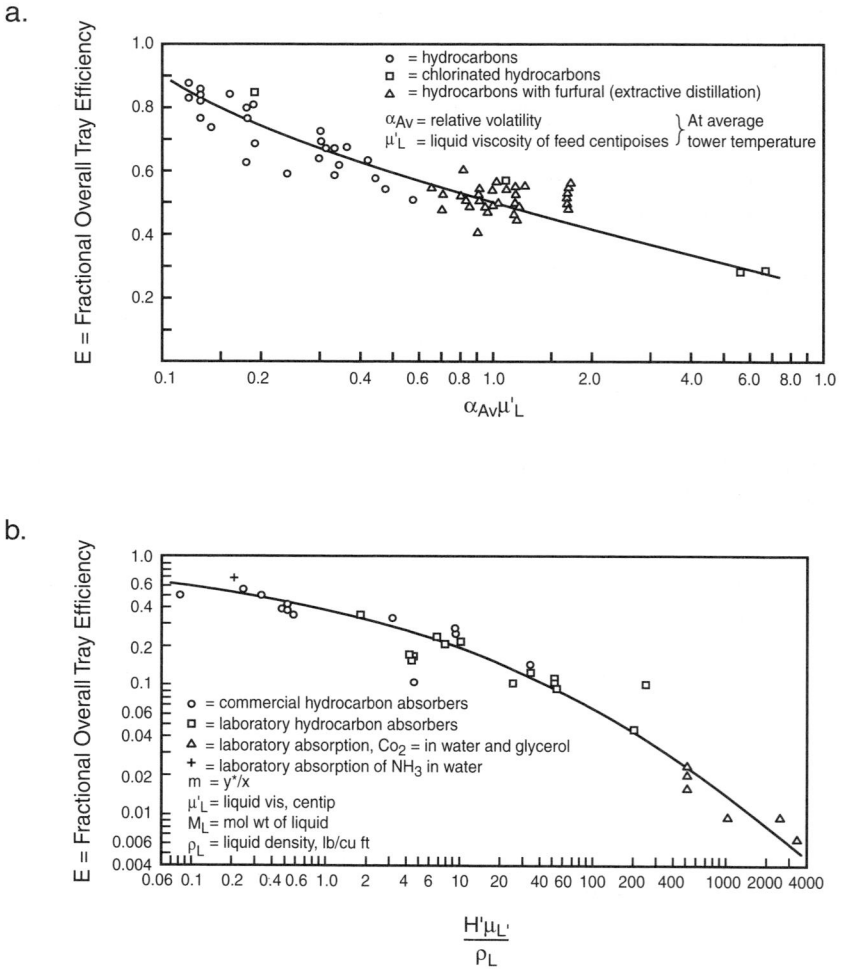

a.

b.

**FIGURE 7.23**
O'Connell's correlations for bubble-cap trays: (a) distillation; (b) absorption. (From O'Connell, H.E., *Trans. Am. Inst. Chem. Engrs.*, 42, 741, 1946. With permission.)

rate of liquid and gas flow, the mechanical design of the trays themselves, as well as physical properties of the systems. For stirred tanks, the rate of stirring and the design of the stirring mechanism itself, the contact time allowed, and the physical properties of the systems all play a role.

There are two ways out of this dilemma. We can draw on information provided by the equipment manufacturers, who frequently have test facilities available for determining stage efficiencies, or we can make use of rough correlations and guidelines for a first estimate in the preliminary design of these units.

### 7.2.1 Distillation and Absorption

Two classical empirical correlations due to O'Connell can be used to obtain a measure of the efficiencies of bubble-cap trays for distillation and absorption (Figure 7.23). The principal correlating parameters in both cases are the viscosity of the liquid $\mu_L$, and the relevant equilibrium constants, $\alpha$ for distillation, and Henry's constant $H$ for absorption. High values of either of these variables adversely affect the tray efficiency.

The correlation for distillation is based on limited data for systems of hydrocarbons and chlorinated hydrocarbons, that for absorption on the performance of hydrocarbon absorbers and the scrubbing of ammonia and carbon dioxide with water. $\mu_L'$ is in units of centipoises, $H'$ in units of mole-fraction ratios, and $\rho_{ML}$ is the molar density of the liquid solvent in lb mol/ft$^3$. Note that gas absorbers generally have much lower plate efficiencies than distillation columns due primarily to their lower operating temperatures.

### 7.2.2 Extraction

The extraction of solutes in agitated vessels is, at the very least, on a par in complexity with that we have seen for gas–liquid contact on trays. Here again the mechanical design of the system, this time that of the impeller, enters the picture, as well as the physical properties to which we must now add the surface tension of the dispersed phase. As a result, no clear-cut correlation has emerged from the host of experimental studies reported in the literature. The studies do, however, provide some guidelines that we summarize for the convenience of the reader in Table 7.2. The parameter values listed represent "safe" lower limits designed to achieve stage efficiencies of more than 80%. Much lower values do materialize on occasion. Thus, contact times of as little as 1 minute have been known to result in efficiencies of more than 75%. The table is therefore to be regarded as providing a comfortable margin of safety for conservative first estimates.

### 7.2.3 Adsorption and Ion-Exchange

In mass transfer operations involving a solid porous phase, the resistance in the majority of cases resides predominantly in the solid phase. This brings

**TABLE 7.2**

Conservative Parameter Values for Batch
Extraction Efficiencies of $E > 0.8$

| | |
|---|---|
| Residence Time | $t > 25$ min |
| Speed of Agitation | rpm $> 200$ |

about a considerable simplification in the estimation of stage efficiency, as neither the mechanical design nor the speed of agitation contributes significantly to the stage efficiency. A minimum rpm must, of course, be maintained to keep the slurry in suspension and keep the liquid film resistance within reasonable bounds. This is usually achieved at a level of 50 to 100 rpm.

In principle, mass transfer in solid particles is distributed in both time and distance, calling for the use of a PDE (Fick's equation) for a rigorous description of the process. In an elegant study conducted in the 1950s, it was shown by Glueckauf that the results of the formal treatment can be approximated by a volumetric solid-phase mass transfer coefficient, given by

$$k_S a(s^{-1}) = 15\frac{D_s}{R^2} \tag{7.29a}$$

where $D_S$ = solid phase diffusivity, $R$ is the particle radius, and $a$ = surface area per unit volume. Use of this transfer coefficient eliminates radial distance as a variable and results in a reduction of the model to the ODE level. We show in the next illustration how this reduced model can be applied to describe the efficiency of a liquid–solid contact in an agitated vessel.

### Illustration 7.12: Stage Efficiencies of Liquid–Solid Systems

We start by noting that the use of a solid-phase mass transfer coefficient calls for a driving force based on the internal concentration $X$ (kg solute/kg solid). A solute mass balance, performed about a single particle assumed to be a sphere, leads to the expression

$$\text{Rate of solute in} - \text{Rate of solute out} = \begin{array}{c}\text{Rate of change} \\ \text{in solute content}\end{array}$$

$$\rho_p k_S a V_p(X^* - X) - 0 = \rho_p V_p \frac{d}{dt} X \tag{7.29b}$$

where $V_p$ is the particle volume and $\rho_p$ its density. Introducing Equation 7.29a we obtain

$$15\frac{D_s}{R^2}\frac{4}{3}\pi R^3(X^* - X) = \frac{4}{3}\pi R^3 \frac{d}{dt} X \tag{7.29c}$$

and after cancellation of terms and separation of variables

$$15\frac{D_s}{R^2}\int_0^t dt = \int_0^q \frac{dX}{X^* - X} \tag{7.29d}$$

Hence,

$$15\frac{D_S}{R^2}t = \ln\frac{X^*}{X^*-X} \tag{7.29e}$$

and

$$1 - X/X^* = \exp(-15D_S t/R^2) \tag{7.29f}$$

Because $X/X^*$ is by definition the fractional stage efficiency $E$, we obtain the following relation between $E$ and the contact time $t$:

$$E = 1 - \exp(-15D_S t/R^2) \tag{7.29g}$$

*Comments:*

The success of Equation 7.29g in predicting stage efficiencies evidently hinges on how well we can estimate the solid-phase diffusivity $D_S$. Although the complex interior geometry of porous particles makes this a difficult task, reasonable first estimates of $D_S$ can be made using a relation previously presented in Chapter 3, which is repeated here:

$$(D_S)_{eff} = \frac{D\varepsilon}{\tau} \tag{3.13a}$$

Here $\varepsilon$ is the interior pore volume fraction, typically of the order 0.3 to 0.4, while the tortuosity $\tau$ is often given a representative value of 4. Thus, with diffusivities in liquids being of the order $10^{-9}$ m$^2$/s (see Chapter 3), we can expect $(D_S)_{eff}$ to have a typical value $10^{-10}$ m$^2$/s. We make use of these considerations in Practice Problem 7.10, which deals with the efficiency of a column adsorption process. In general, however, it is more fruitful to use Equation 7.29g to explore the effect of contact time $t$ or particle radius $R$.

Suppose, for example, that for a given process it is proposed to double radius $R$ to improve the settling rate. We then have from Equation 7.29g

$$\ln(1 - E_{new}) = \left(\frac{R_{old}}{R_{new}}\right)^2 \ln(1 - E_{old}) \tag{7.29h}$$

so that for $E_{old} = 0.9$, $E_{new}$ drops to a value of 0.44. This is a drastic reduction in efficiency for a mere doubling of particle radius.

## 7.2.4  Percolation Processes

In our preceding discussion of this topic, we managed to reduce the underlying model, which consists of two PDE mass balances, to a simple algebraic cumulative balance joined to an appropriate continuous spectrum of equilibrium stages. When this restriction is removed and the mass transfer resistance is brought back into play, no alternative simplifications are possible and we must return to the full PDE model. This model has been solved for a number of different equilibrium relations, most notably the linear case expressed by $X = HY$. The results for the latter can be expressed in terms of the following two dimensionless parameters:

Dimensionless Distance Z

$$Z = k_s a(z/v) \tag{7.30a}$$

Dimensionless Time T

$$T = k_s a \frac{\rho_f}{\rho_b} \frac{t}{H} \tag{7.30b}$$

Solutions of the PDEs as a function of these parameters are given in Table 7.3 at concentration levels of 1% and 10% of the feed concentration at a distance $z$ from the inlet. These tabulated values of $Z$ and $T$ can be used to calculate the time $t$ it takes for a particular concentration level to reach the position $z$ (for example, the outlet of an adsorber) or, conversely, to calculate the height of an adsorber or ion-exchange column needed for it to remain functional over a prescribed period $t$.

**TABLE 7.3**

Parameters for Nonequilibrium Adsorption

| Z | \multicolumn{2}{c}{T} | \multicolumn{2}{c}{Efficiency E = T/Z} |
|---|---|---|---|---|
|   | 1% of $Y_F$ | 10% of $Y_F$ | 1% of $Y_F$ | 10% of $Y_F$ |
| 1000 | 900 | 950 | 0.9 | 0.95 |
| 800 | 700 | 740 | 0.88 | 0.93 |
| 600 | 520 | 550 | 0.87 | 0.92 |
| 400 | 330 | 360 | 0.83 | 0.90 |
| 200 | 150 | 170 | 0.75 | 0.85 |
| 100 | 70 | 83 | 0.70 | 0.83 |
| 80 | 52 | 65 | 0.65 | 0.81 |
| 60 | 37 | 48 | 0.2 | 0.80 |
| 40 | 22 | 30 | 0.55 | 0.75 |
| 20 | 7.8 | 13 | 0.39 | 0.65 |
| 10 | 2.5 | 5.0 | 0.25 | 0.50 |
| 8 | 1.2 | 3.5 | 0.15 | 0.44 |
| 6 | 0.38 | 2.2 | 0.063 | 0.37 |
| 5 | 0.10 | 1.6 | 0.020 | 0.32 |

The following example illustrates the use of the table. Suppose we are required to reduce the level of pollutant in water to 1% of its existing level. It is expected to keep the adsorber on-stream for 100 days before break-through at the 1% level occurs. What should the height of the unit be?

Data: $\qquad H = 10^4 \qquad\qquad v = 1 \text{ cm/s} \qquad\qquad \rho_f/\rho_b \approx 1$

$k_s a$ is estimated from Equation 7.29a to be of the order $10^{-2} \text{ s}^{-1}$. The corre-sponding values of dimensionless time and distance are

$$T = 10^{-2} \times 1 \times \frac{3600 \times 24 \times 100}{10^4} = 8.64$$

$$Z \text{ (Table 7.3)} \cong 21$$

Consequently,

$$z = \frac{v}{k_s a} Z = \frac{10^{-2}}{10^{-2}} 21$$

$$z = 21 \text{ m}$$

This is somewhat excessive, and would suggest the use of two 10-m col-umns in series. Note that when we deal with multicomponent systems, the substance with the lowest Henry constant breaks through first. It is this value of $H$ that must then be used in computing dimensionless time $T$.

Table 7.3 can also be used to calculate an efficiency for the percolation process sometimes referred to as bed utilization and defined as the ratio of the minimum mass of adsorbent required under equilibrium conditions to the mass used in the actual operation ($W_m/W_a$). This is shown in the Illus-tration that follows.

## Illustration 7.13: Efficiency of an Adsorption or Ion-Exchange Column

We start the procedure by composing the ratio of the dimensionless distance $Z$ to dimensionless time $T$. We obtain in the first instance

$$\frac{Z}{T} = \frac{k_s a(z / v)}{k_s a(\rho_f / \rho_b)(t / H)} = \frac{z \rho_b}{v \rho_f t} H \qquad\qquad (7.31a)$$

We now multiply the numerator and denominator of this fraction by the cross-sectional area of the column $A_C$. This has the effect of transforming the ratio into an expression representing the actual mass of the stationary solid per mass of fluid treated, which we term $W_a$.

$$\frac{Z}{T} = \frac{z\rho_b A_C}{v\rho_f t A_C} \qquad H = \left[\frac{\text{kg solid}}{\text{kg fluid treated}}\right]_a \qquad H = W_a H \qquad (7.31b)$$

But as seen in the last chapter, Equation 6.7c, the corresponding *minimum* mass $W_m$ is given by the *inverse* of the Henry constant $H$. Hence, $H = 1/W_m$ and we can write for the efficiency $E$ of the process

$$E = \frac{W_m}{W_a} = \frac{T}{Z} \qquad (7.31c)$$

Thus, the efficiency of a percolation process is given simply by the ratio of dimensionless time to dimensionless distance. For the process considered in the previous section, for example, we have $T = 8.64$ and $Z = 21$. Consequently,

$$E = \frac{8.64}{21} = 0.41 \text{ or } 41\%$$

This implies that 59% of the adsorber is occupied by the mass transfer zone.

As we move up the columns in Table 7.3, efficiency improves dramatically. At $Z = 100$, the efficiency becomes 70% and at $Z = 1000$, it becomes 90%, and the percentage of the bed occupied by the mass transfer zone drops to 30% and 10%, respectively. The penalty to be paid is an increase in bed height.

## Practice Problems

7.1. *The Operating Diagram*

Describe what is represented in an operating diagram. What is the meaning of an operating line or curve? How many operating lines are required in crosscurrent cascades, and how many in a countercurrent operation? What are the requirements for fractional distillation?

7.2. *Single-Stage Adsorption: The Freundlich Isotherm*

The Freundlich adsorption isotherm is a special type of equilibrium relation of the general form:

$$Y = mX^n \tag{7.32a}$$

It is an empirical relation that does not converge to a saturation value at high loadings, nor does it yield the required Henry's constant at low coverages. It does, however, in many cases provide an adequate description of adsorption or ion-exchange equilibria over an intermediate concentration range. Suppose that a particular liquid–solid system is described by the following Freundlich isotherm:

$$Y = 8.91 \times 10^{-5}\, X^{1.66} \tag{7.32b}$$

It is desired to reduce the impurity concentration in a given liquid from $Y_o = 9.6$ units/kg solvent to 10% of this value in a single stage. Determine the minimum mass of adsorbent per 1000 kg solution required to accomplish this.

*Answer*: 32.0 kg

### 7.3. *Leaching of Solids: Extraction of Edible Oils*

Leaching involves the intimate contact of particular solids with an appropriate solvent for the purpose of removing undesirable components or recovering valuable materials contained in the solid. The phase diagram that describes this operation, displayed in Figure 7.24a, involves plots of dimensionless solid mass $N$ against liquid mass fractions $x$ and $y$ in the pore space of the solid and the free liquid extract, respectively. We use the same symbols as used previously in extraction; i.e., $B$ and $C$ denote solvent and solute, and $A$ is the solid component that here takes the place of the raffinate solvent. The liquid equilibrium compositions $x$ and $y$ are located as before, at the ends of a tie-line connecting the curves $(N, x)$ and $(N, y)$, which represent the solid content of the leached phase and residual particulate matter in the otherwise liquid extract phase. $R$ and $E$ denote the mass $(B + C)$ of liquid leaving with the leached solid and in the liquid extract, respectively, and are subject to the same mixing or lever rules we established for triangular diagrams in Chapter 6 (Illustration 6.6). Suppose we wish to extract 1000 kg of an oil-bearing seed containing 20% by weight of oil with 400 kg solvent. The extract is assumed to be free of solids so that the $(N, y)$ curve coincides with the abscissa (Figure 7.24b). The tie-lines are slightly sloped, indicating a higher oil content in the solid phase due to adsorption. Make the appropriate mass balances for this and construct an operating diagram that will yield the concentration of oil in the extract and the fractional degree of recovery. (*Hint*: Locate the mixing point $M$.)

*Answer*: $y = 0.275$, $r = 0.41$

a.

b.

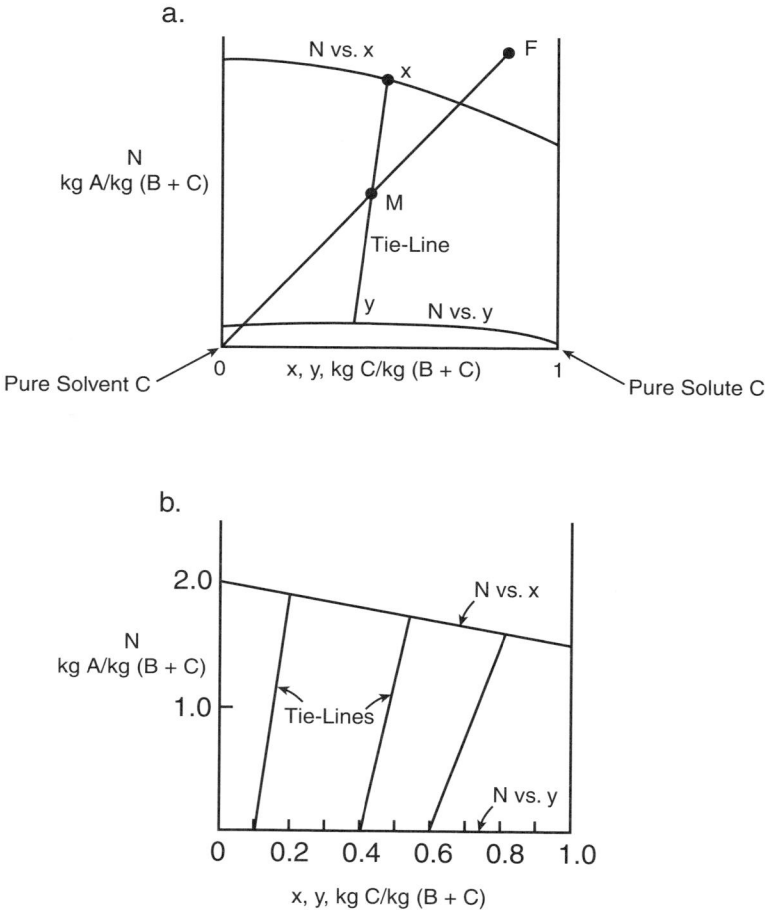

**FIGURE 7.24**
Staged leaching process: (a) phase diagram; (b) phase diagram for oil-bearing seeds.

### 7.4. *The Rayleigh Equation in Biotechnology: Ultrafiltration*

Ultrafiltration is a membrane process in which a solution containing a valuable solute such as a protein is concentrated by applying pressure to it and forcing the solvent across a semipermeable membrane, i.e., a membrane more permeable to the solvent than it is to the solute. Some of the latter will usually leak through as well; that is, the process is not 100% efficient. Efficiency is here defined as $1 - C_p/C_R$, where $C_p$ is the concentration at any instant of the solution passing through — the *permeate* — and $C_R$ denotes the concentration in the enriched solution left behind, termed the *retentate*. In a test run of a batch ultrafiltration unit to determine leakage, it was found that the retentate concentration had doubled after 53.7% of the solution had passed through the membrane. Show in the first instance that the model is represented by the Rayleigh equation and then:

a. Determine the efficiency of the process.

b. Calculate the enrichment obtained, i.e., the ratio of final retentate concentration to the cumulative concentration of the permeate.

*Answers*: a. 0.9; b. 14.5

*Note:* More about ultrafiltration appears in Chapter 8.

7.5. *Countercurrent Washing of Granular Solids*
Consider the case of a steady flow of granular solids that emerges from a leaching operation with a fraction $f$ of the final extract solution still adhering to it ($f$ = mass of adhering solution/mass of solids). If the leached substance is considered sufficiently valuable, or conversely is too objectionable to leave behind, it may become necessary to wash the leached solids for further recovery of solute. A countercurrent stages operation has the advantage of providing continuity of operation while maximizing effluent concentration. Show that the Kremser equation applies and provide the pertinent expressions for residue factor $R$ and extraction ratio $E$.

7.6. *Design of a Countercurrent Cascade in the Linear Regime*
A gas scrubber is to be used to reduce solute content in a dilute gas feed entering at $10^4$ kg/h to 1% of the incoming concentration. A solvent with a Henry constant of 0.1 kg/kg is available.

a. What is the minimum solvent flow rate to be used?

b. What is the *actual* solvent flow rate you would propose, given that the plate efficiency is 30%?

7.7. *The Countercurrent Cascade in Triangular Coordinates*
Figure 7.25 displays the operating diagram in triangular coordinates for a countercurrent cascade. The point 0, which is located exterior to the diagram, is termed the operating point, to which all operating lines, such as $\overline{R_1 E_2}$, $\overline{R_2 E_3}$, etc., converge. In a typical design problem, feed parameters $x_F$ and $F$, the amount and purity of solvent B, and the location of $R_a$, which defines the desired recovery, are known or prescribed. The task is then to establish the number of stages required to attain the reduction in solute content from the feed value $x_F$ to that of the exiting raffinate, $x_n$. This can be done using the construction shown in Figure 7.25. Explain and justify the operating diagram, drawing on the mixing rule and appropriate mass balances. (*Hint:* Compose the differences of the streams $R_{m-1} - E_m$ and show that they all yield the same value 0, which is called the operating point.)

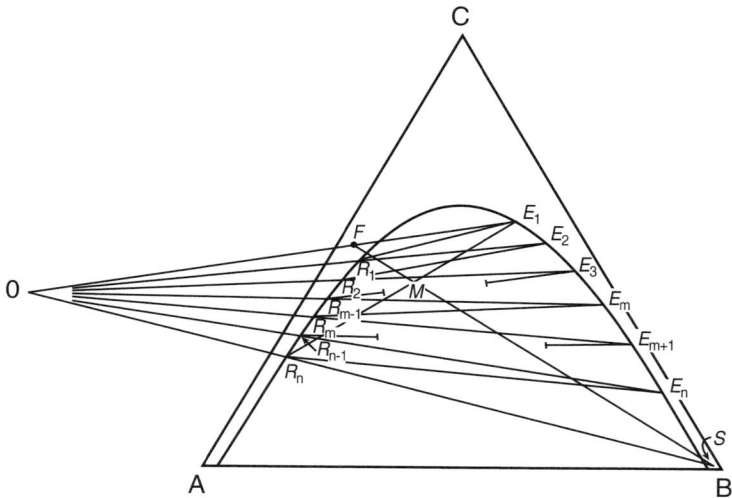

**FIGURE 7.25**
The countercurrent extraction cascade in triangular coordinates (see Practice Problem 7.7).

7.8. *Effect of Feed and Reflux on Column Performance*
Consider the following changes in the operating conditions of the fractionation column designed in Illustration 7.8.

a. Feed rate is doubled.

b. Feed concentration fluctuates ±10%.

c. The reflux rate is doubled.

Which variables are affected by these changes? Which remain unchanged? Support your statements with actual calculations as far as possible.

7.9. *The Use of Open Steam in Fractional Distillation*
When an aqueous solution is fractionated to give an organic component as the distillate and the water is removed as the residue product, the heat required may be provided by direct injection of steam at the bottom of the tower. The reboiler is then dispensed but this is done at the expense of a larger number of trays in the fractionation tower. Revise the normal McCabe–Thiele diagram to reflect this change in operating conditions.

7.10. *Maximum Recovery in Batch Distillation*
A batch still with 10 theoretical plates is used to fractionate a binary mixture $x_W^o = 0.4$ with $\alpha = 2$ into an overhead product of constant composition $x_D = 0.95$.

a. Calculate the maximum recovery.

b. How would you proceed if the column, instead of having trays, contains a high-efficiency packing?

(*Hint:* Use the Fenske equation.)

**7.11.** *Total Boil-Up in Batch Distillation*

Derive an expression that will give the total amount of liquid boiled up in a batch still operating at constant overhead composition $x_D$. (*Hint:* Make an unsteady balance over *part* of the still, then evaluate $\int G dt$ in terms of the varying bottoms composition and reflux.)

$$\textit{Answer: } G_{\text{Tot}} = \int_{w_W{}^o}^{x_W} \frac{W^o(x_D - x_W{}^o)dx_W}{(x_D - x_W)^2\left(1 - \dfrac{R}{R+1}\right)}$$

*Note:* The varying value of $x_W$ and matching reflux ratios have to be read off the McCabe–Thiele diagram.

**7.12.** *Design of an Ion-Exchange Column*

A synthetic ion-exchange resin in bead form is to be used for collecting and concentrating the copper in a dilute waste stream. Velocity of the feed of concentration 13 meq/l is 4 cm/s, bed density $\rho_b$ = 1200 kg/m³, and the specific gravity of the solution can be set at 1.0. Laboratory equilibrium data are available and are as follows:

| meq $Cu^{2+}$/l sol'n | 0.2 | 2.0 | 4.0 | 8.0 | 12 | 16 | 20 |
|---|---|---|---|---|---|---|---|
| meq $Cu^{2+}$/g resin | 2.0 | 4.0 | 4.4 | 4.7 | 4.75 | 4.85 | 4.9 |

If the bed is to remain on stream for 10 h, what is the minimum height of the column required to treat the solution?

*Answer:* 3.3 m

**7.13.** *Adsorption Purification Revisited*

In Illustration 6.4 we addressed the problem of sizing a carbon bed for the removal of benzene from water. The calculations there are limited to determining the minimum requirements in the absence of transport resistance, i.e., under conditions of local equilibrium. We now wish to calculate the efficiency of the process by incorporating the effect of a mass transfer resistance. Carry over the pertinent variables from Illustration 6.4 and make use of Table 7.3 and the Glueckauf relation (Equation 7.29a). Assume $D = 10^{-9}$ m²/s, $R = 2$ mm, $\varepsilon = 0.3$, $\tau = 4$.

Answer: 38%

**7.14.** *Pollution from the Mist over Niagara Falls*

In a 1986 study of pollution in the Niagara Falls region it was found that an estimated 60 tonnes of PCBs, chloroform, and chlorobenzenes are released each year into the atmosphere from the mist over

Niagara Falls. This is entirely due to the high fugacity (or activity coefficients) of these substances, which brings about a near quantitative transfer to the atmosphere when no more than a tiny fraction of the liquid has evaporated. We have shown this to be the case for mercury in Illustration 7.4 but have not established the time frame over which the event occurs. Consider the same process of evaporation of mercury, this time taking place from a droplet 1 mm in diameter. Mass transfer here is almost entirely liquid-film controlled, and a measured value of $k_L = 2.6 \times 10^{-5}$ m/s has been reported. Calculate the time required for the droplet to release 90% of its original mercury content. How does this compare with the lifetime of a droplet over Niagara Falls, given that the drag coefficient of a sphere in the turbulent regime is 0.44?

# 8

## Continuous-Contact Operations

The various staged operations taken up in the previous chapter have several features in common: The two phases involved in the transfer of mass were brought together and mixed intimately in discrete stages, which took the form of stirred tanks or some equivalent device. As a result, the concentrations were generally distributed uniformly in space and any variations they underwent were with respect to time, and not with distance. The operations were not only allowed to go to equilibrium but were actively encouraged to do so by means of agitation and the provision of sufficient contact time. Any departures from equilibrium were lumped into an entity called the stage efficiency. An efficiency of 100% signified the attainment of complete equilibrium and values below that expressed varying degrees of non-equilibrium.

Continuous-contact operations are diametrically different from staged operations in almost every aspect. The two phases are in continuous flow and in continuous contact with each other, rather than repeatedly separated and re-contacted in an array of stages. Second, the attainment of equilibrium is shunned. An active driving force is maintained at all times and its constituent concentrations vary continuously from the point of entry to the exit. The result is that the concentrations are now distributed in space and, assuming a normal steady-state operation, are invariant in time. Thus, while staged operations vary at most with time, but not at all with distance, the exact opposite holds in continuous-contact operations.

A final difference concerns the equipment used in the two cases. Continuous-contact processes are generally carried out in empty or packed columns or in tubular devices of various configurations. Staged operations may use columns, but these are usually subdivided into discrete contact stages in the form of trays. More commonly, staged processes are carried out in agitated vessels of some type, which may be used singly or in suitable arrangements.

In the following, we divide continuous-contact operations into two distinct categories. The first deals with classical packed-column operations in countercurrent flow. We revisit the packed-gas scrubber we first saw in Chapter 2 and provide a general survey of packed-gas absorption operations. Packed-column distillation is addressed next and, in a somewhat unusual departure from the norm, we reexamine coffee decaffeination by supercritical extraction.

The process involves a moving-bed configuration, which although superficially different from the conventions of packed column processes has identical operational parameters and can be analyzed by identical procedures.

The second category involves membrane processes, a contemporary topic of considerable importance with the promise of a bright future. Among the subcategories considered here are reverse osmosis, hemodialysis, and membrane gas separation, and the text provides some useful relations to address problems in these areas.

## 8.1   Packed-Column Operations

Packed columns are used primarily in gas absorption and liquid extraction and in air–water contact operations such as humidification and water cooling, which we take up in Chapter 9. They are found less frequently in distillation operations where their use is confined mostly to small-scale processes involving high-efficiency packing.

The analysis of packed-column operations has the same three basic goals we have seen in similar contexts, i.e., design, prediction of equipment performance, and, to a lesser extent, parameter estimation from experimental data. We may, for example, wish to calculate the height of a column required to achieve a certain degree of separation or level of purification. Or it may be required to assess the effects of changes in feed concentration or flow rate on column performance. In all these problems, the same basic model equations are applied and manipulated in appropriate ways to extract the desired information.

The principal mathematical feature shared by all steady-state packed-column operations is the distribution of the concentration variables in space, principally in the direction of flow. The model must therefore be composed of mass balances taken over a difference element of each phase, which are then converted into ODEs and integrated to obtain concentration profiles and other useful information. Algebraic (integral) mass balances also appear and can often be combined with the ODE balances to obtain important results.

We have already, in Illustration 2.3, alerted the reader to the existence of this multitude of mass balances, and we have occasion now to obtain a broader picture of their derivation and various applications. To do this, we return to the example of the gas scrubber taken up there and reexamine it in greater detail.

### Illustration 8.1: The Countercurrent Gas Scrubber Revisited

The basic model equations for countercurrent continuous mass transfer are the differential mass balances over each phase, which had been derived in

Illustration 2.3 and the companion equilibrium relation (Equation 2.12f). They represent a complete model for the system and are reproduced below.

Gas-Phase Mass Balance

$$G_s \frac{dY}{dz} + K_{OY}a(Y - Y^*) = 0 \tag{2.12d}$$

Liquid-Phase Mass Balance

$$L_s \frac{dX}{dz} + K_{OY}a(Y - Y^*) = 0 \tag{2.12e}$$

Equilibrium Relation

$$Y^* = f(X) \tag{2.12f}$$

Here $X$ and $Y$ are solute concentrations in units of kg solute per kg carrier or solvent, and the mass velocities $G_s$ and $L_s$ have units of kg (carrier or solvent)/m$^2$ s.

The two mass balances can be manipulated in a number of ways to yield specific results of interest. One such operation consists of subtracting Equation 2.12d and Equation 2.12e, which results in the elimination of the mass transfer terms. We obtain

$$G_s \frac{dY}{dz} - L_s \frac{dX}{dz} = 0 \tag{8.1}$$

This equation can be integrated between different limits to yield the algebraic mass balances derived in Illustration 2.3 by performing integral solute balances over the column. They are

$$(L_s X_2 + G_z Y) - (L_s X + G_s Y_2) = 0 \tag{2.11a}$$

and

$$(L_s X_2 + G_s Y_1) - (L_s X_1 + G_s Y_2) = 0 \tag{2.11b}$$

These two expressions are then recast in a form that makes them more suitable for graphical representation. We have termed these alternative formulations *operating lines* and reproduce them below.

$$\frac{Y - Y_2}{X - X_2} = \frac{L_s}{G_s} \tag{2.11c}$$

and

$$\frac{Y_1 - Y_2}{X_1 - X_2} = \frac{L_s}{G_s} \tag{2.11d}$$

We soon show how these expressions can be used to arrive at a graphical solution of the model equation.

The ODE mass balances (Equation 2.12d and Equation 2.12e) can also be tackled separately and in isolation by performing a formal integration. We obtain, for the gas-phase balance,

$$Z = \frac{G_s}{K_{OY}a} \int_{Y_2}^{Y_1} \frac{dY}{Y - Y^*} \tag{8.2a}$$

where $Z$ = height of the scrubber and $K_{OY}$ is an average overall mass transfer coefficient that is obtained by experiment. The integral on the right side is referred to as the number of transfer units (NTU) and the factor preceding it as the height of a transfer unit (HTU). Thus,

$$Z = \text{HTU} \times \text{NTU} \tag{8.2b}$$

On the surface, we do not seem to have gained much by taking this step since the integral contains too many variables and therefore cannot be evaluated. Some thought will reveal, however, that the difference $Y - Y^*$ in the integrand can be read from a joint graph of the operating line and equilibrium curve. This is shown in Figure 8.1, where the difference in question is given by the vertical distance between the two curves.

Let us note some features of this diagram. To begin, it has a familiar air to it. We in fact have seen an identical representation of the operating line and equilibrium curve in the operating diagram, dealing with a countercurrent staged cascade (Figure 7.14a). However, that diagram differs from Figure 8.1 because it uses the staircase construction to establish the number of stages required to achieve a desired separation. In the present case, graphical evaluation of the NTU integral takes the place of the staircase construction, while the HTU represents, in a sense, the inherent mass transfer resistance of the process and can therefore be viewed as the equivalent of a *stage efficiency*.

Another feature that the two operations share is the existence of a *minimum solvent flow rate*. In both cases this flow leads to a pinch at the level of the feed concentration between the operating line and equilibrium curve. This has previously led to the number of stages going to infinity. The result here is similar: The NTU integral increases rapidly as we approach the pinch point and ultimately diverges to infinity. This leads to a column of infinite height $Z$. While operating costs are reduced with diminishing solvent flow

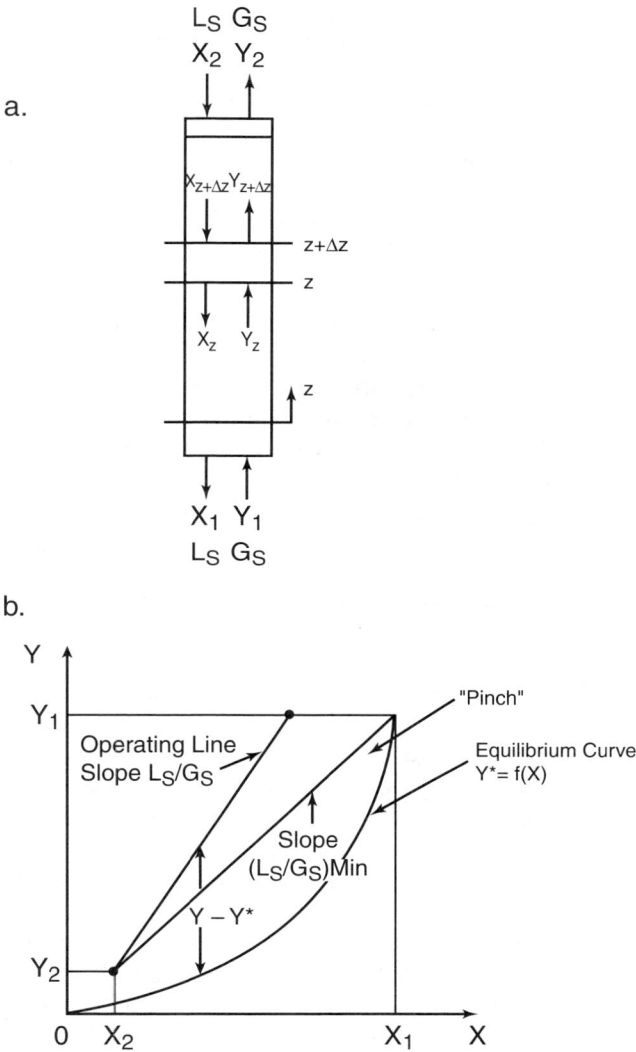

**FIGURE 8.1**
The countercurrent packed-gas scrubber: (a) column variables; (b) operating diagram.

in both cases, the capital cost of the plant, i.e., the cascade or column, rises rapidly and goes to infinity when the minimum solvent flow rate is imposed.

The design procedure for a packed-gas scrubber parallels that of the countercurrent staged cascade. We start by fixing the point $(X_2, Y_2)$ representing the compositions at the top of the column. $Y_2$ denotes the prescribed effluent concentration and $X_2$ the purity of the solvent. A line of slope $L_s/G_s$ is next drawn through that point and extended to the level of the feed concentration $Y_1$. The slope $L_s/G_s$ is usually set at a value in the range 1.2 to 1.5 $(L_s/G_s)_{Min}$.

The vertical difference between the two graphs, the operating line and the equilibrium curve, are then used to evaluate the NTU integral. This can be done either graphically or numerically. Finally, the height of a transfer unit is established using the given carrier flow rate $G_s$ and a volumetric mass transfer coefficient $K_{OY}a$ determined experimentally or drawn from existing correlations or tabulations (see Table 5.6). These values typically vary over the range 10 to 50 cm. With both HTU and NTU in hand, the height of the scrubber can then be determined using Equation 8.2b.

*Comments:*

Let us first return to the basic model equations (Equation 2.12d to Equation 2.12f). This model is quite general, and able to accommodate arbitrary equilibrium relations and system parameter values. Its solution, which is not undertaken here, must be implemented numerically and can be used both for design and prediction of scrubber performance and for parameter estimation. The reduced form of the solution we derived, which rests on the use of the operating diagram Figure 8.1b and Equation 8.2b, is not able to provide direct information on scrubber performance except by a process of trial and error. Suppose, for example, that we wish to establish the effluent concentration that results from doubling the feed rate, i.e., $G_s$, in an existing scrubber of height $Z$. The HTU value would first have to be modified to reflect the new flow rate, but more importantly, the upper limit of the NTU integral would have to be adjusted and the evaluation of the integral repeated by trial and error until the product of HTU and NTU exactly matches the given height $Z$ of the scrubber. The numerical solution suffers from the same dilemma, but can be used, once it is properly programmed and in place, to carry out a wide range of repeated calculations with a minimum of effort.

A second point of note concerns the solvent flow rates to be used. Low amounts of solvent carry the advantage of low solvent inventory but lead to greater scrubber heights, which in the limit of the minimum flow rate leads to an infinitely high tower. If, on the other hand, we allow an unbounded increase in solvent flow, column height and cost will be reduced to a minimum but operating costs will go to infinity. Between these two extremes there must be an optimum flow rate that will minimize the total expenditures composed of capital and operating costs.

This situation is depicted in Figure 8.2, which shows a plot of total cost vs. solvent flow rate. That cost is made up of three component expenditures consisting of the cost of the column, the cost of the solvent, and the cost of pumping the solvent to the top of the column. The column cost starts at infinity when solvent flow is at a minimum, thereafter declines sharply, and ultimately levels off to a near-constant value. The pumping cost also starts at infinity because the column is infinitely high, then declines, passes through a minimum, and gradually increases, going to infinity again as the solvent flow rate becomes unbounded. The cost of the solvent shows a much simpler

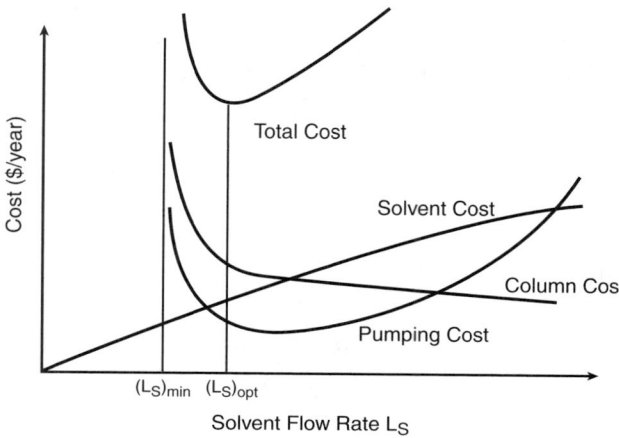

**FIGURE 8.2**
The countercurrent gas scrubber cost vs. solvent flow rate.

relation to flow rate, rising almost in a straight line, with any departures from linearity resulting from volume discounts allowed for the solvent. The total cost curve that results from the three-component expenditures shows a minimum, which, as mentioned, lies in the typical range (1.2 to 1.5) $(L_s/G_s)_{Min}$.

### *Illustration 8.2: The Countercurrent Gas Scrubber Again: Analysis of the Linear Case*

We consider here the case where both the operating line and equilibrium curves are linear, the latter represented by Henry's law:

$$Y^* = HX \qquad (8.3)$$

On introducing this relation into the NTU integral of Equation 8.2a, we obtain

$$NTU = \int_{Y_2}^{Y_1} \frac{dY}{Y - Y^*} = \int_{Y_2}^{Y_1} \frac{dY}{Y - HX} \qquad (8.4a)$$

Drawing on the material balance (Equation 2.11a) to express $X$ as a function of the gas-phase mass ratio $Y$ yields

$$NTU = \int_{Y_2}^{Y_1} \frac{dY}{Y(1 - 1/E) + (1/E)Y_2 - HX_2} \qquad (8.4b)$$

where $E$ is the so-called *absorption factor*, equal to the ratio of the slopes of operating line to equilibrium line, $L/HG$. $E$ has its counterpart in staged operations and is called the extraction ratio there.

A standard evaluation of this integral yields

$$\text{NTU} = \frac{1}{1-1/E} \ln \frac{Y_1(1-1/E)+(1/E)Y_2 - HX_2}{Y_2(1-1/E)+(1/E)Y_2 - HX_2} \tag{8.4c}$$

or equivalently

$$\text{NTU} = \frac{\ln[R(1-1/E)+1/E]}{1-1/E} \tag{8.4d}$$

Here $R$ is given by

$$R = \frac{Y_2 - HX_2}{Y_1 - HX_2} \tag{8.4e}$$

and in the case of pure solvent ($X_2 = 0$) becomes a direct measure of the depletion $Y_2/Y_1$ in the solute content of the gas phase. It is identical to the residue factor previously defined for staged countercurrent processes (see Table 7.1), and a plot of it relating it to the absorption factor $E$ and the number of transfer units appears in Figure 8.3a. The reader will note that the diagram is quite similar in appearance to the plot seen in Figure 7.15, which relates the number of stages in a countercurrent cascade to the same parameters.

We now address a number of features that both plots have in common and that define the behavior of these linear systems.

As briefly noted in Illustration 7.7, the parameter value $E = 1$ represents an important dividing line. Above it, the residue factor $R$ converges asymptotically to a constant value with an increase in the number of transfer units. We cannot, in other words, absorb arbitrarily large fractions of the incoming solute by increasing solvent flow rate, as long as the value of $E$ remains below unity. This becomes possible only when the absorption factor is greater than 1; that is, whenever the slope of the operating line exceeds that of the equilibrium line. The solute fraction remaining in the outgoing gas can now be made arbitrarily small by the simple expedient of increasing the solvent flow rate. A complete reduction to zero solute content, however, becomes possible only in the limiting case of an infinite flow rate of virgin solvent.

A rational explanation of this behavior can be obtained by inspecting the pertinent operating diagrams, shown in Figure 8.3b. We note here that for $E < 1$, operating and equilibrium lines always converge and ultimately intersect in a pinch, making it impossible to set arbitrarily high solute recovery ratios. When $E$ exceeds unity, the two lines diverge and no longer intersect except at the origin. It now becomes possible to fix the effluent gas

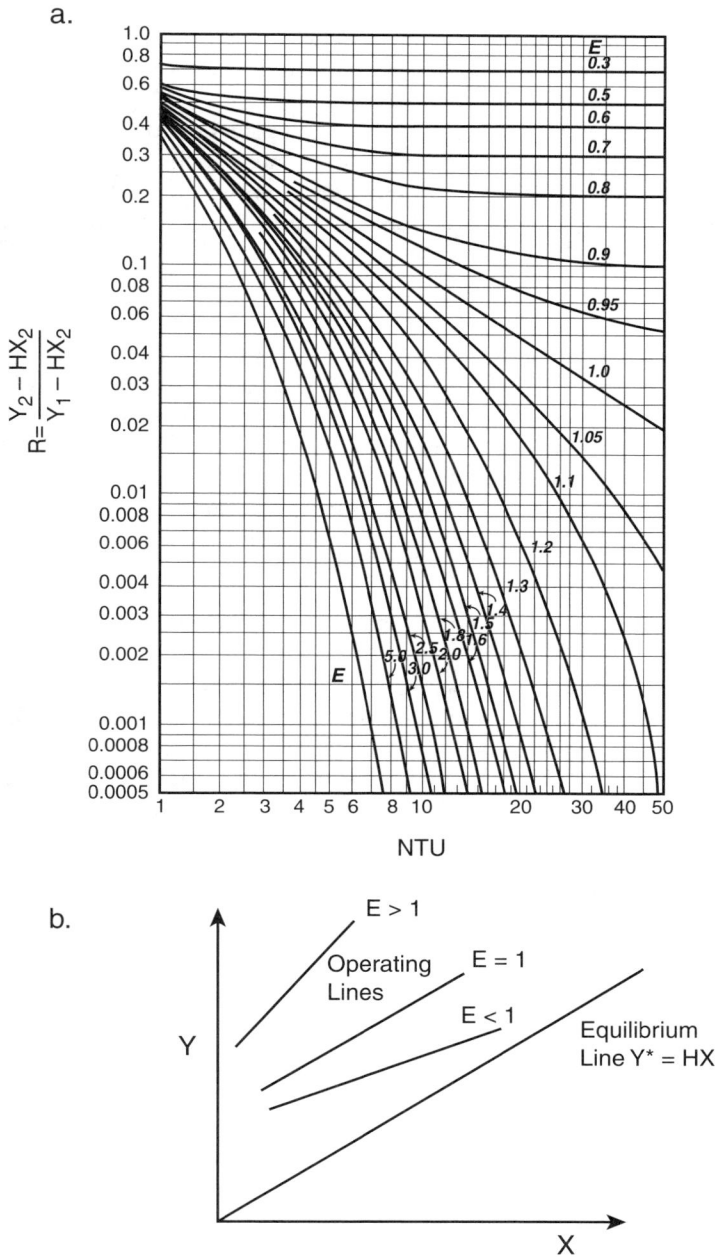

**FIGURE 8.3**
The gas scrubber in the Henry's law region: (a) evaluation of NTU; (b) operating diagram.

concentration $Y_2$ at any level other than zero without risking a pinch. The dividing line for these two types of behavior is the operating line, which runs parallel to the equilibrium line. No intersection occurs in this case and the general behavior is the same as for $E > 1$.

The NTU plot of Figure 8.3a can also be applied to other packed-column processes operating in the linear region. These include gas stripping and extraction, and require the pertinent parameters to be redefined to fit each new operation. This has been done in Table 8.1, which lists the different versions of $R$ and $E$ for each case.

*Comments:*

This brief analysis of gas absorption with systems obeying Henry's law leads to some interesting conclusions. The lower solvent flow-rate limit here is no longer set by graphically locating the operating line, which causes a pinch. We make use, instead, of the criterion that the slope of the operating line must at least equal and preferably exceed that of the equilibrium line, i.e., Henry's constant. This results in a quick and convenient resolution of the problem of finding the lower limit of the solvent flow rate.

### Illustration 8.3: Distillation in a Packed Column: The Case of Constant α at Total Reflux

Packed-column distillation, while practiced much less frequently than gas absorption, still finds considerable use in medium- and small-scale applications. The model equations parallel those for the absorption case and result in the same HTU–NTU relations seen there, with mole fractions taking the place of mass ratios as the pertinent concentration units. We obtain

$$Z = HTU \times NTU = \frac{G}{K_{OY}a} \int_{y_2}^{y_1} \frac{dy}{y^* - y} \tag{8.5a}$$

**TABLE 8.1**

Packed-Column Parameters in the Linear Region

| Process | $R$ | $E$ | Equilibrium Relation |
|---|---|---|---|
| Gas absorption | $\dfrac{Y_2 - HX_2}{Y_1 - HX_2}$ | $L/HG$ | $Y = HX$ |
| Gas stripping | $\dfrac{X_1 - Y_1/H}{X_2 - Y_1/H}$ | $HG/L$ | $Y = HX$ |
| Liquid extraction | $\dfrac{X_2 - Y_1/m}{X_1 - Y_1/m}$ | $mB/A$ | $Y = mX$ |

and supplement this expression with the equilibrium relation given by

$$y^* = f(x) \tag{8.5b}$$

The operating lines needed for the evaluation of the NTU integral are identical to those established in Chapter 7 and can be seen in Figure 7.17 and Figure 7.18. Either graphical or numerical procedures can be used for this purpose.

A case of special interest is that of a system with constant relative volatility $\alpha$ run at total reflux. The equilibrium relation is now given by

$$y^* = \frac{\alpha x}{1 + x(\alpha - 1)} \tag{8.5c}$$

Because at total reflux the operating line coincides with the 45° diagonal, we have equality of the mole fractions, $y = x$, and Equation 8.5c becomes

$$y^* = \frac{\alpha y}{1 + y(\alpha - 1)} \tag{8.5d}$$

Introducing this relation into the NTU integral yields

$$NTU = \int_{y_1}^{y_2} \frac{dy}{\dfrac{\alpha y}{1 + y(\alpha - 1)} - y} \tag{8.5e}$$

or equivalently

$$NTU = \int_{y_1}^{y_2} \frac{1 + y(\alpha - 1)}{y(\alpha - 1)(1 - y)} \, dy \tag{8.5f}$$

This expression can be evaluated analytically using the additional integration formula

$$\int \frac{dx}{x(a + bx)} = -\frac{1}{a} \ln \frac{a + bx}{x} \tag{8.5g}$$

Its application leads to the result

$$NTU = \frac{1}{\alpha - 1} \ln \frac{y_2(1 - y_1)}{y_1(1 - y_2)} + \ln \frac{1 - y_1}{1 - y_2} \tag{8.6}$$

This is a convenient expression to determine NTU and, from it, HTU values in laboratory experiments, which are usually run at total reflux. It also finds use in the distillation of closely boiling mixtures such as isomers or isotopes. Here $\alpha$ values are quite low and constant, and the operating line very nearly coincides with the 45° diagonal so that Equation 8.6 becomes a valid approximation of the actual number of transfer units. An example of its application in the distillation of isotopes is found in Practice Problem 8.5.

### Illustration 8.4: Coffee Decaffeination by Countercurrent Supercritical Fluid Extraction

Section 6.2.5 introduced the reader to the concept of supercritical fluid extraction (SCE), and Illustration 6.7 provided a first simple example of its application by considering the single-stage extraction of caffeine from coffee beans using supercritical carbon dioxide. The amount of $CO_2$ required in this elementary operation was calculated at 45.9 kg $CO_2$/kg coffee, an inordinately high level. Commercial decaffeination processes use a countercurrent operation carried out in tall columns (see Figure 6.15), with the coffee beans entering at the top and discharged intermittently at the bottom. This process is close enough to a continuous operation to allow us to use the principles established for such an operation.

The operating diagram for the process is shown in Figure 8.4. The equilibrium curve is drawn from the relation established in Chapter 6, i.e.,

$$x = 1.24y^{0.316} \qquad (6.9a)$$

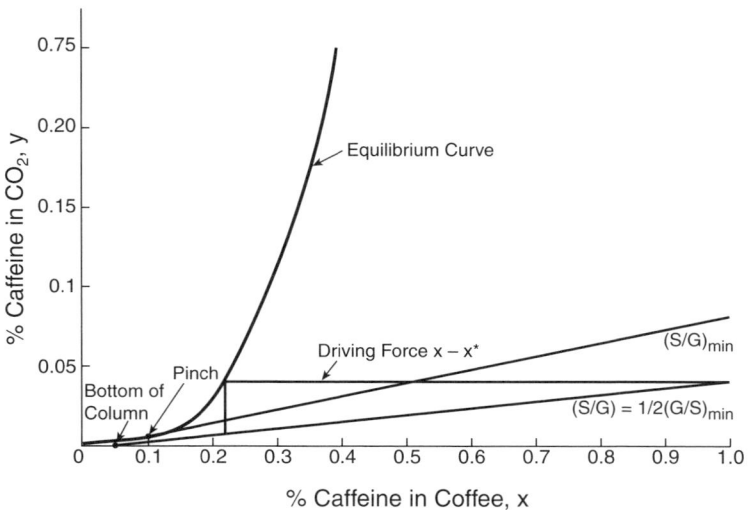

**FIGURE 8.4**
Operating diagram for decaffeination.

The operating line will now be located *below* the equilibrium curve since the process is one of stripping or desorption, i.e., the reverse of that seen in gas absorption. The resistance here would evidently be in the solid phase, with the driving force given by the *horizontal* distance between operating line and equilibrium curve (Figure 8.4). We do not make use of the driving force in this illustration, but address instead the following problem: Suppose that the caffeine content is to be reduced from 1 to 0.05%, a typical goal in commercial decaffeination and one used in the single-stage extraction process in Illustration 6.7. We first wish to establish how much the $CO_2$ consumption is reduced if we replace the single-stage process with a countercurrent mode of contact. Second, given a typical column height of 20 m, what would be the height equivalent to a theoretical stage, HETS?

We start in the usual fashion: by first establishing the minimum amount of "solvent" $CO_2$ required for the operation. The pinch here does not occur at the feed condition, as is usually the case, but much earlier at some intermediate point of the equilibrium curve. That point is located by drawing a tangent to the equilibrium curve starting from the point representing conditions at the bottom of the column (see Figure 8.4).

If we now set the actual $G/S$ ratio at *twice* the minimum amount, we obtain the following result:

$$(S/G) = \tfrac{1}{2}\,(S/G)_{Max} = \Delta y/\Delta x = 0.04/0.95 \qquad (8.7a)$$

The $CO_2$ consumption, which is the inverse of this ratio, is then given by

$$G/S = 0.95/0.04 = 23.8 \text{ kg } CO_2/\text{kg coffee} \qquad (8.7b)$$

The number of stages between feed and effluent is obtained by the usual staircase construction and equals approximately four. The corresponding HETS value is then given by

$$\text{HETS} = \frac{\text{Height of column}}{\text{Number of stages}} = \frac{20}{4} = 5\,\text{m} \qquad (8.7c)$$

*Comments:*

There are a number of features in this example drawn from industry that fall outside the conventional boundaries of textbook problems. The first item of note is the location of the pinch, which occurs at a point between feed and effluent conditions. The resulting minimum $CO_2$ consumption is at the modest level of 11.9 kg $CO_2$/kg beans. Even doubling that amount for actual operating conditions still yields a consumption that is less than that calculated for a single-stage contact (Illustration 6.7). This is a clear-cut vindication of the countercurrent mode of operation.

The second feature of note is the small number of stages required (four), and connected to it, the inordinately high value of the height equivalent to a theoretical stage (5 m). This was partly the result of the relatively generous amount of $CO_2$ we allowed over and above the minimum required. If that number is reduced to 1.5 times the minimum, the number of stages rises to approximately 10, and with it comes a corresponding drop in the HETS to 2 m. It is likely that the actual column operation is in fact in the range of 1.2 to 1.5 times the minimum $CO_2$ requirement and that the HETS is of the order of 1 m, a value much more in line with that expected in operations of this type. The reader is encouraged to examine these cases more closely by enlarging the relevant section of the operating diagram and stepping off the stages for each slope of the operating line.

## 8.2   Membrane Processes

The notion of separating substances by means of membrane barriers that allow the selective passage of one or more species to the exclusion of all others has both a powerful appeal and the appearance of an impossible dream.

Consider the advantages of such a process compared to other operations:

- Elimination of thermal energy as an operating expense
- Ability to process heat-sensitive materials
- No intrusion of solvents, which merely produce an enriched phase and require further processing
- No necessity to regenerate or replace adsorbents or ion-exchange resins

In fact, the only operating costs visible to the casual observer are conveying the fluids to and (usually) through the barriers and the ultimate disposal of the product stream.

The age-old quest for such membranes accelerated in the first half of the 20th century but was initially plagued by a lack of suitable materials, low throughput rates and selectivity, and continuous problems with fouling and plugging. The advent of synthetic polymers brought about a wide new range of materials capable of being altered and modified to suit a particular application. Low throughput rates initially persisted but were overcome with the dramatic development in 1960 of high-flux cellulose acetate membranes by Loeb and Sourirajan. This event signaled the start of a burst of activity in the field of membrane development that continues undiminished to this day.

## 8.2.1 Membrane Structure, Configuration, and Applications

The basic structure of present-day membranes is illustrated in Figure 8.5. It consists of a selective polymeric film and a much thicker but more porous sublayer, which provides the necessary structural support but otherwise does not actively participate in the separation. Such membranes composed of a dual layer of different materials are referred to as *anisotropic* or *asymmetric*.

The performance of a membrane process is influenced by a number of factors, chief among which are the physical structure of the membrane and the nature and composition of the feed. These factors can be used to order and classify the processes into subcategories, each with its own distinctive features. These categories have gradually assumed proportions that have turned them into self-contained operations in their own right. This is shown in Table 8.2, which summarizes the various subprocesses that have emerged, their relation to physical structure expressed through the membrane pore diameter, and the different applications that have come about as a result. Note the impressive range of both the physical parameters and the actual applications. Pore diameters vary over the span of six orders of magnitude capable of barring particles as small as sodium ions (3.7 Å) in reverse osmosis or bacteria with a diameter of 1 µm in microfiltration, while allowing free passage to water molecules. Intermediate-size particles such as proteins (100 Å) or viruses (1000 Å) can either be retained or allowed to permeate, depending on the requirements of the process.

Although fairly distinct in their capabilities, these processes have several features in common . The membrane materials used are generally polymeric in origin and are extensions, both in mode of preparation and in composition, of the cellulose acetate membranes pioneered by Loeb and Sourirajan. Another common feature shared by most of the processes listed is the use of pressure as the driving force; the exceptions are the different forms of dialysis. Thus, in hemodialysis, concentration replaces pressure as the driving force, while electrodialysis is driven by an electrical potential.

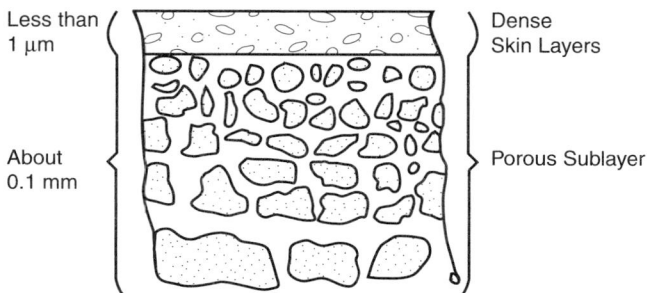

**FIGURE 8.5**
The asymmetric membrane.

**TABLE 8.2**

Membrane Processes and Their Applications

| Process | Pore Diameter | Applications |
|---------|---------------|--------------|
| NF Nanofiltration | 10 Å | High-flux desalination |
| RO Reverse osmosis | 2–30 Å | Desalination |
| HD Hemodialysis | Similar to RO membranes | Removal of urea, creatinine, and other substances from urine |
| ED Electrodialysis | Similar to RO membranes | Desalination in the range 500–2000 ppm salt content |
| GS Gas separation | Similar to RO membranes | Recovery of $H_2$ in ammonia plants, separation of $N_2$ from air, sweetening of natural gas |
| UF Ultrafiltration | 10–1000 Å | Separation of proteins and other macromolecules; treatment of process water and whey; clarification of fruit juices |
| MF Microfiltration | 0.1–10 μm | Removal of microorganisms from water and beverages, or product antibiotics |
| CF Conventional filtration | 10–100 μm | Cell harvesting |

*Note:* 1 nm = 10 Å = $10^{-3}$ μm.

Geometrical configurations of the membrane modules have passed through various stages of development from which two major contenders have emerged: In the spiral-wound configuration, membrane *sheets* flanked by two spacer-supported channels are wound around a central perforated collector tube, which receives the permeate (Figure 8.6). Feed enters axially through one of the channels and selectively permeates radially through the membrane into the second channel, which discharges the permeate into the collector tube. In the hollow-fiber configuration, a *shell and tube* arrangement is used in which the feed enters either through the membrane fibers (core-fed) or is confined to the shell side (shell-fed). These flow geometries are illustrated in Figure 8.7.

Let us now turn to a more detailed examination of these processes: Both *nanofiltration* (NF) and *reverse osmosis* (RO) draw on principles of osmosis for their implementation. The principal features of this phenomenon and its manifestation in reverse-osmosis operations are illustrated in Figure 8.8a. Consider a selective membrane, i.e., one that is freely permeable to water but much less so to salt, separating a salt solution from water, as shown in Part 1. In such an arrangement, water will flow from the pure-water side into the side less concentrated in water, i.e., the saltwater side. This process is referred to as normal osmosis. If, now, a hydrostatic pressure is applied to the salt side, the flow of water will be retarded, and if that pressure is sufficiently high, the flow will ultimately cease completely. At this point we will have reached what is termed *osmotic equilibrium* (Part 2), and the hydrostatic pressure associated with this state is referred to as the *osmotic pressure*. A further increase in applied pressure will act to reverse the flow from the

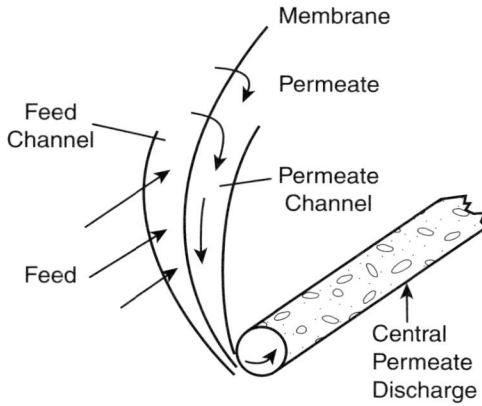

**FIGURE 8.6**
The spiral-wound membrane.

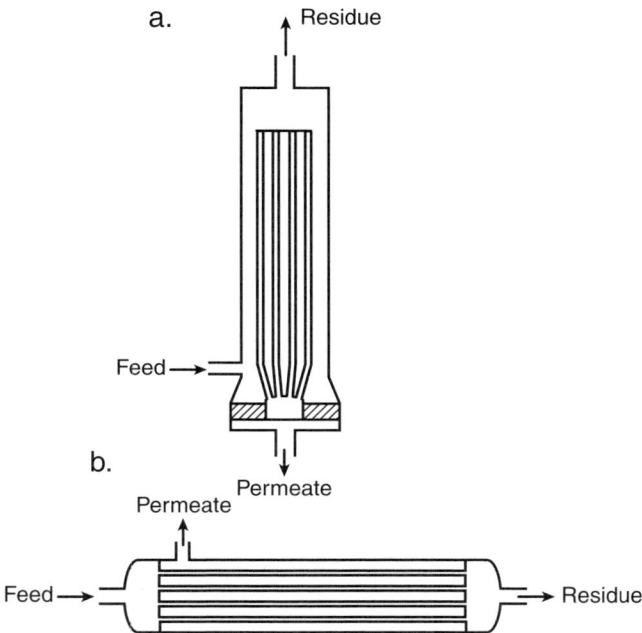

**FIGURE 8.7**
Hollow-fiber reverse-osmosis modules: (a) shell-side feed; (b) bore-side feed.

salt solution to the pure-water side (Part 3), and that flow will rise as the pressure is increased. It is this phenomenon, termed reverse osmosis, that is exploited in all RO operations. The membranes used in these processes are so dense that no discrete pores can be discerned. Permeation proceeds instead by a solution-diffusion mechanism that makes use of a statistical distribution of free volume or holes (see Chapter 3). The daily production of water using RO and NF  amounts to approximately 1 billion gallons, of which 60% involves desalination of brackish and sea water. The remainder of the membrane capacity is in the production of ultrapure water for the electronics industry. Both hollow-fiber and spiral-wound configurations are employed, with pressures in the range 50 to 70 atm for sea water, and 15 to 20 atm for brackish water. Osmotic pressures for these feeds are in the range 1 to 20 atm and are listed, along with other items of interest, in Table 8.3.

*Gas separation* and *hemodialysis* often employ membranes similar to those used in RO, but differ from them in several aspects. Neither involves osmotic pressure as a countervailing driving force. In gas separation, the partial pressure difference of the diffusing species acts as the driving potential, while in hemodialysis the concentration difference is across the membrane to provide the necessary impetus. Hollow fiber configurations are predominant in both applications.

*Membrane gas separation* is a rapidly evolving technology, with a strong presence in hydrogen recovery from plant off-gases, and in the separation of nitrogen from air. Approximately 50% of all nitrogen used in ammonia synthesis and other applications is produced by membrane separation, the rest coming from the cryogenic liquefaction and distillation of air. Hydrogen recovery typically employs pressures as great as 150 atm on the feed or shell side of the module, with correspondingly small hollow-fiber diameters of 50 to 100 μm (Table 8.2). Nitrogen separation employs much lower feed pressures of only up to 10 atm, which can be easily accommodated in the bore-side of the hollow fiber module and can tolerate higher fiber diameters of up to 1 mm. Because of the extreme sensitivity of the separation process to small pinholes in the membrane, it has become customary to cover the basic fiber with a permeable defect-sealing layer of silicone rubber. We address questions of enrichment and selectivity attained in these processes in the next section.

*Hemodialysis* is a special case of a wider class of operations covered by the umbrella term *dialysis*. It involves, as is shown in Figure 8.8b, the simultaneous permeation of solutes from the higher to the lower concentration side of the membrane while water passes  in the opposite direction driven by the

**TABLE 8.3**

Osmotic Pressures of Various Solutions at 25°C

| Solute | Concentration, mg/l | Osmotic Pressure, atm |
|---|---|---|
| NaCl | 35,000 | 27 |
| Sea water | 32,000 | 23 |
| Brackish water | 2,000–5,000 | 1–2.7 |
| Sucrose | 1,000 | 1.1 |
| Dextrose | 1,000 | 2 |
| Blood plasma and intracellular fluid (37°C) | — | 7.2 |

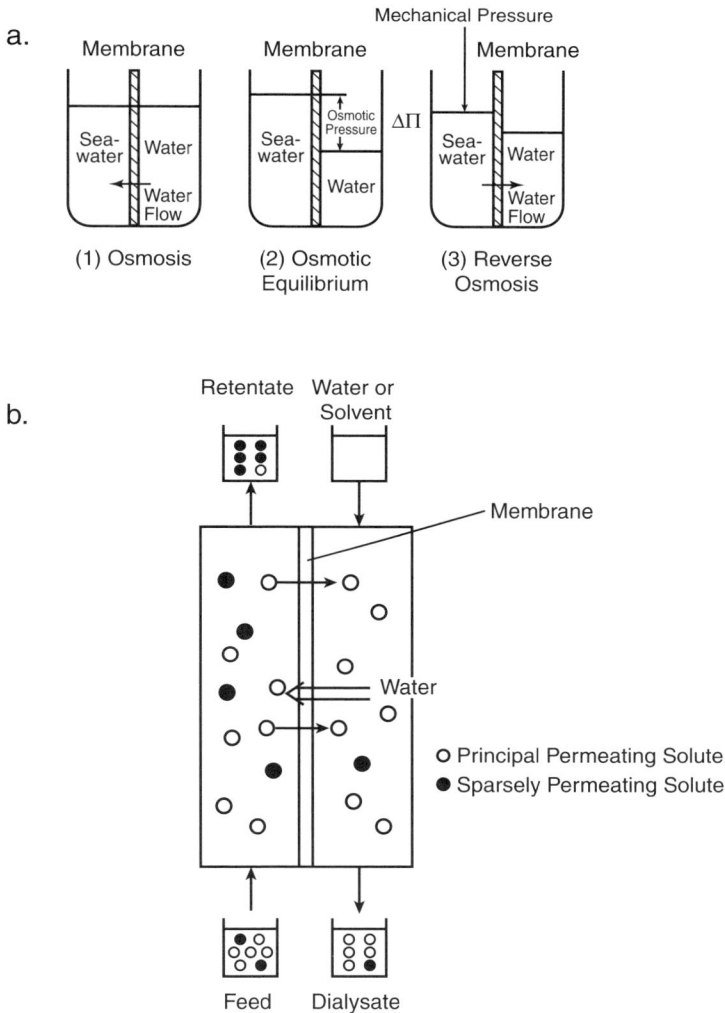

**FIGURE 8.8**
(a) Principles of reverse osmosis; (b) principles of dialysis.

osmotic pressure difference of the two solutions. In hemodialysis, water flux is suppressed by using iso-osmotic solutions on both sides. The nonpermeating solution is referred to as the retentate and contains blood cells, proteins, and other valuable components of the blood. The stream carrying away the toxic solutes, which have permeated through the membrane, is termed the *dialysate* or permeate. The use of hollow-fiber membranes in the dialysis of blood is now widespread and serves close to 1 million patients worldwide involving over 100 million procedures a year. The device, often referred to as an artificial kidney, is illustrated in Figure 8.9. In it, a saline solution with an osmotic pressure nearly equal to that of blood is supplied to the shell

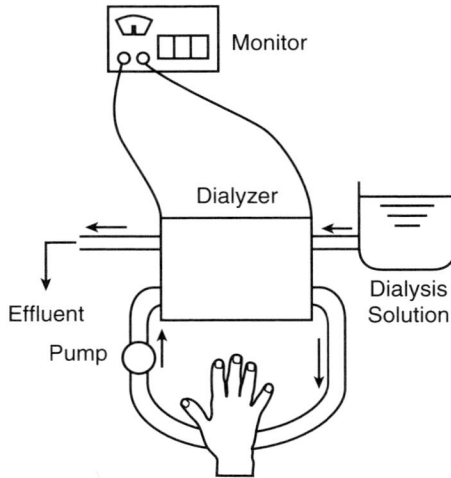

**FIGURE 8.9**
The hemodialyzer. (From Matsuura, T., *Synthetic Membranes and Membrane Separation Processes,* CRC Press, Boca Raton, FL, 1990. With permission.)

side of the module while the blood is pumped countercurrently through the core side of the hollow-fiber module. The device contains several thousand fibers in a 5-cm-diameter tube with a total membrane area of 1 to 2 m². The cumulative area used worldwide for medical purposes exceeds that employed in all industrial applications.

*Ultrafiltration* (UF) is commonly used to strip smaller-sized molecules such as salts and sugars along with water from solutions containing valuable macromolecules. The most prominent applications that have emerged in recent times are the concentration of automotive paint in process rinse water and the concentration of valuable proteins in whey, the supernatant liquid produced in cheese manufacture. In both cases the application of UF technology not only helps to recover valuable materials but also eliminates a bothersome disposal problem. The extension of the process to the treatment of other waste streams has been hampered by the tendency of the membrane to accumulate retained macromolecules at the surface, often in the form of adhering gel layers. This *concentration polarization,* as it is termed, which arises in all liquid-phase membrane processes but is most marked here, can be partially overcome by membrane modification, repeated cleaning, or the use of high fluid velocities. It remains, however, a serious impediment to the flourishing of UF technology. Its likely future course will be an incremental improvement through the development of membranes that discourage the adhesion of gel layers. UF has a potentially huge market in the treatment of wastewaters such as those that arise in the pulp and paper industry but must await a solution of the problem of concentration polarization.

## 8.2.2 Process Considerations and Calculations

The engineer dealing with membrane processes is confronted with the usual set of tasks associated with operations of this type. In the area of design, we wish to establish the type of membrane and the total area required for a specified separation or production rate. Conversely, for a given unit, real or imaginary, we would want to project performance or product concentration and recovery. Finally, it is often desirable to establish the effect of operating parameters such as feed composition, module geometry, or applied pressure on process performance. We provide some guidelines for dealing with these tasks, aided by the survey of fiber size and operating pressure given in Figure 8.10.

Membrane processes in general involve concentration changes in at least two directions, axial and lateral or radial, and therefore inevitably give rise, at least in principle, to PDEs. When both solute and water transport is involved, the equations will be two in number and they will also be coupled. Rigorous numerical solutions of these models were established in the 1970s and 1980s but are generally shunned by process engineers because of their complexities. Certain simplifying assumptions can be introduced that reduce the models to the ODE or even algebraic level while providing reasonably valid estimates of the answers being sought.

In all these endeavors, rigorous or approximate, transport parameters play a key role and must be addressed first. They are dependent on a number of variables, including flow conditions, membrane and module geometry, and the specific membrane process being addressed. Thus, the flow can be

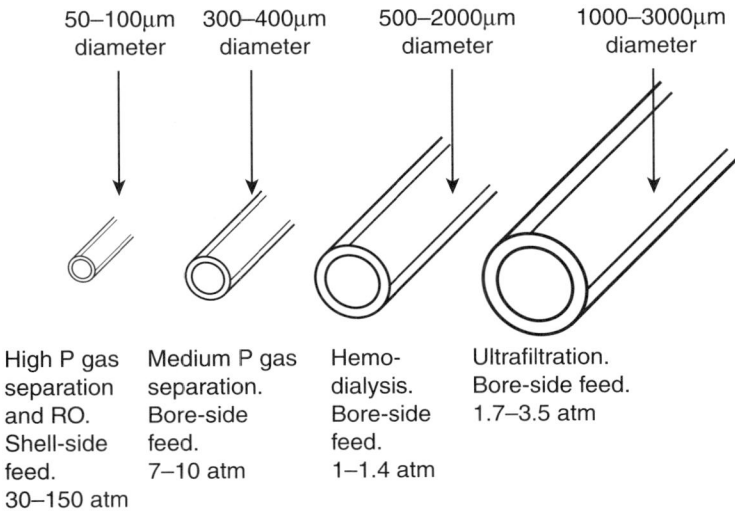

**FIGURE 8.10**
The principal types of hollow-fiber membranes.

**TABLE 8.4**

Membrane Transport Characteristics

| Type of Process and Geometry | Flow Regime | Controlling Resistance | Solute Transport Coefficients | Type of Model |
|---|---|---|---|---|
| Reverse osmosis spiral wound | Laminar (Entry region) | Core-side fluid | $Sh = 2.12\left(Re\,Sc\,\dfrac{d}{L}\right)^{1/3}$ | PDE |
| Reverse osmosis hollow fiber | Turbulent (Shell-side fed) | Shell-side fluid | $k \sim 10^{-5} - 10^{-6}$ m/s | PDE |
| Gas separation hollow fiber | Laminar | Membrane | Permeabilities | ODE |
| Hemodialysis hollow fiber | Laminar (Entry region) | Core-side fluid | $Sh = 1.86\left(Re\,Sc\,\dfrac{d}{L}\right)^{1/3}$ | ODE |
| Reverse osmosis hollow fiber (bore-side fed) | Laminar (Entry region) | Core-side fluid | $Sh = 1.86\left(Re\,Sc\,\dfrac{d}{L}\right)^{1/3}$ | PDE |
| Ultrafiltration hollow fiber | Laminar/ turbulent | Core-side fluid and gel layer | $Sh = 1.86\left(Re\,Sc\,\dfrac{d}{L}\right)^{1/3}$ $Sh = 0.23\,Re^{0.8}\,Sc^{1/3}$ (Turbulent) | PDE or ODE |

*Note:* Water permeability for contemporary RO membranes is in the range $P_w = 10^{-7}$ to $10^{-8}$ mol/m²sPa.

laminar or turbulent, involve gases or liquids, or take place in hollow-fiber or spiral-wound geometries. We have listed some of these features and the resulting transport parameters in Table 8.4.

We note, first of all, that transport resistance is on the fluid side in most cases; the main exception is gas permeation, which, because of the much higher diffusivities involved, is controlled by the membrane resistance. When fluid flow is laminar on the core side, mass transport will be in the entry region (see Chapter 5). This is reflected by the Sherwood number tabulated in Table 8.4. Transport coefficients on the shell side are subject to some uncertainty, which is indicated by listing a suggested range of values. Finally, we note that some of the models can be reduced to the ODE level. This occurs primarily as a result of the absence of water flux and the prevailing resistance residing in an entry-region boundary layer (hemodialysis), or in the membrane itself (gas separation). In the illustration that follows, use is made of these and other simplifications to provide the reader with a set of tools for the preliminary analysis of membrane processes.

## Illustration 8.5: Brian's Equation for Concentration Polarization

Concentration polarization, which has briefly been mentioned in our previous discussions, arises in all processes involving the passage of water or some other solvent through the membrane wall. The resulting flux will initially lead to a depletion of water and a consequent rise in solute concentration in the vicinity of the wall. This sets in motion a countervailing diffusion of solute back into the core fluid, which is exactly balanced by the rate at which solute is transported to the wall by the permeating water. We have indicated these opposing flows and the resulting concentration profile in Figure 8.11, where $C_b$, $C_w$, and $C_p$ are solute concentrations in the bulk fluid, at the wall, and in the permeate, respectively. Note that simultaneously, and in conjunction with this process, an entry-region development of the concentration profile takes place, which is described by the Levêque-type transport relations given in Table 8.2.

This local balance of solute transport was first described by Brian in analytical form as follows:

$$\begin{array}{cc} \text{Rate of transport of} & \text{Rate of back} \\ \text{solute by water flux} & = & \text{diffusion of solute} \end{array}$$

$$J_w C - J_w C_p = D\frac{dC}{dx} \qquad (8.8a)$$

where $J_w$ is the volumetric flux in units of $m^3/m^2\,s$ and $x$ is the distance in the direction of water flux. Note that both $C$ and $x$ increase in the same

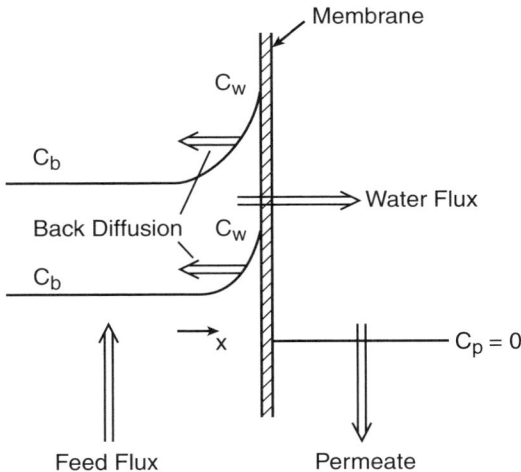

**FIGURE 8.11**
Concentration polarization.

direction so that Fick's law takes a positive sign. This expression can be integrated over the boundary layer thickness $\Delta x$, which yields

$$\int_{C_b}^{C_w} \frac{dC}{C-C_p} = \int_0^{\Delta x} \frac{J_w\,dx}{D} \qquad (8.8b)$$

or

$$\ln \frac{(C_w - C_p)}{(C_b - C_p)} = J_w \Delta x / D \qquad (8.8c)$$

where the right side can be viewed as a Peclet number, given that $J_w$ has the units of velocity.

Recognizing that the ratio $D/\Delta x$ defines the mass transfer coefficient $k_c$, Equation 8.8c can be recast in the form

$$\frac{C_w - C_p}{C_b - C_p} = \exp(J_w / k_c) = \exp(\text{Pe}) \qquad (8.8d)$$

which is the expression derived by Brian. The ratio $C_w/C_b$ is a measure of solute accumulated at the membrane wall and is termed the *concentration polarization modulus* $E_{Cp}$. We have occasion to use this equation in the illustration that follows, and in a modified and revealing form in Practice Problem 8.6.

### Illustration 8.6: A Simple Model of Reverse Osmosis

We set as our task here the replacement of the rigorous PDE model of reverse osmosis by a simple and approximate treatment, which makes use of certain empirical findings. Given the uncertainty of the parameters used in even the most rigorous models, this is no more than what good engineering sense dictates.

Our starting point is the basic flux equation for water transport through membranes driven by hydrostatic and osmotic potentials, a process alluded to in Chapter 1 (see Table 1.2). We have

$$N_w/A = P_w(\Delta p - \Delta \pi) \qquad (8.9a)$$

where $N_w$ is the molar transmembrane water flux (mol/s) and $P_w$ is the water permeability in mol/m$^2$ s Pa. A range of values for $P_w$ is given in Table 8.4.

The osmotic pressure $\pi$ is a colligative property, i.e., a property that depends on the particle molarity and is for infinitely dilute solutions given by the van't Hoff equation:

$$\pi = Cn_pRT \tag{8.9b}$$

where $C$ = molar concentration and $n_p$ = number of particles formed on dissociation.

This equation holds well for the dilute solutions involved in nanofiltration and for brackish water. Even for sea water, with $Cn_p \sim 10^3$ mol/m³, it predicts an osmotic pressure of $\pi = 10^3 \times 8.31 \times 298 = 2.48$ MPa = 24.6 atm, compared to an experimental value of 23 atm (Table 8.3). We adopt this equation to replace osmotic pressure by concentration and make the further assumption that the osmotic pressure in the permeate is zero, i.e., that the membrane is 100% effective in rejecting salt and the bulk concentration $C_b$ is constant, as was deduced in Illustration 5.2 for laminar entry region flow. We can then combine the two equations and factor out the constant $C_b$ to obtain

$$N_w / A = P_w \left( \Delta p - C_b n_p RT \frac{C_w}{C_b} \right) \tag{8.9c}$$

This expression opens a way to introduce Brian's equation (Equation 8.8d) to eliminate the unknown polarization modulus $C_w/C_b$. We have

$$N_w / A = P_w \left[ \Delta p - C_b n_p RT \exp\left( \frac{J_w}{k_c} \right) \right] \tag{8.9d}$$

and noting that the volumetric flux $J_w$ (m³/m² s) equals $1.8 \times 10^{-6}$ times the molar flux $N_w/A$ (mol/m² s), we obtain the following expression in the single variable $J_w$:

$$J_w = 1.8 \times 10^{-6} P_w \left[ \Delta p - C_b n_p RT \exp\left( \frac{J_w}{k_c} \right) \right] \tag{8.9e}$$

This is a non-linear equation in which $J_w$ must be solved numerically. We can use this equation to estimate water flux $J_w$ drawing on $k_c$ values given in Table 8.4 or those for the Peclet number $J_w/k_c$ listed in Table 8.5. Let us

**TABLE 8.5**

Experimental Membrane Process Parameters

| Process | $E_{Cp} = C_w/C_b$ | Pe | $E_e$ |
|---|---|---|---|
| Reverse osmosis | 1–1.5 | 0.3–0.5 | $<10^{-2}$ |
| Ultrafiltration | 70–150 | 5–10 | $\sim 10^{-2}$ |
| Gas separation | $\sim 1$ | $3 \times 10^{-3}$ to $7 \times 10^{-2}$ | $\sim 1$ |

choose an average value for Pe = 0.4, a water permeability of $10^{-7}$ mol/m²s Pa (see Table 8.4) and an applied hydrostatic pressure of $\Delta P$ = 1000 psig = $6.8 \times 10^6$ Pa. The particle molarity $C_b n_p$ for sea water is approximately $10^3$ mol/m³. We obtain from Equation 8.9e:

$$J_w = 1.8 \times 10^{-6} \times 10^{-7} (6.8 \times 10^6 - 10^3 \times 8.314 \times 298 \exp (0.4)) \quad (8.9f)$$

or

$$J_w = 5.6 \times 10^{-7} \text{ m}^3/\text{m}^2 \text{s} = 5.6 \times 10^{-7} \text{ m/s} \quad (8.9g)$$

This compares with a water flux of about $9 \times 10^{-7}$ m/s obtained in commercial desalination processes under identical conditions.

*Comments:*

The degree of agreement obtained with this simple model is quite remarkable but should be seen as being partly due to the judicious choice of experimental Peclet number and water permeability $P_w$. The range of these parameters will result in the answers extending between certain limits, which will differ by a factor of at least 2, introducing uncertainties of the same magnitude in the quantity being sought. A more fruitful application of Equation 8.9d and Equation 8.9e lies in their use as a tool for exploring the effect of changes in the operating variables on the performance of an *existing* unit with a *known* water permeability. Feed rate, water salinity, or operating pressure are the more obvious candidates for such a study. We address the effect of these variables on reverse osmosis performance in Practice Problems 8.7 and 8.8.

### Illustration 8.7: Modeling the Artificial Kidney: Analogy to the External Heat Exchanger

We have previously indicated that transmembrane transport in hemodialysis is dominated by the fluid resistance on the blood (core) side and that this resistance lies entirely in the laminar entry region. It was further seen that the dialysate concentration is adjusted to have the same osmotic pressure as that of the blood and that consequently no water flux is allowed to intrude on the proceedings. The only transport is that of the solutes to be removed from the blood, which is characterized by a core-side mass transfer coefficient listed in Table 8.3. We are, in other words, dealing with a straightforward combination of convective flow in the axial direction and radial transport through a film resistance. This combination can be expressed in terms of simple ODEs. Assuming co-current flow, the following results.

For the core-side mass balance:

$$\text{Rate of solute in} - \text{rate of solute out} = 0$$

$$Q_B C_B \big|_x - \left[ \frac{Q_B C_B \big|_{x+\Delta x}}{K_0 \pi d (C_B - C_D)_{avg} \Delta x} \right] = 0 \tag{8.10a}$$

For the shell-side mass balance:

$$\text{Rate of solute in} - \text{Rate of solute out} = 0$$

$$\left[ \begin{array}{c} Q_D C_D \big|_x \\ + K_0 \pi d (C_B - C_D)_{avg} \Delta x \end{array} \right] - Q_D C_C \big|_{x+\Delta x} = 0 \tag{8.10b}$$

On dividing by $\Delta x$ and letting $\Delta x \to 0$ we obtain the two ODEs

$$Q_B \frac{dC_B}{dz} + K_0 \pi d (C_B - C_D) = 0 \tag{8.10c}$$

and

$$Q_D \frac{dC_D}{dz} + K_0 \pi d (C_B - C_D) = 0 \tag{8.10d}$$

where the subscripts $B$ and $D$ refer to blood and dialysate, respectively, and $K_0$ (m/s) is an overall mass transfer coefficient that includes the effect, usually minor, of the membrane resistance itself.

The two ODEs are solved simultaneously by an extension of the $D$-operator method outlined in the Appendix or by Laplace transformation. The results can be arranged into the following dimensionless form:

For the co-current case:

$$E = \frac{1 - \exp[-N_T(1+Z)]}{(1+Z)]} \tag{8.11a}$$

For the countercurrent case:

$$E = \frac{1 - \exp[N_T(1-Z)]}{Z - \exp[N_T(1-Z)]} \tag{8.11b}$$

where the parameters $Z$, $N_T$, and $E$ are defined as follows:

$$Z = Q_B/Q_D = \text{Flow rate ratio} \qquad (8.11c)$$

$$N_T = K_0 A/Q_B = \text{Number of mass transfer units} \qquad (8.11d)$$

$$E = (C_{Bi} - C_{Bo})/(C_{Bi} - C_{Do}) = \text{Extraction ratio} \qquad (8.11e)$$

and the subscripts $i$ and $o$ refer to inlet and outlet conditions, respectively.

In most practical applications, dialysate flow is large compared to blood flow ($Z \cong 0$) and its solute concentrations much lower than those prevailing in the blood. Under these conditions we obtain for both the co-current and countercurrent cases

$$E = 1 - \exp(-N_T) \qquad (8.12a)$$

or equivalently

$$C_{Bi}/C_{Bo} = \exp(-K_0 A/Q_B) \qquad (8.12b)$$

The expressions above are general-purpose equations that can be used not only for design ($A$ in $N_T$), but also for parameter estimation ($K_0$ in $N_T$) from experimental concentration data, and the calculation of effluent concentration $C_{Bo}$ for different flow rates. We address this type of calculation in Practice Problem 8.10.

If the task is to assess the response of the *body* to dialysis, it becomes necessary to integrate the events in the blood compartment with those taking place in the dialyzer. The configuration then becomes identical to one that is well known in the field of heat transfer. It involves heating the contents $M_t$ of a well-stirred tank by pumping it through an external heat exchanger at a flow rate $F$ and returning it to the tank. The tank represents the blood compartment while the dialyzer takes the place of the heat exchanger. The analogy is illustrated in Figure 8.12.

The model is based on the assumption that the heat exchanger is at a quasi-steady state since flow through it is much faster than the rate at which incoming temperature changes with time. The solution, which is a classical and well-known one, is given by

$$t = \frac{M_t}{F}\left[1 - \exp\left(-\frac{UA}{FC_p}\right)\right]^{-1} \ln\frac{T_s - T_t^o}{T_s - T_t} \qquad (8.13a)$$

where the steam temperature $T_s$ corresponds to the external dialyzer concentration, which is usually taken to be vanishingly small. To convert this expression to represent the duration of the dialysis procedure, we drop the

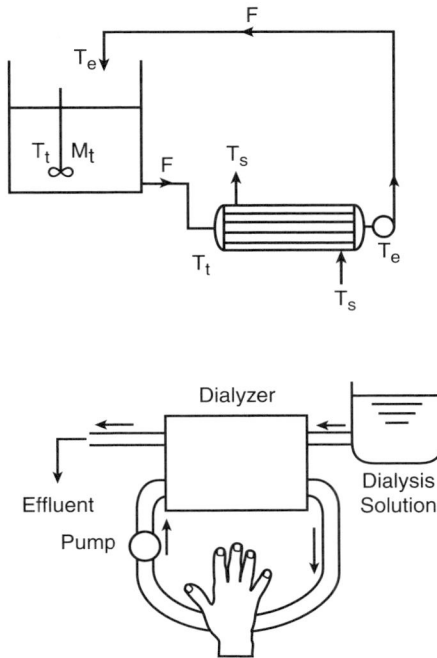

**FIGURE 8.12**
Analogy between an external heat exchanger and the dialyzer.

heat capacity $C_p$ and replace the heat transfer coefficient $U$ by the mass transfer coefficient $K_o$. The time $t$ it takes to reduce the toxin concentration from $C^o$ to $C$ is then given by

$$t = \frac{V}{Q}\left[1 - \exp\left(-\frac{K_o A}{Q}\right)\right]^{-1} \ln\frac{C^o}{C} \qquad (8.13b)$$

where $V$ and $Q$ are now the blood volume and volumetric flow rate, respectively. Derivation of this expression is left to the exercises.

A typical value for $K_o A$ is 1 cm³/s; that for the blood flow rate is 5 cm³/s. Blood volume for an adult is about 5 l. We then have, for a tenfold reduction in toxic concentration,

$$t = \frac{5000}{5}\left[1 - \exp\left(-\frac{1}{5}\right)\right]^{-1} \ln 10 \qquad (8.13c)$$

i.e.,

$$t = 12{,}690 \text{ s} = 3.52 \text{ h} \qquad (8.13d)$$

This is the duration of a typical dialysis procedure.

Equation 8.13b is a highly useful expression to assess the effect of the percent toxin to be removed, of the value $K_o A$ for different dialyzers, and the impact of flow rate $Q$ on the duration of the precedent. Note that $K_o$ varies with $Q$ as well, but the dependence is a weak one-third power one (see Table 8.4).

### Illustration 8.8: Membrane Gas Separation: Selectivity $\alpha$ and the Pressure Ratio $\phi$

It has previously been indicated (see Table 8.2) that membrane-based gas separation processes, while still of considerable complexity, can be modeled at the level of ODEs. This is because the principal transport resistance resides within the membrane itself and because the mass balances need therefore be concerned only with concentration changes in the direction of flow. These equations, however, must be supplemented by force balances to take account of the nonlinear pressure drop associated with permeation processes. The resulting set of ODEs will generally have to be solved numerically. We do not take up this problem here but instead acquaint the reader with two important system parameters that provide useful information for a preliminary assessment of these operations. These parameters are the pressure ratio $\phi$ and the membrane selectivity $\alpha$.

The pressure ratio is an operational parameter and is defined as the ratio of feed pressure to that prevailing on the permeate side, i.e.,

$$\phi = P_F/P_P \tag{8.14a}$$

where the subscripts refer to feed and permeate conditions, respectively.

Selectivity $\alpha$, on the other hand, is a property of the membrane and deals with the relation between *compositions*, usually expressed as mole fractions $y_1$, on either side of the barrier. Its definition, at least in its form, is identical to that of relative volatility, which was seen in connection with vapor–liquid equilibria (see Equation 6.20a). For a two-component system, it is given by the relation

$$\alpha = \frac{y_{1P}(1-y_{1F})}{y_{1F}(1-y_{1P})} \tag{8.14b}$$

where the ratio $E = y_{1P}/y_{1F}$ can be viewed as an enrichment factor of the more permeable component 1 attainable by a particular type of membrane. Both $\alpha$ and $\phi$ are used, singly or in conjunction, to define two important regions of limiting operational behavior. When $\phi$ is small, and the selectivity is far in excess of the pressure ratio, i.e.,

$$\alpha \gg \phi \tag{8.14c}$$

we speak of the process as *pressure-ratio limited*. Conversely, if the pressure ratio substantially exceeds $\alpha$, i.e.,

$$\phi \gg \alpha \tag{8.14d}$$

the process is said to be in the *selectivity-limited region*. Let us consider these two cases, and the information and lessons to be drawn from them.

### 1. Pressure-Ratio-Limited Region

We start by noting that gas permeation is driven by a concentration or partial pressure difference across the membrane, i.e., it is a process based on diffusion, and not on D'Arcy-type bulk flow. For diffusion to take place, the feed partial pressure must be higher than that on the permeate side; i.e., we must have

$$p_F > p_P \tag{8.15a}$$

or equivalently in terms of mole fraction $y$ or total pressure $P$

$$y_F P_F > y_P P_P \tag{8.15b}$$

As a consequence of this inequality, the enrichment attained will generally be smaller than the total pressure ratio and can at most equal it; i.e., we must have

$$E = \frac{y_P}{y_F} < \frac{P_F}{P_P} \tag{8.15c}$$

and in the limit

$$(E)_{\text{Max}} = \frac{y_P}{y_F} = \frac{P_F}{P_P} = \phi \tag{8.15d}$$

This limiting value corresponds to the situation in which the partial pressures on either side of the membrane have attained equality, and diffusion has ceased. Note that while partial pressures are now equal, the total pressures on either side are not, nor are the corresponding mole fractions.

The enrichment given by Equation 8.15d represents the maximum value of $E$ attainable, irrespective of the selectivity of the membranes involved. In other words the process is now dependent on $\phi$ only and is consequently in the pressure-ratio-limited region, with a resulting enrichment given by Equation 8.15d. This condition arises when $\alpha$ exceeds $\phi$ by a factor of more than 5.

## 2. Selectivity-Limited Region

When the pressure ratio is very large, it ceases to be a limiting factor and the process becomes solely dependent on the membrane selectivity $\alpha$. The enrichment $E$ will in this case no longer be dictated by the value of $\phi$ but must instead be extracted from Equation 8.14a. We obtain by simple rearrangement

$$E = \frac{y_{1P}}{y_{1F}} = \frac{\alpha}{1 + y_{1F}(\alpha - 1)} \tag{8.16}$$

which is entirely analogous to the enrichment obtained in vapor–liquid equilibria. This condition holds when $\phi$ exceeds $\alpha$ by a factor of more than 5.

*Comments:*

The limiting expressions (Equation 8.15d and Equation 8.16) are invaluable in arriving at a first assessment of gas-separation processes. Suppose, for example, that $\alpha = 5$, which is a typical membrane selectivity for the separation of nitrogen and oxygen. Then in any practical operation, the pressure ratio will be in excess of $\alpha$ and the operation will consequently take place in the selectivity-limited region. The maximum enrichment of nitrogen attainable is then given by (Equation 8.16)

$$(E_{Max}) = \frac{\alpha}{1 + y_{1F}(\alpha - 1)} = \frac{5}{1 + 0.79(5 - 1)} = 1.2 \tag{8.17}$$

This is still a respectable value and has led to the large-scale adoption of the membrane process for the separation and enrichment of air.

The limiting expressions laid down above are also helpful in setting limits on the usefulness of enhancing membrane selectivity. This comes about as a result of the high cost of compressing feed gas to very high pressures or drawing a hard vacuum on the permeate side, which limits practical pressure ratios $\phi$ to the range 10 to 50. Suppose, for example, that a value of $\phi = 10$ is chosen. Then a membrane selectivity of 50 will locate the process in the pressure-ratio-limited region with the enrichment factor given by $E = \phi$, independent of $\alpha$ (Equation 8.15d). Any further increases in membrane selectivity will therefore have no effect on the performance of the process under these conditions.

## Practice Problems

8.1. *The Operating Line in a Gas Scrubber*
Explain the meaning and consequences of the following special cases of an operating line for a gas scrubber:
a. The line is vertical.
b. The line is horizontal.
c. The line is curved.
d. The line is displaced parallel to itself and away from the equilibrium curve.
e. The line lies *below* the equilibrium curve.
What is the effect on NTU in cases a, b, and d?

8.2. *The Effect of Feed Flow Rate in Gas Absorption*
Analyze how an increase in the gas flow rate would affect the performance of an existing absorber.

8.3. *The Effect of Packing Size in Gas Absorption*
Qualitatively plot the component and total costs that are affected by packing size in a gas absorber. (*Hint:* Use Figure 8.2 as a guide.)

8.4. *Packed Column Liquid Extraction in the Linear Regime*
An aqueous solute is to be extracted in a packed column using an organic solvent with a distribution coefficient of 1.5. The raffinate concentration is to be reduced tenfold from 0.1 to 0.01 wt ratio and it is proposed to use a solvent ratio 1.5 times the minimum required. The entering solvent is essentially solute free and HTU is initially set at 4 m. Such high HTU values of several meters are not unusual in liquid-extraction applications. Establish the required height of the extraction column.

*Answer*: 11.2 m

8.5. *Another Look at Isotope Distillation*
An isotope pair with a separation factor of 1.01 is to be separated into overhead and bottoms product with compositions of 0.99 and 0.10 mole fraction, respectively, by distillation in a packed column. A particular high-efficiency packing has been touted by its manufacturer as having HTU values of no more than 3 cm. The facility considering the process is, for various reasons, constrained to column heights of no more than 10 m. Will the process meet this requirement?

8.6. *Concentration Polarization, Peclet Number, and Enrichment Factor*
Show that Brian's equation (Equation 8.8d) can be recast in the form

$$E_{Cp} = \frac{C_w}{C_b} = \frac{\exp(\text{Pe})}{1 + E_e[\exp(\text{Pe}) - 1]} \qquad (8.8\text{d})$$

where $\text{Pe} = J\Delta x/D$ is the flux Peclet number, and $E_e$ is an enrichment factor defined by the ratio $C_p/C_w$. Table 8.5 conveys an idea of the magnitude of these parameters for various membrane processes.

8.7. *The Effect of Pressure in Reverse Osmosis*
Typical sea water desalination plants using reverse osmosis produce 30 to 50 gfd of water [gal (U.S.) per square ft per day] (1 gfd = 4.72 × 10$^{-7}$ m$^3$/m$^2$ s). Suppose it is decided to double the operating pressure from 50 to 100 atm. Will the production rate increase by less or by more than a factor of 2? Assume 100% salt rejection.

8.8. *The Effect of Feed Rate and Water Salinity in Reverse Osmosis*
Consider the process described in Illustration 8.6. Using the numerical parameter values cited there determine:
a.  The effect of halving salinity on $J_w$.
b.  The effect of doubling the shell-side feed rate on the same quantity. (*Hint:* Consider the effect on $v$ in turbulant flow.)

*Answer:* a. $J_w = 8.9 \times 10^{-5}$ m$^3$/m$^2$ s

8.9. *Concentration of an Isotonic Saline Solution*
Isotonic saline is an NaCl solution that has the same osmotic pressure as blood. It is used in medical treatments, for example, to make up fluid loss, and also as the extracting fluid on the dialysate side of an artificial kidney. Calculate the concentration of this solution. (*Hint:* Consult Table 8.3.)

*Answer:* 0.141 molar

8.10. *Performance of a Hemodialyzer*
Use the following data to calculate the urea concentration exiting from a hemodialyzer with a total surface area of 1 m$^2$ and 100 hollow fibers of length $L = 17$ cm.
   Urea diffusivity   $D = 2 \times 10^{-5}$ cm$^2$/s
   Urea inlet concentration   $C_{Bi} = 100$ ppm
   Blood flow rate   $Q_B = 1$ cm$^3$/s (total)

*Answer:* 560 ppm

8.11. *Effect of Membrane Resistance: The Wall Sherwood Number*
    For the hemodialyzer of Practice Problem 8.10, calculate the percent resistance in the membrane wall, using the wall Sherwood number defined in Illustration 5.1.

*Answer*: 5.1%

8.12. *The Hemodialyzer*
    Derive the solution (Equation 8.13b) describing the performance of a hemodialyzer. (*Hint:* Start with a differential balance for the dialyzer to obtain the effluent blood concentration, and use this value for the balance over the blood compartment.)

8.13. *Performance of a Membrane Process in Gas Separation*
    The membrane selectivity for the removal of $CO_2$ from natural gas is of the order 20. For a feed containing 5% $CO_2$, give an approximate evaluation of the enrichment attainable using a pressure ratio of 20.

# 9

## Simultaneous Heat and Mass Transfer

Our treatment so far has made occasional reference to heat transfer, primarily to draw the reader's attention to the analogies that exist between the transport of heat and mass. For example, in Chapter 1 we highlighted the similarities between the rate laws governing convective and diffusive heat and mass transfer. The analogy between the two phenomena when dealing with co-current or countercurrent operations has been brought out on several occasions, notably Illustration 8.7.

We now turn our attention to processes where heat and mass transfer occur in unison. This is far from an unusual event, but it raises the complexity of the underlying model, a fact that persuaded us to defer its consideration to the final chapter.

Simultaneous heat and mass transfer occur in a natural way whenever the transport of mass is accompanied by the evolution or consumption of heat. An important class of such operations, and one that occupies a considerable portion of this chapter, involves the condensation of water vapor from an airstream and the reverse process of evaporation of liquid water into air. There are a host of important operations in which this type of transfer occurs. The humidification and dehumidification of air, often identified with air conditioning, is practiced within domestic, commercial, and industrial contexts. Process water that was used in a plant for cooling purposes is often cycled through a cooling tower, where it is contacted in countercurrent flow with air and undergoes evaporative cooling before being returned for reuse. The drying of solids, an important class of operations in its own right, also draws on the principles that underlie the transport of water between its liquid phase and air. Indeed, the air–water system and the temperature and concentration changes that arise in air–water contact are of such importance in the physical sciences that they have led to the construction of *psychometric* or *humidity charts*. These charts summarize in convenient fashion the thermal and concentration variables relevant to operations involving the air–water system. In the illustrations and practice problems that follow, we use these charts repeatedly to establish parameters of interest in various operations based on the air–water system.

Simultaneous heat and mass transfer also occurs in exothermic or endo-thermic heterogeneous reacting systems and in the absorption or adsorption from concentrated gas streams. These topics are addressed in separate illus-trations but we retain the air–water system as the central theme of this chapter.

## 9.1   The Air–Water System: Humidification and Dehumidification, Evaporative Cooling

### 9.1.1   The Wet-Bulb Temperature

We start our deliberations by examining the events that occur when a flowing gas comes in contact with a liquid surface. The reader will be aware from personal experience that this process results in a drop in the temperature of the liquid, often referred to as evaporative cooling. The chill we experience when wind blows over our perspiring bodies is one manifestation of this effect.

Let us assume that both the water and the air are initially at the same temperature. During the first stage of evaporation, the energy required for the process, i.e., the latent heat $\Delta H_v$, will come from the liquid itself, which consequently experiences a drop in temperature. That decline, once it is triggered, will cause a corresponding amount of heat transfer to take place from the air to the water. At this intermediate stage, the latent heat of vaporization is provided both by the liquid itself and by heat transfer from the warmer gas.

As the liquid temperature continues to drop, the rate of heat transfer accelerates until a stage is reached where the entire energy load is supplied by the air itself. A steady state is attained in which the rate of evaporation is exactly balanced by the rate at which heat is transferred from the gas to the liquid. The liquid is then said to be at its "wet-bulb temperature," $T_{wb}$, and the corresponding air temperature is referred to as the "dry-bulb tem-perature," $T_{db}$. The difference $(T_{db} - T_{wb})$ constitutes the driving force for the heat being transferred from the gas to the liquid. This is indicated in Figure 9.1, which also shows the associated humidities of the air, $Y_{wb}$ (kg $H_2O$/kg air), the saturation humidity prevailing at the surface of the liquid, and $Y_{db}$, the humidity in the bulk air. The wet-bulb temperature and its associated saturation humidity play a central role in humidification and dehumidifica-tion, in water cooling operations, as well as in drying processes. These are taken up in subsequent illustrations.

The relation among $T_{wb}$, $Y_{wb}$, and the system parameters is established by equating the rate of heat transfer from air to water to the rate of evaporation, i.e., the rate at which moisture is transferred from the water surface to the air. Thus,

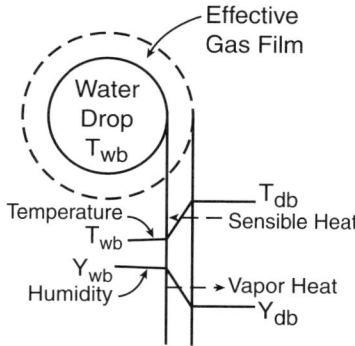

**FIGURE 9.1**
Temperature and humidity distribution around a water drop exposed to a flowing airstream.

Rate of evaporation = Rate of heat transfer

$$k_Y A(Y_{wb} - Y_{db})\Delta H_v = hA(T_{db} - T_{wb}) \qquad (9.1a)$$

where $k_Y$ is the mass transfer coefficient in units of kg $H_2O/m^2$ s $\Delta Y$.
Canceling terms and rearranging we obtain

$$\frac{Y_{wb} - Y_{db}}{T_{wb} - T_{db}} = -\frac{h}{k_Y \Delta H_v} \qquad (9.1b)$$

where the difference $T_{db} - T_{wb}$ is referred to as the wet-bulb depression.
We note from Equation 9.1b that the humidity of the air $Y_{db}$ can, in principle, be established from measured values of $T_{db}$, $T_{wb}$, and $Y_{wb}$, the latter being obtained from the relation

$$Y_{wb} = \frac{\left(P^o_{H_2O}\right)_{wb}}{P_{Tot} - \left(P^o_{H_2O}\right)_{wb}} \frac{M_{H_2O}}{M_{air}} \qquad (9.1c)$$

where $P^o_{H_2O}$ is the vapor pressure of water, available from tables, and $M$ = molar mass. $T_{db}$ is measured by exposing a dry thermometer to the flowing air, while $T_{wb}$ is obtained in similar fashion using a thermometer covered with a moist wick. More recent devices for measuring $Y$ rely on changes in electrical properties of the sensor element with the moisture content of air. Both wet- and dry-bulb properties appear in the humidity charts that are taken up shortly.

## 9.1.2   The Adiabatic Saturation Temperature and the Psychrometric Ratio

Before addressing the properties and construction of the humidity charts, we consider a small variation on the simple contact of water with flowing air, which led to the wet-bulb conditions. In this modified arrangement, shown in Figure 9.2, a stream of air is humidified in contact with constantly recirculated water. Both the water and the exiting gas stream attain adiabatic saturation temperature, $T_{as}$, which is lower than the dry-bulb temperature of the entering air because of evaporative cooling.

If care is taken to introduce the make-up water at the same adiabatic saturation temperature, and the datum temperature is set at $T_{as}$, a simple energy balance will yield

$$\text{Rate of energy in} - \text{Rate of energy out} = 0$$

$$[C_s(T_{db} - T_{as}) + Y_{db}\Delta H_v] - [C_s(T_{as} - T_{as}) + Y_{as}\Delta H_v] = 0 \qquad (9.2a)$$

which on rearrangement leads to the expression

$$\frac{Y_{as} - Y_{db}}{T_{as} - T_{db}} = -\frac{C_s}{\Delta H_v} \qquad (9.2b)$$

where $C_s$ is the specific heat of the air, also termed humid heat, in units of kJ/kg dry air. Plots of this equation appear in the humidity charts discussed in the next illustration.

The striking similarity between the adiabatic saturation and wet-bulb relations, Equation 9.2b and Equation 9.1b, led to a detailed examination of the ratio of the two slopes, $h/k_Y C_s$, also known as the psychrometric ratio. These studies culminated in the finding that for the water–air system, *and only for that system*, its value is approximately unity. Thus,

$$\text{Psychrometric Ratio } h/k_Y C_s \approx 1 \qquad (9.2c)$$

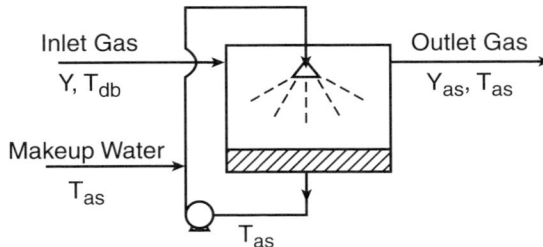

**FIGURE 9.2**
Flow sheet showing the attainment of adiabatic saturation conditions.

This expression, known as the Lewis relation, when used to compare Equation 9.1b and Equation 9.2b, leads to the conclusion that the adiabatic saturation and wet-bulb temperatures are essentially identical. The Lewis relation has other important implications as well, as will become apparent in Illustration 9.4 dealing with the design of water-cooling towers. It is seen there that the underlying model equations can be enormously simplified by making use of the Lewis relation.

### Illustration 9.1: The Humidity Chart

The psychrometric or humidity charts are displayed in Figure 9.3 and Figure 9.4 for the low and high temperature ranges shown. To familiarize ourselves with the properties of these diagrams, we start by defining and deriving a set of variables, which appear implicitly or explicitly in the two figures.

*Absolute Humidity Y*

This quantity was already referred to in connection with the wet-bulb temperature and is redefined here for convenience:

$$Y(\text{kg H}_2\text{O}/\text{kg dry air}) = \frac{18}{29}\frac{p_{H_2O}}{P_T - p_{H_2O}} \qquad (9.3a)$$

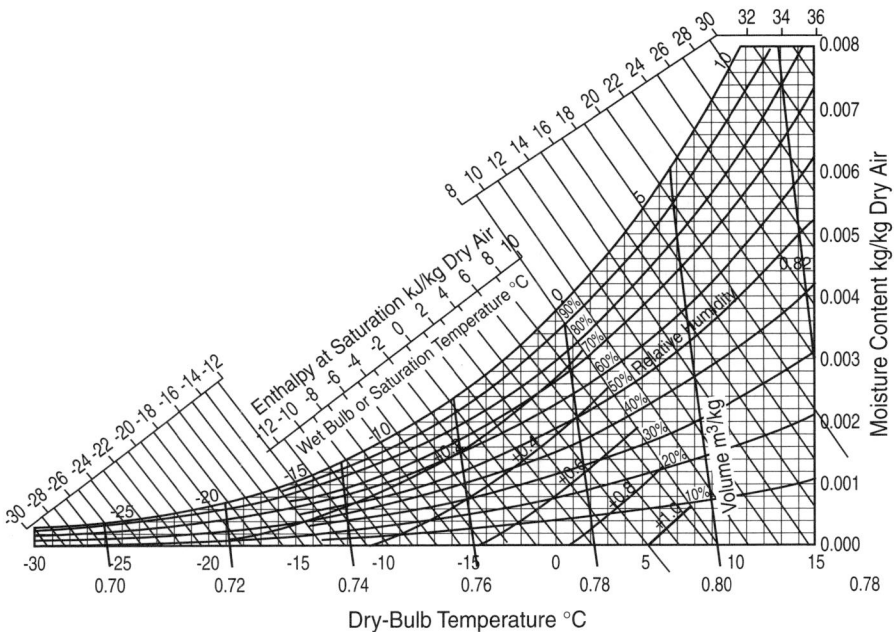

**FIGURE 9.3**
Humidity chart: low temperature range. (From Carrier Corporation. With permission.)

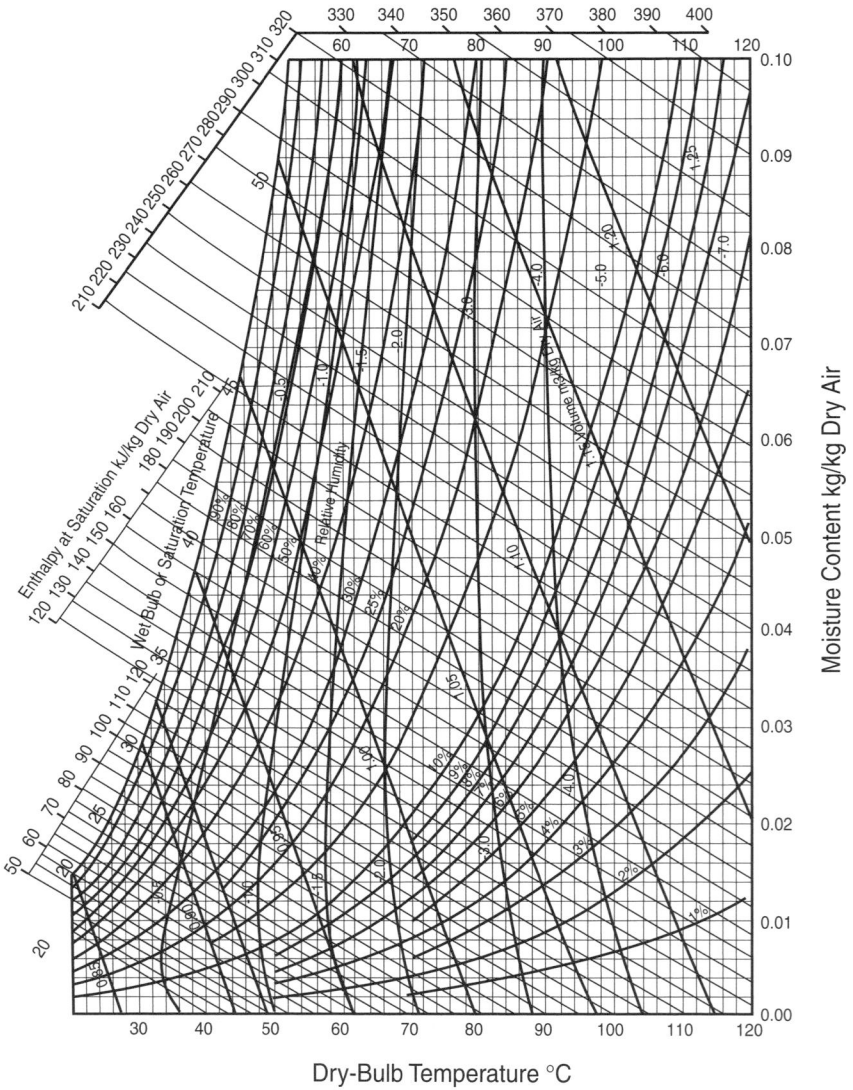

**FIGURE 9.4**
Humidity chart: high temperature range. (From Carrier Corporation. With permission.)

where $p_{H_2O}$ is the partial pressure of water vapor.

$Y$ appears as the right-hand side ordinate in the humidity charts.

*Relative Humidity* RH

To obtain a sense of the relative degree of saturation of the air, we define

$$\%RH = \frac{P_{H_2O}}{P^o_{H_2O}} 100 \qquad (9.3b)$$

where $P^o_{H_2O}$ is the saturation vapor pressure of water at the temperature in question.

RH varies over the range 0% (dry air) to 100% (fully saturated air) and appears in the humidity charts as a set of parametric curves that rise smoothly from left to right.

### Dew Point $T_{dp}$

This is the temperature at which moist air, cooled at constant $P_{Tot}$ and $Y$, becomes saturated, i.e., attains 100% relative humidity. Its value is established by moving from the initial defining point of a given air–water mixture on the humidity chart along a horizontal line to the eventual intersection with the curve of 100% relative humidity.

### Humid Volume $V_H$

The humid volume of moist air is the specific volume in m³/kg dry air measured at $P_{Tot}$ = 101.3 kPa (1 atm) and the temperature $T$ of the mixture. Values of $V_H$ appear in the humidity charts as a set of lines of negative slope.

### Humid Heat $C_s$

This quantity, which has already been encountered in connection with the adiabatic saturation temperature, is the specific heat of moist air expressed in units of kJ/kg dry air.

Humid heat does not usually appear explicitly in the charts but is contained in the enthalpies shown there. It can be calculated from the following equation:

$$C_s(\text{kJ/kg dry air}) = 1.005 + 1.88Y \qquad (9.3c)$$

### Enthalpy H

With the humid heat in hand, we are in a position to formulate the enthalpy of an air–water mixture. With $T°$ chosen as the datum temperature for both components and adding sensible and latent heats we obtain

$$H(\text{kJ/kg dry air}) = C_s(T - T°) + Y\Delta H_v° \qquad (9.3d)$$

$$\underset{\substack{\text{Sensible} \\ \text{heat}}}{} \quad \underset{\substack{\text{Latent} \\ \text{heat}}}{}$$

where the datum temperature is usually set equal to 0°C for both liquid water and dry air.

Values of the enthalpies of various air–water mixtures are read from the left-hand oblique ordinate.

*Adiabatic Saturation Temperature* $T_{as}$ *and Wet-Bulb Temperature* $T_{dp}$

Plots of the adiabatic saturation line, Equation 9.2b, appear in the humidity charts as lines extending from the abscissa to the 100% relative humidity curve. The point of intersection with that curve defines the wet-bulb temperature $T_{wb}$, which is also the adiabatic saturation temperature $T_{as}$.

*Example*

This concrete example illustrates the various uses to which the humidity charts may be put: We choose moist air with a relative humidity of 25% and a (dry-bulb) temperature of 50°C and proceed to calculate various properties of interest using the chart shown in Figure 9.4.

*Absolute Humidity* Y. This value is read from the right side rectangular ordinate, which yields

$$Y = 0.0195 \text{ kg H}_2\text{O/kg dry air} \tag{9.4a}$$

*Dew Point* $T_{dp}$. We follow the horizontal line through the point $Y = 0.0195$, $T = 50°C$ to its intersection with the 100% relative humidity curve and obtain

$$T_{dp} = 24.5°C \tag{9.4b}$$

This corresponds to the temperature at which, on isobaric cooling of the moist air, the first condensation of water occurs.

*Wet-Bulb Temperature* $T_{wb}$. Here the procedure is to follow the adiabatic saturation line to its intersection with the 100% relative humidity curve. We obtain

$$T_{wb} = 30.4°C \tag{9.4c}$$

Note that the wet-bulb temperature is not identical to the dew point.

*Water Partial Pressure* $p_{H_2O}$. This quantity can be obtained directly from the absolute humidity and Equation 9.3a. Solving it for $p_{H_2O}$ yields

$$p_{H_2O} = \frac{29\,Y\,P_{Tot}}{18+29\,Y} = \frac{29\times 0.0195 \times 101.3}{18+29\times 0.0195} = 2.09 \text{ kPa} \tag{9.4d}$$

*Humid Volume* $V_H$. The plots for $V_H$ are shown as steep lines of negative slope. The point $Y = 0.1095$, $T = 50∞C$ is located between the lines for $V_H = 0.90$ and $0.95$. Linear interpolation yields the value

$$V_H = 0.945 \text{ m}^3/\text{kg dry air} \qquad (9.4\text{e})$$

*Enthalpy* H. This value is read from the oblique left-hand ordinate of Figure 9.4 and comes to

$$H = 103 \text{ kJ/kg dry air} \qquad (9.4\text{f})$$

*Water Removal.* Suppose the air mixture considered here is to be cooled and dehumidified to $T = 15°C$ and $RH = 20\%$. The water to be removed can then be calculated as follows:

$$(Y)_{\text{initial}} = 0.1095 \qquad (Y)_{\text{final}} = 0.0021 \qquad (9.4\text{g})$$

Water to be removed:

$$(Y)_{\text{initial}} - (Y)_{\text{final}} = 0.0195 - 0.0021 = 0.0174 \text{ kg H}_2\text{O/kg air} \qquad (9.4\text{h})$$

Alternatively, the result may be expressed in volumetric units by dividing by the humid volume of the original mixture: $0.0174/0.945 = 0.0184 \text{ kg H}_2\text{O/m}^3$ initial mixture.

*Water Removal Heat Load.* In addition to the amount of water to be removed, an important parameter in the design of a dehumidification unit is the associated heat load. That quantity can be computed from the relevant enthalpies read from the humidity chart. We have for the case cited

$$(H)_{\text{initial}} = 103 \text{ kJ/kg dry air, } (H)_{\text{final}} = 20.3 \text{ kJ/kg dry air} \qquad (9.4\text{i})$$

$$\text{Heat removed} = (H)_{\text{initial}} - (H)_{\text{final}} = 103 - 20.3 = 82.7 \text{ kJ/kg dry air} \qquad (9.4\text{j})$$

Alternatively, using volumetric units

$$\text{Heat removed} = [(H)_{\text{initial}} - (H)_{\text{final}}]/V_H = 82.7/0.945 = 87.5 \text{ kJ/m}^3 \qquad (9.4\text{k})$$

### Illustration 9.2: Operation of a Water-Cooling Tower

As previously mentioned, warm process water that was used in a plant for cooling purposes can be restored to its original temperature by contacting it with an airstream, which causes it to undergo evaporative cooling. The operation is generally carried out in cooling towers containing stacked packings of large size and voidage to minimize pressure drop. We propose here to model the operation of such a tower and, in the course of the model development, introduce the reader to some ingenious simplifications based on the Lewis relation (Equation 9.2c).

As in all packed-column operations, the fundamental model equations consist of differential balances taken over each phase; the principal novelty here is the simultaneous use of mass and energy balances.

The pertinent variables and the differential elements around which the balances are taken are displayed in Figure 9.5a.

*Water Balance over Gas Phase (kg $H_2O/ml$ s)*

This balance is no different from similar mass balances used in packed-gas absorbers and distillation columns (see Chapter 8) and takes the form

Rate of water vapor in − Rate of water vapor out = 0

$$\left[ \begin{array}{c} G_s Y|_z \\ +N_{avg} \end{array} \right] - \left[ G_S Y|_{z+\Delta z} \right] = 0 \tag{9.5a}$$

which upon introduction of the auxiliary mass transfer rate equation, division by $\Delta z$, and letting $\Delta z \to 0$ yields the usual form of ODE applicable to these cases:

$$G_s \frac{dY}{dz} - K_Y a(Y^* - Y) \tag{9.5b}$$

where $Y^* - Y$ is the humidity driving force.

*Water Balance over Water Phase*

This balance is omitted since the water losses are usually less than 1%.

*Gas Phase Energy Balance (kJ/m² s)*

Here we must be careful to include both sensible heat transfer as well as the latent heat brought into the air by the water vapor. We obtain

Rate of energy in − Rate of energy out = 0

$$\left[ \begin{array}{c} G_s H|_z + q_{avg} \\ +\Delta H_v N_{avg} \end{array} \right] - \left[ G_s H|_{z+\Delta z} \right] = 0 \tag{9.5c}$$

which, after applying the same procedure as before, yields

## a. Column Variables

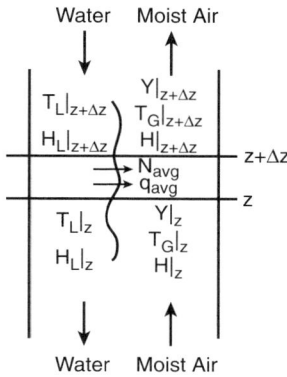

## b. Operating Diagram

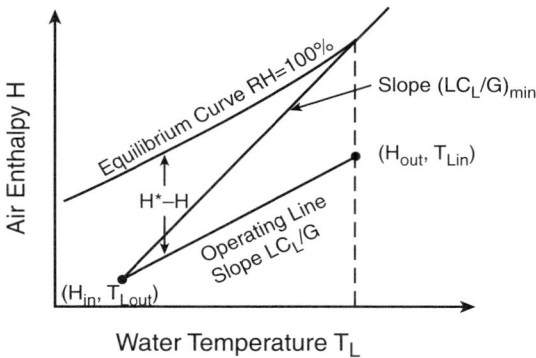

**FIGURE 9.5**
Variables and operating diagram for a packed cooling tower.

$$G_s \frac{dH}{dz} - Ua(T_L - T_G) - \Delta H_v K_Y a(Y^* - Y) = 0 \qquad (9.5d)$$

Here $T_L$ and $T_G$ are the water and air temperatures, respectively, and $H$ equals the enthalpy of the moist air at a given point in the tower.

*Liquid Phase Energy Balance (kJ/m² s)*

A completely analogous derivation to the gas-phase energy balance yields

$$LCp_L \frac{dT_L}{dz} - Ua(T_L - T_G) - \Delta H_v K_Y a(Y^* - Y) = 0 \qquad (9.5e)$$

where we have replaced the liquid enthalpy $H_L$ by $Cp_L (T_L - T°)$. The model is completed by adding the relevant equilibrium relation, which coincides with the 100% RH curve in the humidity charts and is also available in analytical form. Thus, for the equilibrium relation

$$Y^* = f(T_L) \tag{9.5f}$$

Equation 9.5b, Equation 9.5d, Equation 9.5e, and Equation 9.5f together with the previously given expression for $H$ (Equation 9.3d) constitute a set of five equations in the five state variables $Y$, $Y^*$, $T_G$, $T_L$, and $H$.

Although a numerical solution of these equations is today easily accomplished, early workers in the field had to cast about for alternative means of solving the model. To do this, they used the ingenious device of introducing the Lewis relation into the gas-phase energy balance, which has the effect of combining $T_G$ and $Y$ into a single variable, the air enthalpy $H$. We sketch the procedure below, using interfacial values in place of $Y^*$ and $T_L$ to accommodate the film coefficients $h$ and $k_Y$ used in the Lewis relation. We use that relation to replace $h$ by $k_Y C_s$ and obtain in the first instance

$$G_s \frac{dH}{dz} - k_Y a \left[ (C_s T_i + \Delta H_v Y_i^*) - (C_s T_G + \Delta H_v Y) \right] = 0 \tag{9.5g}$$

where, as seen from Equation 9.3d, the bracketed terms ( ) represent enthalpies of air–water mixtures. We therefore can write

$$G_s \frac{dH}{dz} - k_Y a (H_i - H) = 0 \tag{9.5h}$$

where $H_i - H$ can be considered an enthalpy driving force, which replaces and combines the temperature and humidity driving forces in the original model.

We now assume that the two-film theory can be applied to this system, with the result that Equation 9.5h can be cast in the form

$$G_s \frac{dH}{dz} - K_Y a (H^* - H) = 0 \tag{9.5i}$$

where $K_Y a$ is now the overall mass transfer coefficient and $H^*$ the gas enthalpy in equilibrium with the bulk water temperature $T_L$.

This equation is of the same form as gas-phase differential balances encountered in gas absorption and distillation, so that the design procedures used there can be replicated, provided an appropriate operating line can be constructed. That line is obtained from an overall two-phase integral heat balance and takes the form

$$G_s(H_1 - H) = LC_L(T_{L1} - T_L)$$ (9.5j)

for part of the column, and for the entire tower

$$G_s(H_1 - H_2) = LC_L(T_{L1} - T_{L2})$$ (9.5k)

The gas-phase energy balance (Equation 9.5i) can in turn be formally integrated to yield the familiar HTU–NTU relation:

$$Z = \frac{G_s}{K_Y a} \int_{H_{in}}^{H_{out}} \frac{dH}{H*-H} = \text{HTU} \times \text{NTU}$$ (9.5l)

The model is completed with the addition of the equilibrium relation:

$$H* = f(T_L)$$ (9.5m)

which is constructed from the 100% RH curve of the psychrometric charts.

The original set of five equations, three of which are ODEs, have thus been reduced to the three relations (Equation 9.5k, Equation 9.5l, and Equation 9.5m). What is more, the set is now cast in the familiar form of an HTU–NTU expression, joined to an operating line and equilibrium relation. The graphical procedure used to solve this much simpler set is outlined in Figure 9.5b and follows the usual routine of drawing an operating line, this one of slope $LC_L/G_s$, through the point $(H_1, T_{L1})$, and evaluating the NTU integral using the enthalpy driving force read from the operating diagram. Note that it is now $G_{Min}$, not $L_{Min}$, which corresponds to an infinitely high tower. In Practice Problem 9.3, the reader is asked to apply this method to the design of a cooling tower.

## 9.2 Drying Operations

The drying of solids is a topic of considerable proportions that merits an entire monograph for its proper treatment. Our purpose here is to give the reader a brief survey of the operation and to provide practice in carrying out simple calculations.

Drying can be carried out in a variety of physical configurations in which the solids can be stationary, conveyed on a moving belt, or allowed to tumble through an inclined rotary kiln. What these operations have in common is the use of heated air to assist in the drying process. When air passes through a stationary mass of solids, which may be contained in a set of perforated trays or in a fixed bed or column, the process is referred to as *through-flow drying*. *Cross-flow drying* occurs when air passes through a perforated conveyor belt at right angles to the solids being conveyed. Rotary kilns, on the other hand, usually make use of a countercurrent mode of contact, with hot combustion gases flowing upward into the inclined kilns while the tumbling solids make their way downward in the opposite direction.

The drying of solids in general is a highly complex process involving both heat and mass transfer. If the solid is porous, moisture content and the temperature will vary internally, as well as externally in the direction of airflow. Thus we could be dealing with at least two coupled PDEs (mass and energy balance) in time and two dimensions.

Early studies of drying processes revealed that considerable simplifications result by recognizing three distinct drying periods, shown in Figure 9.6. During an initial adjustment period, the surface moisture quickly drops to the wet-bulb temperature. If the moisture content in the air is either negligible or otherwise constant due to steady-state conditions, the humidity driving force will assume a constant value $Y^* - Y$ and the drying process will consequently proceed at a constant rate. During this constant-rate period, as it is called, the process can be modeled algebraically if we assume negligible change in air humidity in the direction of flow, or by an ODE if the latter varies. Thereafter, the drying process becomes more complex as moisture removal now must take place from the interior porous structure and a continually receding water interface. A lengthy drying period results as the moisture becomes increasingly inaccessible. It is this interval, called the *falling-rate period*, that leads to the aforementioned PDEs.

In the following illustration, which deals with the drying of a steamed activated carbon bed, we assume that the operation takes place entirely in the constant-rate period. This is based on the fact that carbon is a hydrophobic substance that allows little penetration of its porous structure by the condensate produced during the steaming process. In other words, the condensate is assumed to be present entirely as surface moisture, which is removed by a constant-rate drying mechanism.

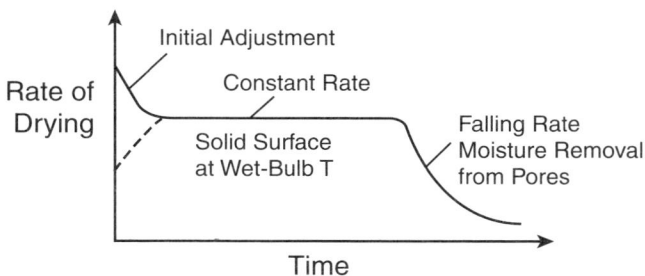

**FIGURE 9.6**
The various drying periods.

## Illustration 9.3: Debugging of a Vinyl Chloride Recovery Unit

Vinyl chloride monomer (VCM) is a volatile substance (boiling point 14°C) used as a starting material for the production of polyvinyl chloride (PVC). It has been identified as a potential carcinogen, and occupational health regulations now call for an upper limit of 1 to 5 ppm VCM in factory air.

A preferred method of air purification is to pass VCM-laden air through beds of activated carbon. These beds operate on a four-step cycle:

1. Saturation with VCM to 1 ppm breakthrough
2. Stripping of the adsorbed VCM with steam, which is subsequently condensed, leaving essentially pure VCM that is dried and recycled
3. Drying of the carbon bed with hot air
4. Cooling of the regenerated bed with cold purified air

A dual-bed system is commonly employed, so that while one bed is "on-stream," the second bed can be regenerated and prepared for the adsorption step. The time period allowed for each step is typically as follows:

1. Saturation: 4 h
2. Steaming: 2 h
3. Drying: 1 ½ h
4. Cooling: ½ h

A schematic diagram of the cycle appears in Figure 9.7.

A major producer of PVC experienced difficulties with a newly installed adsorption system. The unit performed satisfactorily with fresh carbon, but during subsequent cycling the VCM level in the effluent rose to unacceptable levels. Inspection of the beds after drying revealed considerable residual moisture. Inadequate drying during step 3 was therefore considered to be a possible reason for the malfunctioning of the bed. The total amount of condensate that needed to be evaporated was estimated at 100 kg.

When queried about their choice of airblower and its delivery rate (0.1 kg/s at 49°C), plant personnel responded that it was based on the assumption that the condensate was at or near 100°C. This was a major conceptual error. Basic knowledge of the psychrometric chart should have led to the realization that evaporative cooling would reduce the temperature of the condensate well below 100°C, i.e., to the wet-bulb level. This in turn would dramatically reduce the evaporation rate since the vapor pressure, and hence the driving saturation humidity, is an exponential function of temperature.

If we assume that the entire bed attains the wet-bulb temperature after a brief start-up period, the model reduces to an ODE, which must be complemented by a cumulative mass balance to determine the drying time. We consider instead a limiting version of the full model in which the exiting airstream is assumed to be fully saturated. This asymptotic case provides us

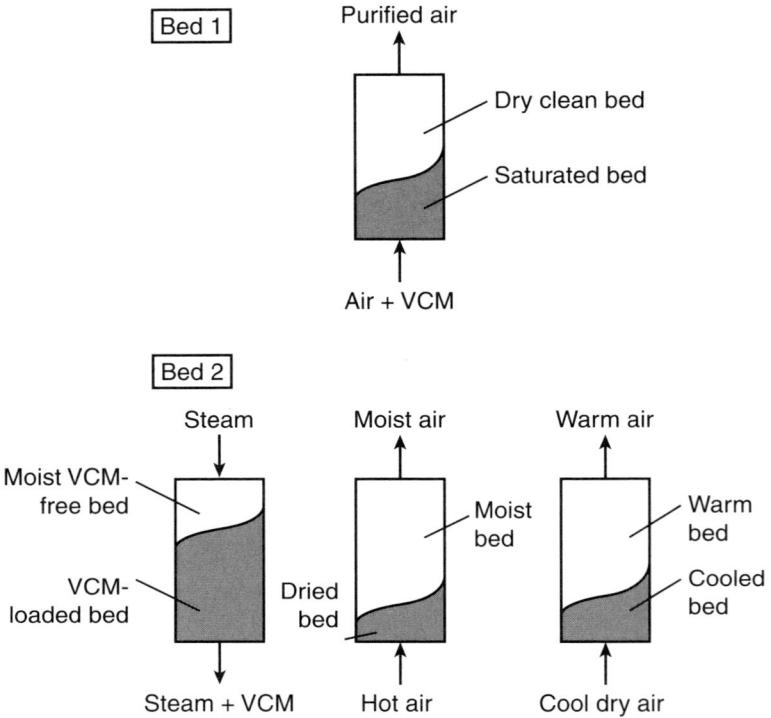

**FIGURE 9.7**
Removal of VCM from air: operation of the adsorption purifier.

with an estimate of the *minimum* airflow rate required to evaporate the charge of 100 kg water in the assigned time of $1\frac{1}{2}$ h.

Let us proceed along these lines. To obtain the minimum airflow rate we compose a cumulative water balance, which is given by

$$
\begin{array}{cc}
\text{Water removed} & \text{Amount of} \\
\text{by exiting gas} & = \text{water to be} \\
\text{in } 1\frac{1}{2} \text{ hr} & \text{evaporated}
\end{array}
$$

$$Y_{wb}(G_s)_{Min}t = 100 \text{ kg} \tag{9.6a}$$

from which, with $t = 1\frac{1}{2}$ h = 5400 s, we obtain

$$(G_s)_{Min} = \frac{100}{Y_{wb}\,5400} = \frac{1}{54\,Y_{wb}} \tag{9.6b}$$

**TABLE 9.1**

Minimum Air Flow Rates for Drying a Carbon Bed Containing 100 kg Moisture

| Air Temperature (°C) | 25 | 50 | 75 | 100 | 125 |
|---|---|---|---|---|---|
| $Y_{wb}$ (kg H$_2$O/kg air) | 0.006 | 0.013 | 0.021 | 0.028 | 0.038 |
| $(G_s)_{Min}$ (kg/s) | 3.1 | 1.4 | 0.88 | 0.66 | 0.49 |

The humidity chart, Figure 9.4, is now used in Equation 9.6b to establish values of $(G_s)_{Min}$ for various levels of incoming air temperature $T_{air}$. This is done in the usual fashion by first locating the incoming air temperature and humidity, assumed to be zero, on the chart, and then moving from that point upward and to the left along the adiabatic saturation line to a point of intersection with the 100% RH curve. The right-hand ordinate of that point yields the value of $Y_{wb}$ to be used in Equation 9.6c. The results obtained are summarized in Table 9.1.

Examination of Table 9.1 shows that the air provided by the plant ($T_{air}$ = 50°C, $G_s$ = 0.1 kg/s) underestimated the minimum requirement by a factor of 14. Even at a temperature of 125°C, the minimum flow required was still five times that actually provided. Clearly, a combination of both higher temperature and greater blower capacity would be required to meet the drying specifications. A reasonable recommendation would be for a flow rate of 1 kg/s at 125°C at the point of delivery. This provides a safety factor of 2 over the tabulated $(G_s)_{Min}$ of 0.49 kg/s.

*Comments:*

Note here that we were able, without resorting to elaborate calculations, to pinpoint the cause of malfunction and to make realistic recommendations for its rectification. The rapid way in which this was achieved and the simple remedies proposed would please industrial clients who value quick results and simple solutions above all else. We should not be blind to the fact, however, that this success was based on a good understanding of the physical process involved. The situation required not so much an extensive expertise in drying operations, but rather the simple recognition that we were dealing with evaporative cooling. This realization led to a rapid resolution of the problem.

## 9.3 Heat Effects in a Catalyst Pellet: The Nonisothermal Effectiveness Factor

Illustration 4.9 considered the model that describes the isothermal diffusion and reaction in a catalyst pellet. Solution of that model yields the reactant concentration profile within the pellet, which is then converted by integration into the so-called catalyst effectiveness factor $E$. Such isothermal effectiveness factors apply to small particles with high thermal conductivities and relatively low reaction rates.

In general the heat of reaction, which is of the order of 100 kJ/mol, cannot be ignored, and the mass balance must then be complemented by an appropriate shell energy balance. That balance must consider the heat conducted in and out of the shell, as well as the heat generated or consumed within the pellet.

We assume the same slab geometry and first-order reaction as before and consider the reaction to be exothermic, which is the more common case. The following formulation is then obtained:

$$\text{Rate of energy in} - \text{Rate of energy out} = 0$$

$$\left[ -k_e A \frac{dT}{dx}\bigg|_x \right] - \left[ \begin{array}{l} -k_e A \frac{dT}{dz}\bigg|_{x+\Delta x} \\ +k_r(T)C_A \Delta H_r A \Delta x \end{array} \right] = 0 \tag{9.7a}$$

Dividing by $A\Delta x$ and letting $\Delta x \to 0$ we obtain the second-order ODE

$$\frac{d^2T}{dx^2} - k_r(T)C_A \Delta H_r / k_e = 0 \tag{9.7b}$$

where $k_e$ is the effective thermal conductivity of the pellet.

This expression is supplemented by the mass balance given in Illustration 4.9 in which the rate constant $k_r$ is now a function of temperature. We repeat it here for completeness:

$$\frac{d^2C}{dx^2} - k_r(T)C_A / D_e = 0 \tag{4.16b}$$

The two ODEs, which are coupled by the two state variables $C_A$ and $T$, generally have to be solved numerically. The resulting concentration profile $C_A(x)$ can then be integrated over the pellet volume $V_p$ as was done in the isothermal case to obtain the nonisothermal effectiveness factor $E_{ni}$:

$$E_{ni} = \frac{\int_0^L k_r(T)C_A(x)dV_p}{k_r(T)C_A V_p} \tag{9.7c}$$

where $k_r(T)$ is given by the familiar Arrhenius relation:

$$k_r(T) = A \exp(-E_a/Rt) \tag{9.7d}$$

Here the reference state is taken to be the surface concentration $C_{AS}$ and the surface temperature $T_s$, i.e., the conditions that would prevail within the pellet in the absence of transport resistances.

A typical, unscaled plot of $E_{ni}$ vs. the nonisothermal Thiele modulus is shown in Figure 9.8. Two additional parameters that contain the thermal factors make their appearance here: the Arrhenius number $E_a/RT_s$, which contains the important activation energy $E_a$; and the dimensionless parameter $\beta$, which reflects the effect due to the heat of reaction and the transport resistances. For $\beta = 0$, i.e., for a vanishing heat of reaction or infinite thermal conductivity, the effectiveness factor reduces to that of the isothermal case. $\beta > 0$ denotes an exothermic reaction and here the rise in temperature in the interior of the pellet is seen to have a significant impact on $E_{ni}$ which may rise above unity and reach values as high as 100. This means that the overall reaction rate in the pellet is as much as 100 times faster than would be the case at the prevailing surface conditions. This is due to the strong exponential dependence of reaction rate on temperature, as expressed by the Arrhenius relation (Equation 9.7d). As expected, the effect varies directly with the heat generated $(\Delta H_r)$ and inversely with the rate of heat removal. Thus, exothermicity, far from being undesirable, actually has a beneficial effect on catalytic conversion. We must guard, however, against an excessive rise in temperature, which might adversely affect catalyst structure, causing a decline or

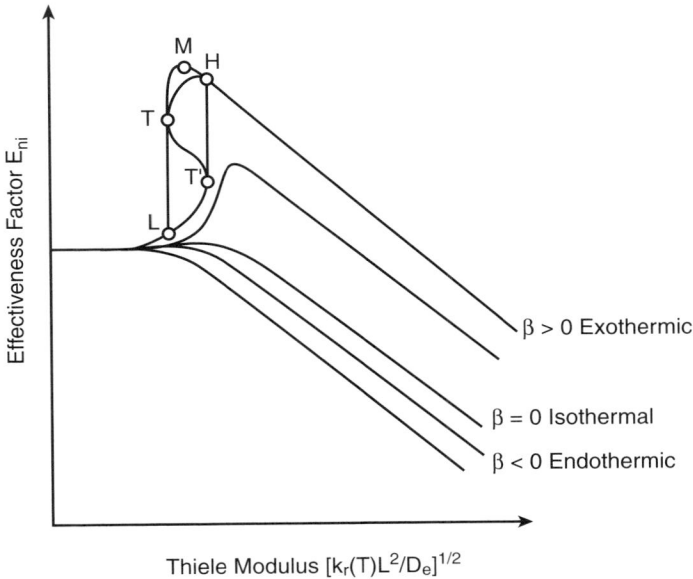

**FIGURE 9.8**
Nonisothermal effectiveness factor and its jump transitions (constant Arrhenius number).

even cessation of catalytic activity. The problem of estimating the maximum temperature that may occur in a catalyst pellet is addressed in Practice Problem 9.4.

For endothermic reactions, $\beta < 0$ applies; $\Delta H_r$ is now a positive number. The effectiveness factor is now *below* the value seen in the isothermal case, due once again to the dependence of $k_r$ on temperature. The effect is shown in the plots of Figure 9.8.

An additional point needs to be noted in connection with the inflections shown by some of the effectiveness curves. It is seen that in this region of inflections $E$ becomes a multivalued function of the Thiele modulus. The question then arises which of the three values actually materializes in practice. A mathematical analysis of such systems, which goes beyond the scope of this book, shows that of the three states, only the upper and lower ones are stable, and that a jump transition occurs from one to the other as one passes through this region. This transition occurs at different locations, depending on the direction from which the region is approached. Suppose, for example, that the Thiele modulus is gradually diminished, by reducing surface temperature $T_s$. Effectiveness then undergoes a gradual increase, reaches a maximum value $M$ shown in Figure 9.8, and then begins a decline until the tangent point $T$ is reached. Here the effectiveness factor experiences a sudden jump decrease to the lower value $L$, after which it continues a smooth decline with diminishing modulus value toward the limiting value of unity. A similar jump transition occurs when one approaches from the opposite direction, but this time it is a jump *increase*, and it occurs at a different location, from the tangent point $T'$ to the location $H$. This phenomenon of obtaining different ordinate values, depending on the direction in which a curve is traversed, is referred to as a hysteresis effect, and the surface temperatures $T_s$ at which the jump transition occurs are known as the ignition and extinction temperatures. In other words, when these temperatures are reached, the reaction rate either undergoes a sudden increase (ignition, point $T'$) or it experiences a sudden drop (extinction, point $T$). These interesting features can be used to control the course of a particular catalytic reaction.

*Comments:*

Although the system we have considered here is a relatively simple one involving a first-order reaction, it has revealed the existence of some fascinating and exotic phenomena. Such phenomena are not limited to catalytic reactions but arise in other nonisothermal systems, for example, in continuous-flow stirred tank reactors, and even in isothermal systems. Their common feature is that the performance curve describing the system has to exhibit an *inflection*. Such inflections have also been observed in a biological context, where they play the role of a *biological switch*, which is activated in response to particular stimuli.

## Illustration 9.4: Design of a Gas Scrubber: The Adiabatic Case

In Illustration 8.1 and Illustration 8.2, we gave a detailed account of the characteristics and performance of a gas scrubber. The tacit assumption was made that the operation was isothermal and that the entire equilibrium curve applied to a single temperature, that of the incoming feed and solvent. Above concentrations of a few mole percent, heat effects make themselves increasingly felt and a corresponding shift in the equilibrium curve to lower gas solubilities takes place. In principle, the temperature rises in both phases, but due to the low volumetric heat capacities of the gas, temperature equilibration is rapid and the enthalpy change lies preponderantly in the liquid phase. This has led to the concept of an adiabatic equilibrium curve, which is constructed by using the predicted temperature rise in the liquid to calculate the local gas solubilities as a function of the solute concentration in the gas phase. The relevant equations for the construction of the adiabatic operating diagram, Figure 9.9a, are as follows:

- The solute mass balance and the resulting operating line are unaffected by the heat effects and remain unchanged. We have, as before in the isothermal case,

$$(Y_1 - Y_2)G_s = (X_1 - X_2)L_s \qquad (9.8a)$$

- To calculate the temperature rise, we draw on an integral enthalpy balance over both phases (Figure 9.9b). Choosing as a reference state the temperature of the incoming solvent and setting $H_{ref} = H_{L2} = 0$, we obtain

$$H_L \text{ (kJ/kg solution)} \times L \text{ (kg solution/s)} =$$

$$G_s \text{ (kg carrier/s)} \Delta H_{sol'n} \text{ (kJ/kg solute)} \times (Y - Y_2) \qquad (9.8b)$$

where the enthalpy of the liquid $H_L$ is given by the auxiliary relation

$$H_L = C_L(\text{kJ/kg K}) \times (T_L - T_{L2}) - \Delta H_{sol'n} \qquad (9.8c)$$

- Combining these equations, we obtain for the local temperature rise

$$T_L - T_{L,2} = \frac{G_s}{L_s} \frac{X}{1+X} \frac{\Delta H_{sol'n}}{C_L}(Y - Y_2) + \frac{\Delta H_{sol'n}}{C_L} \qquad (9.8d)$$

where the factor $X/(1 + X)$ is used to convert total liquid flow rate $L$ to solvent flow rate $L_s$.

To construct the adiabatic equilibrium curve, one chooses a pair of values $(X, Y)$ on the operating line, calculates $T_L$ from Equation 9.8d, and with the

## a. Operating Diagram

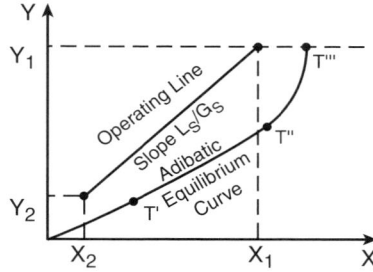

## b. Enthalpy Balance

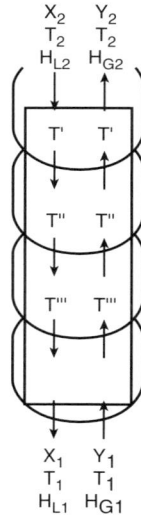

**FIGURE 9.9**
Adiabatic gas scrubber.

values $(Y_1, T_L)$ in hand, establishes $X$ from available equilibrium isotherms. $X, Y^*$ values obtained in this fashion determine a point on the adiabatic equilibrium curve.

The remainder of the calculations proceed as in the isothermal case, i.e., the adiabatic operating diagram (Figure 9.9a) is used to compute NTU, which is then multiplied by HTU to obtain the design height $Z = \text{HTU} \times \text{NTU}$.

*Comments:*

Note the simplification that results from confining the heat effects to the liquid phase and assuming thermal equilibrium between the two phases.

Had this not been done, an additional enthalpy balance for the gas phase would have been required, and the concept of the adiabatic equilibrium curve would have been jeopardized. Once the temperature rise has been accounted for and used in the construction of the adiabatic equilibrium curve, the entire procedure reverts to the familiar territory of the isothermal case. The comfort and value of familiarity is not to be underestimated.

---

## Practice Problems

9.1. *Meteorology*
   a. Using the humidity charts, and given a daytime temperature and relative humidity, indicate how you would proceed in order to predict whether dew will form.
   b. Again using the psychrometric charts, and assuming a strong wind to be blowing, indicate under what conditions frost may form, even though the temperature remains above 0°C. What is the maximum temperature that will allow this to happen?

9.2. *Air Supply to a Dryer*
   An air dryer requires 1 kg/min (dry base) of air at 80°C and RH = 20%. The available air is at 25°C and RH = 50%, and is to be brought to the desired conditions by direct injection of steam. What is the minimum rate at which steam must be supplied, given that its latent heat is 2450 kJ/kg?

*Answer:* 0.08 kg/min

9.3. *Design of a Cooling Tower*
   Water is to be cooled from 43.3 to 29.7°C in a packed column using air entering countercurrently at 29.5°C and a wet-bulb temperature of 23.3°C. The water flow rate is 2.71 kg/m² s and airflow is to be set at 1.5 times the minimum value. The overall mass transfer coefficient for the column is estimated at $K_y a = 0.90$ kg/m³ s $\Delta Y$. Calculate the height of the tower.

*Answer:* 6.6 m

9.4. *Maximum Temperature in a Catalyst Pellet*
   Derive an expression for the maximum possible temperature in a catalyst pellet. (*Hint:* Eliminate the reaction term by dividing mass and energy balances.)

# Selected References

## General Mass Transfer Texts

The earliest major text on mass transfer process is:

R.E. Treybal. *Mass Transfer Operations.* McGraw-Hill, New York, 1952. See also 3rd edition, 1979.

More recent texts on mass transfer, often presented in combination with heat transfer, include the following:

F.M. White. *Heat and Mass Transfer.* Addison-Wesley, Reading, MA, 1988.

S. Middleman. *Introduction to Mass and Heat Transfer.* John Wiley, New York, 1997.

E.L. Cussler. *Diffusion: Mass Transfer in Fluid Systems.* 2nd ed., Cambridge University Press, New York, 1997.

H.D. Baehr. *Heat and Mass Transfer* (translated from the German). Springer, New York, 1998.

A.F. Mills. *Basic Heat and Mass Transfer.* 2nd ed., Prentice Hall, Englewood Cliffs, NJ, 1999.

Separation processes, which make extensive use of mass transfer concepts, are well presented in the recent text:

J.D. Seader and E.J. Henley. *Separation Process Principles.* John Wiley, New York, 2000.

## Transport Phenomena

The classical text by Bird, Stewart, and Lightfoot (1960) has recently been updated:

R.B. Bird, W.R. Stewart and E.N. Lightfoot. *Transport Phenomena*. 2nd ed., John Wiley, New York, 2000.

Transport theory applied to biomedical and materials engineering is discussed in:

D.R. Poirier and G.H. Geiger. *Transport Phenomena in Materials Processing*. Minerals, Metals and Materials Society, Warrendale, PA, 1994.

R.L. Fournier. *Transport Phenomena in Biomedical Engineering*. Taylor & Francis, London, 1999.

## Diffusion

Authoritative compilations of solutions to Fick's and Fourier's equations can be found in the classical monographs:

H.S. Carslaw and J.C. Jaeger. *Conduction of Heat in Solids*. Oxford University Press, Oxford, U.K., 1959.

J. Crank. *Mathematics of Diffusion*. 2nd ed., Oxford University Press, Oxford, U.K., 1978.

They include solutions of source problems.
Diffusivities, permeabilities and solubilities in polymers are compiled in:

J. Brandrup, E.H. Immergut, and E.A. Grulke, Eds. *Polymer Handbook*, 4th ed., John Wiley, New York, 1999.

See also:

J. Crank and G.S. Park, Eds. *Diffusion in Polymers*. Academic Press, New York, 1968.

Tables of diffusivities in various solids and liquids, including metals, molten salts and semiconductors appears in Poirier and Geiger cited above. Similar information on diffusion in porous catalysts can be found in:

C.N. Satterfield. *Mass Transfer in Heterogeneous Catalysts*. MIT Press, Cambridge, MA, 1970.

## Diffusion and Reaction

The twin topics of diffusion and reaction are taken up in most texts on reactor engineering, including:

R.W. Missen, C.A. Mims, and B.A. Saville. *Introduction to Chemical Reaction Engineering and Kinetics*. John Wiley, New York, 1999.

O. Levenspiel. *Chemical Reactor Engineering*. John Wiley, New York, 1999.

## Phase Equilibrium

There is a plethora of handbooks and other compilations of phase equilibrium of relevance to mass transfer operations.

For gas, liquid and solid solubilities in water it is best to draw on environmental source books, among which the monumental treatise by Mackay et al. stands out:

D. Mackay, W.-Y. Shiu, and K.C. Ma. *Illustrated Handbook of Physico-Chemical Properties and Environmental Fate for Organic Chemicals*. Vols. 1 to 5, Lewis Publishers, Boca Raton, FL, 1991–1997.

The treatise also reports vapor pressures and bioconcentration factors of the cited substances.

For vapor–liquid and liquid–liquid equilibria, the reader is referred to the equally monumental compilations:

D. Behrens and R. Eckermann, Eds. *Vapor–Liquid Equilibrium Collection* (25 volumes). *Liquid–Liquid Equilibrium Collection* (5 volumes). DECHEMA Chemistry Data Series, DECHEMA, Frankfurt, 1977–1982.

Equilibria of relevance to supercritical fluid extraction can be found in:

M. McHugh and V. Krukonis. *Supercritical Fluid Extraction.* 2nd ed., Butterworth-Heinemann, Oxford, U.K., 1994.

Compilations of adsorption equilibria appear in:

D.P. Valenzuela and A.L. Myers. *Adsorption Equilibria Handbook,* Prentice Hall, Englewood Cliffs, NJ, 1989.

Equilibria involving metals and systems of metals can be found in:

E. Brandes and G.H. Brooks, Eds. *Smithell's Metals Reference Book.* 7th ed., Butterworth-Heinemann, Oxford, U.K., 1992.

---

## Separation Processes

Equilibrium stage separations with emphasis on distillation and gas absorption, are exhaustively treated in:

E.J. Henley and J.D. Seader. *Equilibrium Stage Separation Operations in Chemical Engineering.* John Wiley, New York, 1981.

See also the text on separation processes by the same authors cited above. The reader will find an up-to-date treatment of distillation in:

J.G. Stichlmair and J.R. Fair. *Distillation: Principles and Practice.* Wiley/VCH, New York, 1998.

Treatments of liquid-liquid extraction appear in:

R.E. Treybal. *Liquid Extraction.* 2nd ed., McGraw-Hill, New York, 1963.
T.C. Lo, M.H.I. Baird, and C. Hanson, Eds. *Handbook of Solvent Extraction.* John Wiley, New York, 1983.

and in:

J. Thornton. *Science and Practice of Liquid–Liquid Extraction,* Vol. 1 and 2. Oxford University Press, Oxford, U.K., 1992.

The definitive and up-to-date monographs on membrane separation are:

T. Matsuura. *Synthetic Membranes and Membrane Separation Processes.* CRC Press, Boca Raton, FL, 1994.

R.W. Baker. *Membrane Technology and Applications.* McGraw-Hill, New York, 2000.

Fundamentals of adsorption, separation, and purification are discussed in a slim volume by the author:

D. Basmadjian. *The Little Adsorption Book.* CRC Press, Boca Raton, FL, 1996.

## Other

The illustration and practice problems in Chapter 3, which deal with transport in plants, used the following as a source:

P.S. Nobel. *Biophysical Plant Physiology and Ecology.* W.H. Freeman, San Francisco, 1987.

See also by the same author:

P.S. Nobel. *Physicochemical and Environmental Plant Physiology.* Academic Press, New York, 1999.

# Appendix A1

## *The D-Operator Method*

The basis of the $D$-operator method consists of replacing the operational part of a derivative, i.e., $d/dx$, by the operator symbol $D$, and treating that symbol as an *algebraic entity*. Thus, the second derivative is written in the form

$$\frac{d}{dx}\left(\frac{d}{dx}\right) + (D)(D) = D^2 \qquad (A.1)$$

and in its full form

$$\frac{d}{dx}\left(\frac{dy}{dx}\right) = D^2 y \qquad (A.2)$$

where $D^2 y$ is considered to be the algebraic product of $D^2$ and $y$. It follows that the quantity $y$ can be separated from $D^2 y$ by factoring it out, just as one would an algebraic quantity. Thus, the ODE

$$\frac{d^2 y}{dx^2} - y = 0 \qquad (A.3)$$

can be written in the equivalent form

$$(D^2 - 1)y = 0 \qquad (A.4)$$

from which it follows that

$$D^2 - 1 = 0 \qquad (A.5)$$

with the solutions

$$D_1 = 1 \qquad D_2 = 1 \qquad (A.6)$$

Equation A.5 is termed the *characteristic equation* of the ODE (Equation A.3) and its solution (Equation A.6) is referred to as the characteristic roots of the ODE.

Consider now the general second-order ODE

$$ay'' + by' + cy = 0 \tag{A.7}$$

Then it can be shown that its solution takes the form

$$y = C_1 \exp(D_1 x) + C_2 \exp(D_2 x) \tag{A.8}$$

where $D_1$ and $D_2$ are the characteristic roots of the ODE, i.e., the solutions of the characteristic equation

$$aD^2 + bD + c = 0 \tag{A.9}$$

When the roots are complex, the exponential terms in Equation A.8 are converted to a trigonometric form using Euler's formula:

$$e^{ix} = \cos x + i \sin x \tag{A.10}$$

We note in addition that the exponential terms can also be expressed in equivalent hyperbolic form and that, when the roots are identical, one of the two solutions is premultiplied by $x$. This follows from the appropriate theory. Table A.1 summarizes the results.

# Appendix A2

*Hyperbolic Functions and ODEs*

**TABLE A.1**

Short Table of Hyperbolic Functions

| | |
|---|---|
| $\sinh x = \dfrac{e^x - e^{-x}}{2}$ | $\tanh x = \dfrac{\sinh x}{\cosh x}$ |
| $\cosh x = \dfrac{e^x + e^{-x}}{2}$ | $\coth x = \dfrac{\cosh x}{\sinh x}$ |

**TABLE A.2**

Solutions of the Second-Order ODE $ay'' \pm by' \pm cy = 0$

| Characteristic Roots | Solutions |
|---|---|
| 1. Distinct and real | $y = C_1 e^{D_1 x} + C_2 e^{D_2 x}$ |
| | or $y = C_1 \sinh D_1 x + C_2 \cosh D_2 x$ |
| 2. Identical and real | $y = C_1 e^{Dx} + C_2 x e^{Dx}$ |
| 3. Imaginary $D_{1,2} = \pm bi$ | $y = C_1 \cos bx + C_2 \sin bx$ |
| 4. Complex conjugate $D_{1,2} = a \pm bi$ | $y = e^{ax}(C_1 \cos bx + C_2 \sin bx)$ |

# Subject Index

## A

Absorption, see Gas–liquid absorption
Activity coefficients, 229
    calculation, from solubilities, 236
    prediction, by UNIQUAC equation, 230
    variation with concentration, 229
Additivity of resistances
    in carbon dioxide uptake by leaf, 119
    in diffusion through composite cylinders, 9
    in heat transfer, 26
    in mass transfer, 26, 35
    two-film theory and, 26, 27
Adsorption (see also Percolation processes)
    batch, of trace substance, 154
    countercurrent cascade for, 265
    crosscurrent cascade for, 257
    desorption from bed in, 298
    efficiency in single stage, 301
    Freundlich isotherm for, 241, 306
    Henry's constants for, 204
    Langmuir isotherm for, 201
    minimum adsorbent inventory in, 260
    minimum bed size in, 207
    moisture isotherms in, 205
    of vinyl chloride monomer, 363
    single-stage, 247, 248, 306
    Toth isotherm for, 241
Agitated vessels
    dissolution time in, 179, 187
    efficiency of, 301, 302
    mass transfer correlations for, 178
Air–water system
    enthalpy of, 355
    humidity charts for, 353, 354
    in drying operations, 361, 371
    in water cooling, 357
Antoine equation, 193
    table of constants for, 194
Artificial kidney, see Hemodialyzer
Azeotropes, 231
    diagrams for, 233
    table of, 234

## B

Batch distillation
    at constant overhead composition, 291
    at constant reflux, 294
    differential, 251
    Rayleigh equation for, 252
    recovery in, 293, 296, 310
    separation factors from, 254
    total boil-up in, 311
Bioconcentration factor (BCF), 182, 183, 187
Biology, Biomedical engineering, and
        Biotechnology
    bioconcentration, 182
    blood coagulation, 37
    controlled-release devices, 35, 90
    diffusion in living cell, 128
    diffusivities of biological substances in
        water, 97
    drug administration, 86
    effective therapeutic concentration (ETC),
        35, 87
    hemodialysis, 330, 332, 333, 334, 338, 346,
        347
    mass transfer in blood, 23, 185, 334
    mass transfer in kidney, 185
    mass transfer in leaf, 112, 119
    nicotine patch, model for, 153
    partition coefficients, 57, 220
    pharmacokinetics, 85
    protein concentration by ultrafiltration,
        308
    toxin uptake and elimination in animals,
        180
    vascular grafts, 23, 37
Blood and blood flow
    anticoagulant release into, 186
    coagulation trigger in, 37
    critical vessel diameter in, 185
    determination by dye dilution, method of,
        88
    hemodialysis of, 330, 332, 333, 334, 338
    isotonic solution for, 346
    mass transfer between tissue and, 57

mass transfer coefficients in, 23
mass transfer regimes in, 185
osmotic pressure of, 229
Boundary and initial conditions for
differential equations, 6, 69, 72,
75, 144, 151
Breathing losses in storage tank, 193
Buckingham $\pi$ theorem, 166, 168

## C

Carbon dioxide
absorption in packed tower, 176
and global warming, 129
caffeine equilibrium in supercritical, 219
compensation of emissions by plant life,
120
emission from car, 120
in carbonation of soft drink, 197
in supercritical extraction of caffeine, 218
net global emissions of, 129
removal from natural gas of, 347
uptake by leaves, 119
Casting of alloys
microsegregation in, 73
modeling of, 75
Rayleigh's equation in, 78
Catalyst pellet
design of, 143
diffusivity in, 115
effectiveness factor for, 147, 365
Raschig Ring form of, 154
reaction and diffusion in, 7, 143
temperature effect on performance of,
154, 365, 371
Coffee decaffeination, 218, 324
Compartments, 40, 241
in animals and humans, 51, 85, 187
in environment, 220
Concentration polarization
Brian's equation for, 335
in alloy casting, 73
in electrorefining, 98
in membrane separation, 332
in reverse osmosis, 335
Conduction of heat, 2, 3
Conservation laws
continuity equation, 80, 89
generalized vectorial of mass, 79
Continuous-contact operations
distillation, 322
gas absorption, 53, 314
liquid extraction, 322, 345
membrane processes, 326
minimum solvent requirement in, 317

supercritical fluid extraction, 324
water cooling, 357
Controlled-release drug delivery, 35, 90, 186
Cooling tower
design equation for, 361
operating diagram for, 359

## D

D'Arcy's law, 2
Dialysis, 330, 331
Diffusion
and Fick's law, 2
and reaction, in liquids, 140, 143, 150, 155
and reaction, in solids, 88, 89, 140, 144
equimolar counter, 18
from sources, 123
from spherical cavity, 8, 10
from well-stirred solution, 89, 138
in animal tissue, 139
in catalysts, 115, 141, 145
in cylinder, 136
in gases, 91, 117
in gas–solid reactions, 142
in hollow cylinder, 8
in leaves, 112, 119
in liquids, 95
in metals, 102
in plane sheet or slab, 67, 136
in polymers, 102
in porous media, 110, 115, 145
in semi-infinite medium, 83, 124, 125, 133,
135
in solids, 101
in sphere, 35
mechanisms of, 92, 95, 101, 102, 111
of dopant in silicon chip, 117
of solids in solids, 116, 120
steady-state multidimensional, 81
through stagnant film, 18, 34
transient, 121, 133
Diffusivities
effective, in porous media, 110, 115, 302
equations for, 93, 96
in air, 92
in liquids, 95, 217
in metals, 96
in molten salts, 96
in polymers, 105, 118
in solids, 106, 116
in supercritical fluids, 217
in water, 97, 118
Knudsen, 111, 118
Dimensional analysis, 166

Dimensionless groups for mass and heat
      transfer, 159
Dissolution of solids, 179, 187
Distillation
   at total reflux, 322
   batch (differential), 62, 251
   batch-column, 290
   construction of trays for, 266
   continuous fractional, 273
   effect of feed and reflux on, 310
   Fenske equation for, 289
   effect of open stream, 310
   isotope, 288, 345
   McCabe–Thiele diagrams for, 276, 278,
      280, 285, 292
   minimum number of trays for, 282, 284
   minimum reflux for, 282, 284, 286
   O'Connell's correlation for tray
      efficiencies, 300
   packed-column, 322
   packing for, 174
   recovery in, 288, 292, 293, 296, 310
   steam, 242
Distribution coefficients in liquid–liquid
      equilibria, 213
*D*-operator method, 379
Driving force
   linear, 3, 5, 34
   overall, 27
Drying
   air supply for, 262, 371
   freeze, 155
   of carbon bed, 362
   of plastic sheets, 172
   periods, 362
   time of, 88, 362
   with air blower, 31

**E**

Effective therapeutic concentration (ETC), 35,
      87
Effectiveness factors for catalyst particles,
      115, 141
   derivation, 147
   plot as function of Thiele modulus, 147
   use in design of catalysts, 147
Electrorefining of copper
   model, 100
   size of plant, 118
Emissions
   concentration histories and profiles, 127
   continuous, 124
   effect of wind on, 131
   from chimney, 153

   from embedded sources, 10, 83, 84, 90
   from plane source, 124
   from point source, 123, 128
   from solvent spill, 153
   from storage tank, 196
   instantaneous, 124, 128
   into infinite medium, 123, 126
   into semi-infinite medium, 125, 128
   net global carbon dioxide, 129
   table of solutions for concentrations of,
      124
Enhancement factor in gas–liquid mass
      transfer with reaction, 152, 156
Environmental topics (see also Emissions)
   adsorption of pollutants in carbon bed,
      205, 311, 362
   attenuation of mercury pollution of water
      basin, 254
   bioconcentration factors for toxins, 183
   carbon dioxide uptake by plant life, 119
   clearance of river bed and soils, 87, 299
   DDT uptake by fish, 222
   discharge of plant effluent into river, 200
   evaporation of pollutant from mist over
      Niagara Falls, 311
   evaporation of pollutant from water
      basin, 42, 242, 254
   global warming, 112, 120, 129
   Henry's constants for adsorption of
      pollutants onto soil, 209
   mass transfer between oceans and
      atmosphere, 36
   mass transfer in leaf, 112
   octanol–water partition coefficient, 221
   partitioning, 220, 225
   pollutant release from buried dumps, 83,
      84
   pollutant release from groundwater onto
      soils, 208
   reaeration of river, 47
   uptake and clearance of toxins in animals,
      180
Error function
   table of numerical values of, 126
   table of properties of, 126
ETC, see Effective therapeutic concentration
Eutectic, 89

**F**

Fenske equation, 289
Fermi problems, 31, 33, 117
Fick's equation, 67, 121, 131
Fick's law, 2
Film theory, 14, 24

Film thickness
   effective, 14, 35
   estimation of, 22
   in entry region, 162, 164
Fish
   bioconcentration in, 183
   uptake of toxin by, 183, 221
   water intake by, 239
Fourier's law, 2
Freeze-drying of food, 155

# G

Gas–liquid absorption, Gas scrubbing
   adiabatic, 369
   countercurrent, continuous contact, 314
   countercurrent, staged, 265
   countercurrent, with linear equilibrium, 270, 309
   design of packed columns for, 317
   diameter for packed column, 187
   Henry's constants for, 197
   HETP for, 176
   Kremser equation for staged and linear equilibrium, 270
   mass balances in, 53
   mass transfer coefficients for packings used in, 175
   minimum solvent requirements in, 267, 317
   NTUs for linear equilibrium, 320, 321
   O'Connell's correlation for plate efficiencies in, 300
   operating diagram for countercurrent, 267, 317, 321
   optimum packing size for, 345
   optimum solvent flow rate for, 319
   trays for, 266
   use of reactive solvent in, 151
Gas–solid reactions and diffusion, 142
Glueckauf equation, 302
Gradient-driven processes, 2, 33
Graetz problem for mass transfer, 70, 81

# H

Hatta number, 152
Heat exchangers, 88, 338
Heat transfer
   additivity of resistances in, 26
   analogy to mass transfer, 14, 28, 159, 338
   convective, 3, 16, 26, 159
Helium, underground storage of, 10, 84
Hemodialyzer, 330, 332
   analogy to external heat exchange, 341

calculation of performance, 338, 346, 347
   mass transfer coefficient for, 341
Henry's constants
   for absorption equilibria, 197
   for adsorption from water onto soil, 209
   for gas–water equilibria, 197
   in Langmuir isotherm, 203
Henry's law, 196
HETP (height equivalent to a theoretical plate), 176, 290
   estimation of, 176
HETS (height equivalent to a theoretical stage)
   in coffee decaffeination, 325
HTU (height of a transfer unit), 316, 345, 361
Humidity
   absolute, 353, 356
   and humid heat, 355
   and humid volume, 355, 357
   charts, 353, 354
   relative, 354

# I

Ice, evaporation of, 155
Ideal solutions, 226
   Raoult's law for, 226
   table of separation factors for, 235
Intalox Saddles, 174, 175
Ion-exchange (see also Percolation processes)
   Efficiency of column, for linear equilibrium, 305
   equilibrium isotherm for, 241
   minimum bed size for, 311
   structure of resins for, 239
Isotonic solution, 346
Isotopes
   $CH_4$–$CH_3D$, 236
   $C^{12}O$–$C^{13}O$, 288
   distillation of, 288, 345
   $H_2O$–HDO, 235, 241
   separation factors for, 235
   use of Fenske equation in distillation of, 288

# K

Kremser or Kremser–Souders–Brown equation, 269, 272

# L

Laminar boundary layer, 13
Laminar flow

entry (Lévêque) region for mass transfer in, 162, 185
fully developed region for mass transfer in, 162, 185
mass transfer coefficients for, 162, 164
release of a substance into, 70, 80
Langmuir isotherm, 201
Laplace's equation, 81
Leaching
countercurrent staged, 270
Kremser equation for stage calculations in, 272
of oil-bearing seeds, 140, 307
of ore, 154
phase diagram for staged, 308
Lewis relation, 353
Linear driving force, 3, 16
Linear phase equilibria
countercurrent cascades of systems with, 267, 309
gas scrubbing in systems with, 272
Kremser equation for staged operations with, 270
minimum solvent or adsorbent inventory for crosscurrent cascades with, 260
NTUs for systems with, 320, 321
Liquid–liquid extraction
calculations in triangular diagram for, 213, 249, 263
continuous contact with linear equilibrium, 345
countercurrent cascade for, 270, 309
crosscurrent cascade for, 261, 263
distribution coefficients for, 213
efficiency in, 301
Kremser equation, use in, 270
minimum solvent inventory in, 260
operating diagrams for, 249, 263
phase equilibria, for, 210, 212, 213, 241
single-stage, 249
Log-mean differences, 9, 17, 34
Loop of Henle, 185

# M

Mass balances
classification of, 49, 51
cumulative, 51, 62, 76, 135, 293, 295
differential, 50
integral, 50, 53
setting up of, 39, 53
steady-state, 41, 53
unsteady, 41, 50, 53
unsteady differential, 51, 67

Mass transfer
analogies with heat transfer, 14, 28, 73, 185, 340, 341
by diffusion, see Diffusion
convective, 15, 70, 87
driving force for, 15
film theory for, 15
rate laws for, 2, 3
resistance to, 3
simultaneous with heat transfer, 349
Mass transfer coefficients
conversion of, 17, 21
definitions of, 17
estimation of, 22, 31
film, 15
for adsorption, 302
for column packings, 175
in agitated vessels, 177
in blood flow, 23, 185, 334
in kidney, 185
in laminar flow around simple geometries, 163
in laminar tubular flow, 162, 164
in membrane processes, 35, 334
in toxin uptake and clearance in animals, 183
in turbulent flow around simple geometries, 171, 172
in turbulent tubular flow, 171
overall, 25
units of, 16, 21
volumetric, 48, 55, 177, 181
Materials science topics
binary liquid–solid equilibria, 74, 78, 89
casting of alloys, microsegregation in, 73
diffusion in metals and molten salts, 96, 106, 116, 120
diffusion in polymers, 102, 105, 118
doping of silicon chips, 117, 154
eutectic, 89
gas–solid reaction with diffusion, 142
membranes, separation by, 326
membranes, structure of, 327
Sievert's law, 106
transformer steel, manufacture of, 138
McCabe–Thiele diagram
in batch fractionation, 292
in continuous fractionation, 285
location of feed plate in, 286
minimum number of plates from, 284
minimum reflux ratio from, 284
$q$-line in, 283
Membranes
asymmetric, 327
hollow-fiber, 328

Loeb–Sourirajan, 326
permeabilities in, 334
spiral-wound, 328, 329
structure of, 327
Membrane gas separation, 330
nitrogen production by, 344
pressure ratio, 342
pressure ratio limited, 343
selectivity, 342
selectivity limited, 344
Membrane processes, 326
for removal of $CO_2$ from natural gas, 347
hollow-fiber dimensions for, 333
mass transfer coefficients for, 334
table of, 328
Models, information from, 64
Moisture adsorption isotherms, 205
Momentum transport, 5
Moving boundary problems
and freeze-drying of food, 155
and reacting particle, 142
shrinking core model for, 148

## N

Newton's viscosity law, 2, 5
NTU (number of transfer units), 316, 345, 361
plot for calculation of, 321

## O

O'Connell's correlations for tray efficiencies, 300
Ohm's law, 3, 12
Operating diagrams
for continuous contact operations, 317, 345
for countercurrent cascades, 267
for crosscurrent cascades, 260, 263
for fractionation (McCabe–Thiele), 276, 278, 280, 286, 292
for percolation processes, 298
for single-stage operations, 247, 250
for supercritical extraction, 324
Osmosis, 331
equation for, 3
Osmotic pressures, table of, 330

## P

Packaging materials, design of, 108, 118, 120
Packings for packed-column operations, 174
Pall Rings, 174, 175, 176
Partial differential equations
how to avoid, 57, 60, 73, 142, 155, 336, 340

setting up of, 66
vectorial form of, 79
Partitioning
blood-tissue, 57
in biology and environment, 220, 241, 255
octanol–water partition as measure of, 221
Peclet number, 159, 170, 336, 338, 345
Percolation processes, 296
as staged operation, 296
bed size for, actual, 305
bed size for, minimum, 207, 305, 311
efficiency of, 305
in adsorption from groundwater onto soil, 298
in clearance of soils and river beds, 298
parameters for design and analysis of, 303
Permeability, 61, 103, 334
Phase diagrams
for binary liquid–solid systems, 74
for binary vapor–liquid systems, 225, 228, 233
for liquid–liquid systems, 209, 241
for pure substances, 191
in triangular coordinates, 212
lever rule in, 213, 226
Pharmacokinetics, 85
Phase equilibria
binary liquid–solid, 74, 78
binary vapor–liquid, 224
fluid–solid, 201
gas–liquid, 196
Henry's constants for, 196, 197, 209, 230
in supercritical $CO_2$, 217, 219, 324
liquid–liquid, 209, 241
of water vapor on adsorbents, 201
Phase rule, 222
Photosynthesis, 119
Poiseuille equation, 2
Pollutants
adsorption onto soils, 208
clearance from river beds and soils, 299
emissions from sources, equations for, 124
evaporation from water basins, 42, 237, 254
removal by adsorption in carbon beds, 205, 311, 362
solubilities in water of, 199
Psychrometric charts, 353, 354
Psychrometric ratio, 352

## Q

Quasi-steady-state assumption, 57, 60, 142, 155, 340

# R

Raoult's law, 226
Raschig Rings, 154, 174
Rate laws, tables of, 2, 3
Rayleigh equation
    determination of separation factor from,
        254
    in batch distillation, 252
    in casting of alloys, 78
    in environment, 254
    in ultrafiltration, 308
Reactive solvent, selection of, 151
Reflux ratio, 278
Relative volatility, see Separation factors
Reverse osmosis (RO)
    concentration polarization in, 335
    effect of pressure in, 346
    effect of salinity in, 346
    flux Peclet number for, 336, 338
    hollow-fiber modules, 329
    mass transfer coefficients for, 334
    production rates of water by, 338
    simple design equation for, 336

# S

Schmidt number, 158
Sea water
    desalination of, 328, 336
    osmotic pressure of, 230
    sea salt from, 239
Separation factors (relative volatility), 232
    effect of total pressure on, 236
    for ideal solution, 235, 241
    for liquid–solid systems, 79
    table of, 235
Shape factors in 3-D diffusion, 83, 84
Shear rate, 185
Shear stress, 5
Sherwood number, 158, 159, 171, 178, 347
Sievert's law, 106
Silicon
    diffusivity in solids, 116
    doping of chips, 117, 154
    use in manufacture of transformer steel,
        138
Solubilities of liquids and solids in water,
    table of, 199
Solubility of gases in water, table of, 197
Solution mining, 186
Stage efficiencies, 299
    in adsorption, 301
    in distillation, 301
    in gas–liquid absorption, 300

    in liquid–liquid extraction, 301
    in percolation processes, 304
Staged operations, 243
    co-current, 244
        countercurrent, 244, 264
    crosscurrent, 244, 257, 263
    differential, 251
    efficiencies of, 244
    single-stage, 245, 249
    with linear equilibria, 260
Stanton number, 158, 170
Supercritical fluid (SCF), 215
    caffeine extraction with, 219, 324
    carbon dioxide as, 217
    diffusivities in, 217
    equilibrium relation, for caffeine
        extraction with, 219, 241
    region of existence, 216
    solubility of naphthalene in, 217
Supercritical fluid extraction (SCE), 192, 210,
    215
    applications, 216
    decaffeination by, 219, 324
    operating diagram for, 324
    plant, at Houston, Texas, 218
    size of extraction vessel, 218

# T

Temperature
    adiabatic saturation, 352, 356
    critical, 191
    dew-point, 355, 356
    dry-bulb, 350
    effect on catalyst effectiveness factor, 365
    effect on gas absorption, 368
    maximum in catalyst pellet, 371
    wet-bulb, 350, 356
Thiele modulus, 146
Triangular diagram, 209, 212
Turbulent flow
    mass transfer coefficients in, 171, 172
Two-film theory, 24

# U

Ultrafiltration, 332, 334
    protein concentration by, 308

# V

van't Hoff equation, 336
Vector operators
    in formulation of conservation laws, 79
    table of, 82

# W

Washing of granular solids, 309
Water purification
    by activated carbon, 205, 311
    by ion-exchange, 240, 311
    by reverse osmosis, 331, 334, 336

# Z

Zero gradients, 6